AF389485

Commemorative Issue to Celebrate the Life and Work of Prof. Roger W.H. Sargent

Commemorative Issue to Celebrate the Life and Work of Prof. Roger W.H. Sargent

Special Issue Editors

Rafiqul Gani
Ian Cameron

MDPI • Basel • Beijing • Wuhan • Barcelona • Belgrade • Manchester • Tokyo • Cluj • Tianjin

Special Issue Editors

Rafiqul Gani
PSE for SPEED Company Ltd. 294/65 RK Office Park
Thailand

Ian Cameron
The University of Queensland
Australia

Editorial Office
MDPI
St. Alban-Anlage 66
4052 Basel, Switzerland

This is a reprint of articles from the Special Issue published online in the open access journal *Processes* (ISSN 2227-9717) (available at: https://www.mdpi.com/journal/processes/special_issues/Roger_Sargent).

For citation purposes, cite each article independently as indicated on the article page online and as indicated below:

LastName, A.A.; LastName, B.B.; LastName, C.C. Article Title. *Journal Name* **Year**, *Article Number*, Page Range.

ISBN 978-3-03928-134-2 (Pbk)
ISBN 978-3-03928-135-9 (PDF)

Cover image courtesy of shutterstock.com.

Contents

About the Special Issue Editors

Rafiqul Gani is the co-founder and CEO of 'PSE for SPEED', a company located in Bangkok, Thailand. He was formerly Professor of Systems Design at The Technical University of Denmark and Head of the Computer Aided Design Center (CAPEC). He has the position of Visiting Professor at Texas A&M University as well as Zhejiang University, China. He received his PhD from the University of London (Imperial College), having also received numerous other academic and professional awards. Rafiqul was also Editor-in-Chief of the Computers in Chemical Engineering journal from 2009–2015 and on the editorial advisory boards of other journals. He served as President of the European Federation of Chemical Engineering for two terms between 2014 and 2018. He continues to pursue the research and application of computer-aided methods and tools for modelling, property estimation, process–product synthesis and design, and process–tools integration.

Ian Cameron is Professor Emeritus at The University of Queensland and a Fellow of the Australian Academy of Technology and Engineering. He is also co-founder, director and principal consultant with Daesim Technologies, Brisbane. He worked for 10 years with the CSR Group in industry sectors such as sugar, building materials and petrochemicals, having roles in process and control system design, plant commissioning, operations, risk management and environmental protection. After completing a PhD at Imperial College London in the area of Process Systems Engineering (PSE), he worked for 9 years as a United Nations (UNIDO) engineering consultant in Argentina and Turkey. His research has focused on intelligent systems for risk management and process diagnosis, advanced control and system modelling, and the development and application of virtual reality systems in industry and innovation in engineering education design and delivery. He has received numerous professional awards that include the Australian Prime Minister's Award for university teaching.

 *processes* 

Editorial

In Memoriam of Professor Roger W.H. Sargent, the Founder of "Process Systems Engineering"

Rafiqul Gani [1],* and Ian Cameron [2],*

[1] PSE for SPEED Company Ltd., Skyttemosen 6, DK-3450 Allerod, Denmark
[2] School of Chemical Engineering, Faculty of Engineering, Architecture and Information Technology, The University of Queensland, Brisbane St Lucia 4072, Australia
* Correspondence: rgani2018@gmail.com (R.G.); i.cameron@uq.edu.au (I.C.)

Received: 20 March 2020; Accepted: 20 March 2020; Published: 30 March 2020

In Memoriam

Roger W.H. Sargent

1926–2018

In September 2018, the global chemical engineering community lost a true pioneer in the field. This Special Issue is in celebration of Roger Sargent's influence, contributions and legacy as a founder of the field of 'Process Systems Engineering' (PSE). His many years, initially at Air Liquide in Paris, and subsequently at Imperial College London laid a foundation that has been built upon by many who did not personally know him, but have benefited through his insights and sustained vision.

Besides his research and industry focus, he was also influential within education and the engineering profession. He held leadership positions such as head of the Department of Chemical Engineering Imperial College, Dean of the City and Guilds College and President of The Institution of Chemical Engineers (UK) and director of the Centre for Process Systems Engineering.

The papers in this issue represent areas that, in many ways, were pioneered by Roger as he and his many students developed new approaches to the integrated design, mathematical modelling and the optimization of complex industrial processes. These developments were facilitated by computer-based tools and a wide range of advanced numerical methods. A number of his direct PhD students and some of their academic descendants are authors of these contributions.

In remembering his impact, there follows a personal tribute from Arthur Westerberg, one of Roger's PhD students in the early 1960s. Professor Westerberg himself continued ground-breaking contributions in PSE at Carnegie Mellon University (USA).

In the late 1970s, both of us were privileged to complete our PhD degrees under Roger's expert and constructive guidance. We were thankful for his time, patience and encouragement during that period and his ongoing friendship.

He will be missed by family, friends and a host of those within his academic family tree. However, he will not be forgotten.

Rafiqul Gani and Ian Cameron.

Tribute to Roger Sargent

by

Arthur W. Westerberg

I am going to steal part of the following tribute to Roger Sargent from a message I sent to his son, Philip, just after I learned, sadly, that Roger had passed away.

Many have and will summarize Roger's many contributions to the Process Systems Engineering area. This tribute will be more personal. I was Roger's sixth PhD student, from 1961 to 1964, and, of course, he was a lifelong friend.

Having done a two month summer tour organized for college students in 1959 with Shel Thompson, my roommate at the University of Minnesota, and having Neal Amundson as my undergraduate advisor, who knew just about everyone in academia in the UK, pushed me to apply to go to the UK for my PhD. Negotiations with Kenneth Denbigh, then chairman of the teaching side of the department at Imperial College, garnered me an Assistant Lectureship offer to provide support. After spending a year at Princeton to obtain first an MS degree, I stepped off the ship (on which, incidentally, I met my wife Barbara) knowing no one in London in September of 1961. I remember first visiting the main office of the department at Imperial College and meeting Kenneth Denbigh. After first presenting me his topics, he sent me around to the rest of the department. I entered Roger Sargent's office shortly thereafter, and, of course, Roger had a project for someone who had on his resume experience in using the computer, on projects both at Minnesota and at Princeton. No one in those days would likely have had any computer experience. Furthermore, as we now know, Roger was into computers and their use in process design and control, with his several years at Air Liquide in Paris in the 1950s. Mind you the computers anyone would have had then were toys by today's standards, but he could see the potential and wanted to be in the business of developing that area as a research area. As I came with an Assistant Lectureship, an entry level faculty position, he insisted I join the faculty for morning coffee, lunch and coffee, and afternoon tea. Thus I saw him at least three times almost every day and had lots of time to interact with him on my project. Our paper from that work was really two parts: a core dump of his ideas on how computers would contribute to design and control in our field and a presentation of the ideas in my work on how to shoehorn a flowsheeting calculation onto a vastly undersized computer. Many people in industry told me they were in the modeling departments of their companies because they mentioned they had read that paper in their interviews.

I worked for two years with Control Data in San Diego before joining the Chemical Engineering Faculty at the University of Florida in 1967. By then Roger had given a well appreciated talk at the invitation of the AIChE and then published as a paper in CEP. This talk and the resulting paper again presented his views on PSE. If memory serves, it had about 50 references to relevant papers from many different fields. I remember going on a hunt to collect and review all those papers. It got me off to a running start on my career when I joined the faculty in Florida.

A bit of humor: In 1990 Roger received the CAST Division Computing in Chemical Engineering Award and gave his acceptance speech at the November AIChE meeting. As all know who have heard Roger give a presentation, his talks were seldom if ever done within the prescribed time limit. We were all sitting at the tables listening and, by this time, consuming our desserts and drinking coffee. I remember clearly lifting my coffee cup for a sip. When I looked down, the wait staff has taken away the saucer. Our table lit up with laughter. The wait staff were ready to go home.

Over the years I had many occasions to interact with Roger. We were at many meetings together. I visited Imperial College to give the annual Sargent lecture, Roger came to Carnegie Mellon University to give a departmental seminar (we have three former and current faculty members who are his students: the oldest: me, Ignacio Grossmann and Erik Ydstie). In the late 1980s, as we were just starting our Engineering Design Research Center at CMU, Roger led a group at Imperial College to create a related organization aimed at process systems engineering. As director of our center, I reviewed his proposal and others, and clearly this was the project to fund among those I saw. After, I had many occasions to visit his centre. I have to admit, he seemed to tolerate administration (certainly much more eagerly that I did) and held many significant such positions at Imperial College and in the Institute of Chemical Engineers, as well as running this centre. So his legacy is much more than his research.

Finally, he is the top node of a huge academic family tree (http://titan.engr.tamu.edu/Sargent_tree/index.php), that is "the tree" for all of us lucky enough to have been on his.

We will all miss him.

Author Contributions: R.G. and I.C. wrote the In Memoriam preface. A.W. wrote the personal tribute to Professor Roger Sargent. All authors have read and agreed to the published version of the manuscript.

Conflicts of Interest: The authors declare no conflict of interest.

Article

Towards the Grand Unification of Process Design, Scheduling, and Control—Utopia or Reality?

Baris Burnak [1,2], Nikolaos A. Diangelakis [1,2] and Efstratios N. Pistikopoulos [1,2,*]

[1] Artie McFerrin Department of Chemical Engineering, Texas A&M University, College Station, TX 77845, USA
[2] Texas A&M Energy Institute, Texas A&M University, College Station, TX 77845, USA
* Correspondence: stratos@tamu.edu

Received: 3 June 2019; Accepted: 17 July 2019; Published: 18 July 2019

Abstract: As a founder of the Process Systems Engineering (PSE) discipline, Professor Roger W.H. Sargent had set ambitious goals for a systematic new generation of a process design paradigm based on optimization techniques with the consideration of future uncertainties and operational decisions. In this paper, we present a historical perspective on the milestones in model-based design optimization techniques and the developed tools to solve the resulting complex problems. We examine the progress spanning more than five decades, from the early flexibility analysis and optimal process design under uncertainty to more recent developments on the simultaneous consideration of process design, scheduling, and control. This formidable target towards the grand unification poses unique challenges due to multiple time scales and conflicting objectives. Here, we review the recent progress and propose future research directions.

Keywords: process design; scheduling; process control; integration

1. Introduction

It has been over half a century since Professor Roger W.H. Sargent envisioned a paradigm shift in chemical process design methodologies, from ad hoc engineering judgment for specific problems to fully computerized systematic approaches based on complex mathematical models [1]. He conceived the notions of "explicitly formulating the techniques" and "precisely defining the objectives" for engineering design problems, which used to be considered to be "an activity not worthy of higher minds" because of the lack of scientific and systematic tools and methodologies. With the advent of computers, he further emphasized the opportunity to expand the process design problem to account for foreseeable variations in the plant environment over its life cycle to achieve more reliable and robust operations. "(During the process design phase) Many parameters are left available for adjustment during plant operation, such as flow rates, tank levels, operating pressures, etc., but here also *the design places limits on the range of variation possible*." stated Professor Sargent to underpin the interdependence between the design and the uncertainty of future operational decisions.

The Process Systems Engineering (PSE) community has been accumulating formidable knowledge and know-how on mathematical modeling techniques in the fields of process design and operations, and developed efficient tools to solve these advanced models since Professor Sargent had outlined the future of PSE in his 1967 perspective article [1]. Moreover, it has been long established that the early design problem should be studied simultaneously with the operational time-variant decisions to improve the operability and flexibility of the process under variable internal and external plant conditions, and consequently to achieve more reliable, economically more favorable, and inherently safer processes. The most recent efforts towards simultaneous consideration of design and operational decisions explore effective methodologies to integrate the short-term process regulatory decisions (process control) and longer-term economical decisions (scheduling) through mixed-integer dynamic optimization (MIDO) formulations. The proposed solution tools and techniques for this class of integrated problems include (i) discretizing the dynamic high-fidelity representation of the process through orthogonal collocation on finite elements followed by solving a mixed-integer nonlinear programming problem [2], (ii) "back-off" approach to ensure constraint satisfaction under some assumed worst-case scenario [3–5], and (iii) multiparametric programming to explicitly represent the operational strategies to derive tractable and equivalent MIDO formulations [6].

In this paper, we present a historical perspective on the development and progress of modern process design techniques that account for the dynamic variability introduced by the process control and scheduling decisions. In retrospect, we observe the evolution of methodologies from fundamental analyses on design and process uncertainty at steady state to dynamic complex models that explicitly encapsulate the scheduling and control decisions, as illustrated in Figure 1, and summarized as follows.

i   *Flexibility analysis and flexibility index.* The early stages for design optimization under uncertainty. The studies here analyze the steady-state feasibility of a nominal process design under a set of unknown process parameters and unrealized operating decisions, as we will discuss in Section 2.

ii *Dynamic resilience and controllability analysis.* Here, the researchers investigate the dynamic response of a system in closed loop, its interdependence with process design, and attempt to develop the "perfect controller" simultaneously the process that the controller can act on. Such attempts will be demonstrated in Section 3.

iii *Complete integration of design, control, and operational policies.* The focus of the most recent studies in the field. The goal is to model tractable dynamic design optimization problems that account for the scheduling and control decisions to guarantee the operability and even profitability of the operation under all foreseeable conditions. These approaches will be discussed in Section 4.

Clearly, it would be inaccurate and redundant trying to reduce the individual research efforts to a single category. The literature is noticeably diverse in this field with numerous different approaches. However, we find it useful to classify into certain schools of thought that are also in alignment with the historical progress of the field. In Section 5, we further seek to pose the pivotal questions on future challenges and opportunities for the seamless integration of the design, scheduling, and control problems based on the cumulative knowledge of the PSE community and the current trends in the academia.

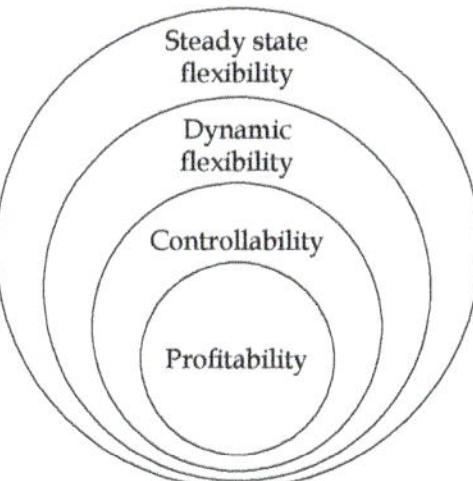

Figure 1. A Venn diagram representation of major operability indices and their relationship with process economics. It is interesting to note that the design optimization approaches started from the outermost layer, and with the advance of modeling techniques, they have progressed towards the center for guaranteed operability, which delivers the optimal process economics.

2. Early Efforts in Design Optimization under Uncertainty

The ongoing collective efforts towards the grand unification of design, scheduling, and control was inaugurated through steady-state design under uncertainty in plant conditions. Takamatsu et al. (1970) [7] estimated the undesirable effects of variations in system parameters, measured process disturbances, and manipulated variables on plant performance by sensitivity analysis on a linearized model. Nishida et al. (1974) [8] adopted the notion of sensitivity analysis to structure a min-max problem for design optimization, presented by Equation (1).

$$\min_{des} \max_{\theta} \quad C(x, des, \theta)$$
$$s.t. \quad h(x, des, \theta) = 0$$
$$g(x, des, \theta) \leq 0 \tag{1}$$
$$\underline{\theta} \leq \theta \leq \overline{\theta}$$

where x is the vector of states of the system, des is the vector of design variables including the steady-state manipulated variables, θ is the vector of parameters that agglomerates the system uncertainties and process disturbances. Equation (1) is one of the first notable attempts to systematically assess the trade-off between minimizing the investment cost and improving the flexibility of the process design. However, this strategy yields conservative solutions since it does not distinguish the time-invariant design variables and time-variant manipulated variables. Grossmann and Sargent (1978) [9] remedied this issue by treating the time-sensitive variables (i.e., manipulated actions and design variables that can be modified in the future) and fixed design variables separately. They further adopted the parametric optimal design problem proposed by Kwak and Haug (1976) [10], and formulated an objective function to minimize the average cost over the expected range of parametric uncertainty, as presented by Equation (2).

$$\min_{u,des} \quad E\{C(x, u, des, \theta)\}$$
$$s.t. \quad \max_{\theta \in \Theta} \quad g_i(x, u, des, \theta) \leq 0, \quad i = 1, 2, \ldots, t \tag{2}$$

where the expected cost function is defined the joint probability distribution of the parameter set θ. Equation (2) requires solving infinite nonlinear programming (NLP) problems. Grossmann and Sargent (1978) [9] proposed an efficient solution procedure for a special case of Equation (2), where each constraint g_i is monotonic in θ, through discretization of the problem over the parameter space. However, solving the NLP problem at a finite number of θ realizations does not ensure the feasibility of the design. This issue was addressed by Halemane and Grossmann (1983) [11] through reformulating an equivalent design feasibility constraint as presented by Equation (3).

$$\max_{\theta \in \Theta} \min_{u \in U} \max_{i \in I} g_i(x, u, des, \theta) \leq 0 \tag{3}$$

The max-min-max problem in Equation (3) mathematically expresses the feasibility question "For all the uncertainty realizations Θ, does there exist a control action u such that the constraint set g is feasible?". Equation (3) was employed in a multiperiod design optimization problem, where the deterministic uncertain parameter θ was allowed to vary within a prespecified range [11]. The feasibility constraint then laid the foundation for the concept of feasibility index, F, proposed by Swaney and Grossmann (1985) [12], as given by Equation (4).

$$F = \max \quad \delta$$

$$s.t. \quad \max_{\theta \in \Theta} \min_{u \in U} \max_{i \in I} g_i(x, u, des, \theta) \leq 0 \tag{4}$$

$$T(\delta) = \{\theta \mid (\theta^{nom} - \delta \Delta \theta^-) \leq \theta \leq \theta \mid (\theta^{nom} + \delta \Delta \theta^+)\}$$

where T is the hyperrectangle for the uncertain parameters, δ is the scaled parameter deviation, and the superscript *nom* denotes nominal conditions. Equation (4) is the first significant attempt to quantify the degree of flexibility of a process design, and has been exploited by numerous studies on design optimization and process operability. However, Equation (4) constitutes a nondifferentiable global optimization problem and is still quite challenging to solve. Therefore, it requires simplifying assumptions and approximations to maintain a tractable problem. Swaney and Grossmann (1985) [13] introduced a heuristic vertex search method and an implicit enumeration scheme for the special case where the critical uncertainty realizations are assumed to lie at the vertices of the hyperrectangle $T(\delta)$. Clearly, this assumption fails to hold when the feasible space of the design problem is non-convex. Grossmann and Floudas (1987) [14] relaxed this assumption by developing a mixed-integer nonlinear programming (MINLP) problem for the feasibility test presented by Equation (3). They further proposed an active constraint strategy for the solution of the resulting MINLP. The mixed-integer formulation also provides a systematic approach to consider all possible critical uncertainty realizations without exhaustive enumeration. The proposed formulation was used for synthesis of a heat exchanger network with uncertain stream flow rates and temperatures [15]. The case of linear constraints reduces to an MILP problem, for which global solution is attainable by standard branch and bound enumeration techniques [14,16,17]. Bansal et al. (2000) [18] developed a computationally efficient theory and algorithm based on multiparametric programming techniques for this special case of flexibility analysis problems. The authors derived explicit expressions for the flexibility index as explicit functions of the continuous design variables. Pistikopoulos and Grossmann (1988a, 1988b, 1988c) used the flexibility test with linear constraints for optimal retrofit design [19–22] and redesign under infeasible nominal uncertainties [23]. Although these approaches are effective and promising to handle the design uncertainty, they require solving nested optimization problems, which poses a major challenge to solve complex and large-scale problems in a reasonable time. Raspanti et al. (2000) [24] proposed replacing the complementarity conditions of the lower level optimization problems with a well-behaved, smoothed nonlinear equality constraints, namely Kreisselmeier and Steinhauser function [25] and Chen and Mangasarian smoothing function [26].

One of the common assumptions in these approaches is the known bounds of the uncertainties, which is rarely the case in real world industrial applications. Pistikopoulos and Mazzuchi (1990) [27] and Straub and Grossmann (1990, 1993) [28,29] extended the flexibility test by assuming a probability distribution model for the parameter uncertainty, which improved the economic performance of the design optimization problem by addressing the "conservativeness" of the solution.

Another common assumption of these approaches is the steady-state operation of the plant design, which creates a significant limitation on the applicability of the methodologies. Although steady-state assumption holds true for the dominant life cycle of the plant operation, design optimization problem may fail to ensure the operability under transient behaviors such as startup or shutdown and transitions between different operating conditions. Dimitriadis and Pistikopoulos (1995) [30] proposed a dynamic feasibility index for the systems that are described by differential algebraic equations (DAE) subject to time-varying constraints. However, the time-dependent uncertainty in their formulation dictates to solve infinitely many dynamic optimization problems. Therefore, the authors assumed that the critical scenarios of uncertainties are known and lie on the vertices of the time-varying uncertainty space, similar to Swaney and Grossmann (1985) [12]. The simplifying assumption reduced the problem to the form given by Equation (5).

$$DF(des) = \max_{\delta, u(t), t} \quad \delta$$

$$\begin{aligned}
s.t. \quad &\dot{x} = f(x(t), u(t), des, \theta(t), t), \quad x(0) = x_0 \\
&g(x(t), u(t), des, \theta(t), t) \leq 0 \\
&\theta(t) = \theta^N(t) + \delta \Delta \theta^c(t) \\
&\delta \geq 0, \quad \underline{u}(t) \leq u(t) \leq \overline{u}(t)
\end{aligned} \tag{5}$$

where the time dependence of the variables constitutes a dynamic optimization problem, and the solution was determined by control vector parameterization techniques [30]. Dynamic flexibility has been widely used in numerous design optimization applications including batch processes [31], separation systems [32–36], reaction systems [37], and heat exchanger network synthesis [38–40].

The dynamic assessment of the plant feasibility under uncertainty has been also studied through exploiting the multiperiod design optimization formulation proposed by Halemane and Grossmann (1983) [11]. Varvarezos et al. (1992) [41] implemented an outer-approximation approach to solve the multiperiod multiproduct batch plant problems operating with single product campaigns, which was formulated as an MINLP. Pistikopoulos and Ierapetritou (1995) [42] considered stochastic process uncertainty and proposed a two-stage decomposition that can handle convex nonlinear problems.

As presented in this section, the early studies on integrated design optimization have primarily focused on (i) investigating the range of operation (flexibility) of a nominal design configuration under foreseeable conditions, and (ii) determining the "best" possible trade-off between the investment cost and the capability of handling variations in the internal and external operating conditions. These studies mostly considered open loop processes, under the traditional assumption that controller design is a sequential task to process design. However, most processes in industry are operated in closed loop, and the controller schemes inherently alter the process dynamics, rendering the open loop flexibility analyses of lesser relevance. In other words, an "attainable" operating point according to open loop flexibility analysis may actually be an infeasible point in closed loop. Realizing the shortcomings of open loop flexibility analyses, researchers began investigating the "controllability" of process systems, and the interdependence of process control and design decisions. In the following section, we present a retrospective background on the integration of process control in the design optimization problem.

3. Integration of Process Control in Design Optimization

The initial efforts towards the integration of process control and design problems established a fundamental understanding on the interdependence of the two decision making mechanisms. The most pronounced school of thought in the early years to evaluate the controllability of the process design is "dynamic resilience", as conceptually defined by Morari (1983a, 1983b) [43,44].

Morari (1983) [43] described dynamic resilience as "the ability of the plant to move fast and smoothly from one operating condition to another and to deal effectively with disturbances". This depiction implies that there is not a clear-cut distinction between flexibility, which was discussed in Section 2, and resilience. However, Grossmann and Morari (1983) [45] pointed out the main difference as "resiliency refers to the maintenance of satisfactory performance despite adverse conditions while flexibility is the ability to handle alternate (desirable) operating conditions". This distinction is the primary motive for most of the flexibility analyses to study steady-state operations, while the resilience deals with the dynamic operations, as we will discuss in this section.

Dynamic resilience, as described by Morari (1983) [43], aims to find the "perfect controller" that is allowed by the physical limitations of the system to assess the controllability of the process by using the internal model control (IMC) structure. The proposed technique decomposes the system transfer function $\tilde{G}$ into (i) a non-singular matrix $\tilde{G}_-$ to design the perfect controller $\tilde{G}_-^{-1}$, and (ii) a singular matrix $\tilde{G}_+$ to generate dynamic resilience indices based on (i) bounds on control variables, (ii) presence of right half plane transmission zeroes, (iii) presence of time delays, and (iv) plant-model mismatch. The proposed indices were used to improve the operability of numerous process, including heat integrated reactor networks [46–48], separation systems [49], heat exchanger networks [50].

Among the four aforementioned resilience indices, Perkins and Wong (1985) [51] studied the last two by adapting the "functional controllability" theorem proposed by Rosenbrock (1970) [52]. The authors further define a system to be functionally controllable if there exists a manipulated action $u(t)$ that can generate any process output $y(t)$ at any time t. Psarris and Floudas studied the dynamic resilience and functional controllability of multiple input multiple output (MIMO) closed-loop systems with time delays [53–55], and transmission zeroes [54,55]. Barton et al. (1991) [56] investigated the open loop process indicators, namely minimum singular value and right half plane zeroes, to assess the interactions between different design configurations and their operability with the best possible control configurations.

In the context of simultaneously assessing the process controllability in process design, one of the first significant contributions is the "back-off approach" introduced by Narraway et al. (1991) [57]. Narraway and Perkins (1994) [58] used this approach to systematically assess the trade-offs between all possible controlled and manipulated variable pairs in a mixed-integer formulation. Bahri et al. (1995) [59] employed the back-off approach to handle process uncertainties in an optimal control problem. The proposed approach is applicable to design linear and mildly nonlinear processes, and relies on three key steps, namely (i) perform a steady-state nonlinear process optimization, (ii) linearize the process at the optimum point, and (iii) "back-off" from the optimal solution by some distance to ensure the feasibility of the operation under some structured disturbance profile. The proposed approach was shown to be effective to select between alternative flowsheets as well as alternative control structures.

With the burgeoning interest in exploring the simultaneous design and control problem, the International Federation of Automatic Control (IFAC) organized the first workshop on "Interactions between Process Design and Process Control" in the Center for Process Systems Engineering at Imperial College London in 1992. The workshop laid the groundwork for a plethora of approaches with a wide range of diversity. Walsh and Perkins (1992) [60] implemented a PI loop in the flexibility analysis, where the input–output loop is selected by an exhaustive screening procedure. Luyben and Floudas (1992) [61] formulated a multiobjective MINLP problem to simultaneously consider the disturbance rejection capacity of the control loop through disturbance condition number and relative gain array to evaluate the interactions between the inputs and outputs of a MIMO system, while designing the process. Shah et al. (1992) [62] used the State-Task Network (STN) representation [63] to simultaneously consider the scheduling and design problems in a batch plant. Thomaidis and Pistikopoulos (1992) [64] introduced a framework to consider the design problem simultaneously with (i) the process flexibility through stochastic flexibility index, (ii) the effect of equipment failures to the overall performance by combined flexibility-reliability index, and (iii) the impact of equipment availability by combined flexibility-reliability index. These aforementioned novel approaches were shown to be promising concepts and techniques to address multiple facets of operational decisions simultaneously with the process design problem. As a result, succeeding studies after this workshop expanded these techniques and branched out to explore further opportunities.

Integrating PI controllers in the design optimization problem was one of the prominent outcomes of the workshop and became the most attractive option for the following research. The literature on PI controllers was already abundant and well-established by the time. Moreover, the explicit form of the controller structure made the integration relatively easy and intuitive, which significantly accelerated the research in closed-loop design optimization. Walsh and Perkins (1994) [65] presented an integrated

PI control scheme and process design for wastewater neutralization. Although the proposed approach was effective for the SISO process, it was reported that it entails further challenges for more complex processes. One major drawback of PI control is its inability to tackle MIMO systems without any advanced modifications in the feedback loop structure. Narraway and Perkins (1993, 1994) [58,66] developed an MILP-based formulation to systematically evaluate the economic performance of every input–output pair combination. Luyben and Floudas (1994a, 1994b) [67,68] adapted a similar approach in a multiobjective framework to determine the best performing input–output pair based on the controllability indices introduced by them, earlier (1992) [61]. The proposed framework was showcased on the design of a heat integrated distillation system [67] and a reactor-separator-recycle system [68]. Mohideen et al. (1996) [32] formulated a multiperiod design and control problem to account for the dynamic variations in the operation, while including the input–output pairing superstructure in the problem. Moreover, the authors used the flexibility index to account for the uncertain parameters in the model and presented a decomposition algorithm for the resulting complex problem. Bansal et al. (2000) [69] constructed a similar formulation as a mixed-integer dynamic optimization (MIDO) problem, which was solved by a Generalized Benders Decomposition (GBD)-based algorithm. The MIDO formulation was presented as follows.

$$
\begin{aligned}
\min_{u,des} \quad & \sum_{i \in NS} w_i C\left(\dot{x}^i(t), x^i(t), u^i(t), des^i\right) \\
s.t. \quad & \dot{x}^i(t) = h_d\left(x^i(t), u^i(t), des^i, \theta^i, t\right), x(t) = x_0 \\
& y^i(t) = h_a\left(x^i(t), u^i(t), des^i, \theta^i, t\right) \\
& g\left(\dot{x}^i(t), x^i(t), y^i(t), u^i(t), des^i, \theta^i, t\right) \leq 0
\end{aligned}
\tag{6}
$$

where w_i is the discrete probability of a scenario i and NS is the discretized set of scenarios. The discretization of uncertainty in the process was first proposed by Grossmann and Sargent (1978) [9].

Although the aforementioned PI-based design and control frameworks are applicable on nonlinear processes, the range of operability is usually limited due to the mismatch between the nonlinear process model and the linearized control model. Ricardez-Sandoval et al. (2008, 2009) [70,71] used robust control tools and the back-off approach to integrate PI control and ensure its stability while solving the design optimization problem. The proposed approach was also tested against the Tennessee Eastman Process [72]. The back-off approach was later generalized for control structure selection in nonlinear processes by Kookos and Perkins (2016) [73]. Ricardez-Sandoval & co-workers have extensively studied back-off approach for simultaneous process design and control under uncertainty [74–76].

One main limitation of integrating PI control in the design optimization in a dynamic formulation is the increasing problem size and complexity. Kookos and Perkins (2001) [77] developed an algorithm for the integrated PI control and design optimization problem, where the size of the search space was reduced systematically in each successive iteration. Malcolm et al. (2007) [78] proposed an "embedded control optimization" procedure, where the authors introduced a two-stage decomposition scheme that approximates the complete integrated problem. The proposed approach reduced the problem size and complexity, and was showcased on larger scale problems including a reactor-separator system [79].

Apart from the inability to naturally handle MIMO systems, PI controllers do not explicitly account for any process constraints stemming from operational, environmental, and safety limitations. Model predictive control (MPC) overcomes these shortcomings by postulating a constrained dynamic optimization problem subject to an explicit model of the process [80]. One of the first remarkable efforts to integrate an MPC scheme in a nonlinear design problem was published by Brengel and Seider (1992) [81].

Here, the authors postulate a bi-level optimization problem, where the *leader* has an economic objective, while the *follower* is the MPC formulation, as presented by Equation (7).

$$
\begin{aligned}
\min_{des} \quad & C_{des}(des) + \kappa C_\kappa(x(t), y(t), u(t), des, \theta(t)) \\
s.t. \quad & f_{des}(des, \theta(t)) = 0 \\
& g_{des}(des, \theta(t)) \leq 0 \\
\min_{u(t)} \quad & C_u(x(t), y(t), u(t), des, \theta(t)) \\
s.t. \quad & \dot{x} = f_u(x(t), y(t), u(t), des, \theta(t)) \\
& g_u(x(t), y(t), u(t), des, \theta(t)) = 0 \\
& h_u(x(t), y(t), u(t), des, \theta(t)) \leq 0
\end{aligned}
\tag{7}
$$

where κ is the design and control integration parameter that scales the trade-off between the controllability of the system and the investment cost. The bi-level problem presented in Equation (7) is challenging to solve without appealing to simplifications. Therefore, the authors proposed replacing the follower problem by complementary slackness equations. However, the solution strategy was still intractable for more complex systems due to the numerical calculation of the second derivatives [81]. As a consequence, integration of the MPC scheme in the design optimization had been rather limited in the literature for almost a decade, until the invention of multiparametric MPC (mpMPC/explicit MPC).

Bemporad et al. (2002) [82] proposed formulating the MPC problem as an explicit function of the initial conditions of the system. This novel strategy allowed for deriving piecewise affine explicit control laws by treating the initial conditions as parameters. The proposed approach formulated the explicit MPC problem as presented by Equation (8).

$$
\begin{aligned}
u_t(\theta) = \quad & \arg\min_{u_t} \|x_N\|_P^2 + \sum_{t=1}^{N-1} \|x_t\|_Q^2 + \sum_{t=1}^{N-1} \|y_t - y_t^{sp}\|_{QR}^2 + \sum_{t=0}^{M-1} \|u_t - u_t^{sp}\|_R^2 + \sum_{t=0}^{M-1} \|\Delta u_t\|_{R1}^2 \\
s.t. \quad & x_{t+1} = Ax_t + Bu_t + Cd_t, \quad y_t = Dx_t + Eu_t + Fd_t \\
& \underline{x}_t \leq x_t \leq \overline{x}_t, \quad \underline{y}_t \leq y_t \leq \overline{y}_t, \quad \underline{u}_t \leq u_t \leq \overline{u}_t, \quad \underline{\Delta u}_t \leq \Delta u_t \leq \overline{\Delta u}_t, \quad \underline{d}_t \leq d_t \leq \overline{d}_t \\
& \theta = [x_{t=0}, u_{t=-1}, d_t, y_t^{sp}, u_t^{sp}]^T
\end{aligned}
\tag{8}
$$

where N is the prediction horizon, M is the output horizon, superscript sp denotes set point, Q, QR, R, and $R1$ are the corresponding weight matrices determined by tuning, P is calculated by discrete algebraic Riccati equation, and $\|\cdot\|_\psi$ denotes weighted vector norm with a weight matrix ψ. Different than conventional MPC, Equation (8) formulates the optimal control problem exactly and completely offline as a function of the set of parameters θ. The solution of this problem can be determined by multiparametric programming techniques, which express the solution space as a piecewise affine function, as presented by Equation (9).

$$
\begin{aligned}
u_t(\theta) &= K_n\theta + r_n, \quad \forall \theta \in CR_n \\
CR_n &:= \{\theta \in \Theta \mid CR^A\theta \leq CR^b\}, \quad \forall n \in \{1, 2, \ldots, NC\}
\end{aligned}
\tag{9}
$$

where CR_n is referred as a critical region and it is the active polyhedral partition of the feasible parameter space, Θ is a closed and bounded set, and NC is the number of critical regions.

The control law given by Equation (9) reduces the complexity of solving an online optimization problem to a simple look-up table algorithm (also known as point location problem) and function evaluation, all of which are affine operations. Hence, the complexity of implementing an MPC scheme is similar to that of a PI controller.

Sakizlis et al. (2003) [83] exploited the explicit nature of the mpMPC solution in the context of design and control integration. The authors formulated a bi-level mixed-integer dynamic optimization problem similar to Equation (7), where the leader accounted for the investment and operating costs in the objective function subject to the dynamic high-fidelity model, and the follower MPC problem was substituted by the affine control law Equation (9). The proposed formulation offered an elegant and systematic methodology to reduce the complexity of the bi-level Equation (7) into a single level dynamic optimization problem. However, the solution strategy still required repetitive linearizations and solving a multiparametric programming problem at every iteration, which can be restrictive for large-scale complex problems. Diangelakis et al. (2017) [84] alleviated that limitation by deriving a "design dependent offline controller", which allowed for solving a single MIDO problem after integrating the control law in the high-fidelity model. Eliminating the linearization step and formulating a single synergistic design and control problem also improved the economic performance of the resulting process compared to the approach proposed by Sakizlis et al. (2003) [83]. The proposed formulation was also showcased on a tank, a continuous stirred tank reactor, and a residential scale combined heat and power unit. The cost effectiveness of the MPC integrated optimal design was also reported to be superior than PI integrated approaches in the literature. Diangelakis and Pistikopoulos (2017) [85] reported that the mpMPC integrated optimal combined heat and power unit operated more fuel efficient in closed loop than PI integrated design. Similarly, Sanchez-Sanchez and Ricardez-Sandoval (2013) [86] showcased a system of CSTRs, where the MPC integrated framework reduced both the operating and the investment costs compared to the PI control integrated approach.

One common aspect of the studies on simultaneous design and control optimization is the assumption that the process will be operated around the same steady-state point throughout the entire life cycle of the plant. However, the external plant conditions, such as market conditions, may dictate a considerably wider operating region with multiple steady-state points [1]. The increasing competition among the businesses impacts the volatility of the market, which creates rapid fluctuations in the energy and raw material prices as well as their availability. Moreover, the demand rate on the product is also subject to considerable variations during the plant operation. Therefore, it is clear that there exists a "best" operating strategy under the knowledge available to the operator, which necessitates the operability of the plant across a wider range. For example, high production rates may be less profitable during the night time because of increased energy prices and hence, operating the energy intensive processes during the daytime may reduce the operating costs. This indicates that the operating level of a processing unit might vary drastically by the choice of the operator. However, the integrated design and control frameworks discussed in this section usually assume a single operating point around which a controllability and flexibility analysis is conducted. Consequently, these frameworks do not attempt to provide any means of guaranteeing the operability of the process at different regions. In the next section, we will discuss several approaches that account for multiple operating regions in a plant, and their scheduling during the operational optimization.

4. Towards the Grand Unification of Process Design, Scheduling, and Control

Process design, scheduling, and control problems are traditionally constructed to address different objectives and they span widely different time scales. In a nutshell, the plant design problem dictates the capacity of processing and it usually comprises the most uncertainty due to its years long lifecycle. The scheduling problem addresses the allocation of the resources and time, as well as the operating level of processing units and their maintenance based on some economic criteria over days/months long horizons. Lastly, the control problem maintains the performance of the plant, while satisfying any physical limitations such as the environmental and safety constraints. The discrepancy in the objectives and time scales creates a challenging problem to systematically evaluate and determine the optimal trade-off between different decision makers.

Process scheduling is more critical in batch operations than continuous operations, as the former are inherently dynamically operated. Accordingly, the initial efforts focused primarily on the batch processes for the integration of the operational optimization and design problems. Birewar and Grossmann (1989) [87] formulated NLP models to incorporate the scheduling decisions in the batch sizing and timing problem in a multiproduct plant for unlimited intermediate storage and zero wait policies. Shah et al. (1992) [62] tackled a similar problem by using the STN representation. White et al. (1996) [88] investigated the switchability of continuous processes between different operating points through formulating an optimal control problem that accounts for the terminal criteria and path constraints within a range of design parameters. Bhatia and Biegler (1996, 1997) [89,90] formulated a dynamic optimization problem, where an economic objective function was subject to a dynamic high-fidelity model of the process described by differential algebraic system of equations. The authors proposed a solution strategy based on discretizing the process model by orthogonal collocation over finite elements, followed by solving the resulting NLP by using a standard solver. The proposed modeling and solution strategy was shown to be promising to satisfy the path constraints, which is a crucial benefit for dynamic systems. Terrazas-Moreno et al. (2008) [2] extended this integration approach to account for the binary decisions in the scheduling problem by formulating a MIDO. Similar to Bhatia and Biegler (1996, 1997) [89,90], the authors first discretized the problem by orthogonal collocation, followed by solving the resulting MINLP.

The early studies that explore the interactions between the scheduling and process control decisions have a significant role in shaping today's approaches for the integrated design optimization problem. In their excellent review article, Baldea and Harjunkoski (2014) [91] classified these attempts to integrate the scheduling and control decisions as (i) "top-down approaches", where the process dynamics and control elements are incorporated in a scheduling skeleton, and (ii) "bottom-up approaches", where the process economics are implemented in the plant-wide control decisions.

In terms of characterizing the transitions between different products in a single operating unit, Mahadevan et al. (2002) [92] introduced a unique "top-down" perspective on the operational optimization problem, revealing that a simultaneous approach on the scheduling and control problem can identify and eliminate the fundamental limiting behavior during the transitions, as showcased on a polymer grade transition process. However, the presented approach requires case specific heuristic decisions to select the "best" fitting scheduling and control configuration and hence, it is not suitable for different applications in the general sense. Chatzidoukas et al. (2003) [93] studied a similar polymerization reactor, and formulated a MIDO problem to determine the time optimal transition between different polymer grades and best performing control structure simultaneously. Flores-Tlacuahuac and Grossmann (2006) [94] introduced a monolithic approach on a multiproduct cyclic CSTR, where the profit was maximized by manipulating the production sequence, transition times, production rates, length of processing times, and amounts manufactured of each product. In contrast to the earlier studies [92,93], the authors focused on the manipulated actions rather than the optimal control configuration. They formulated a MIDO problem, which was solved by discretization of the differential algebraic equations by orthogonal collocation on finite elements followed by solving the resulting MINLP. The presented approach has been extensively studied in the following years to broaden its scope and effectiveness. Terrazas-Moreno et al. (2007) [95] applied this approach on two industrial polymerization reactors. Terrazas-Moreno et al. (2008) [2] formulated a design optimization problem accounting for the scheduling and open loop control trajectories using this approach. Flores-Tlacuahuac and Grossmann (2010, 2011) extended the formulation to partial differential equation systems, and showcased on tubular reactors with single [96] and multiple production lines [97].

This monolithic approach usually generates open loop control trajectories, i.e., no feedback loop is assumed to develop the input and output profiles. However, the processing units are subject to internal process disturbances, and the mismatch between the process and the model leads to deviations in the targeted operations. Zhuge and Ierapetritou (2012) [98] implemented the monolithic approach in closed loop, where the authors initiate a readjustment procedure to solve the integrated problem online if the states deviate from their reference trajectories. This approach does not completely resolve the issue of handling the process disturbances or the process/model mismatch; however, it was shown to mitigate these concerns to a great extent. Gutiérrez-Limón et al. (2014) [99] also implemented a similar closed-loop strategy with a nonlinear model predictive control scheme, while extending the scope of the problem statement to account for an extended horizon production policy. However, both approaches require solving a complex and large-scale MINLP problem at the time steps of the controller, which makes it unsuitable for the processes with fast dynamics.

Low-order representation of fast process dynamics in the scheduling problem has been an effective approach to reduce the computational burden of solving complex optimization problems. Du et al. (2015) [100] proposed a time scale-bridging model that describes the closed-loop input–output behavior of a process in the scheduling formulation, postulated as a MIDO problem. The low-order representation also maintains the stability of the process in the existence of process/model mismatch and handles disturbances. Baldea et al. (2015) [101] extended this approach to MPC governed systems.

Burnak et al. (2018) [102] also addressed the online computational burden of "top-down" approaches by developing a multiparametric programming-based approach, where the authors explicitly mapped (i) the closed-loop dynamic process behavior in a "control-aware" scheduling problem, and (ii) the continuous and binary scheduling level decisions such as the operating level and operational mode of the system in a "schedule-aware" MPC scheme (iii) to yield the optimal operational decisions. The offline nature of the integrated scheduling and control scheme allows for determining the feasible operating space prior to actualizing the operation. Furthermore, reducing the problem complexity from solving online optimization problems to a simple look-up table and affine function evaluation, the framework is well-suited for fast process dynamics. Charitopoulos et al. (2019) [103] employed a similar multiparametric programming approach to include the planning decisions in their framework.

In the "bottom-up" approaches, on the other hand, incorporating the economic objectives in the plant control structures has been perceived as the key for seamless integration of scheduling and control. For this purpose, MPC formulations provide the flexibility to account for a spectrum of objectives in the control level due to their optimization-based structures. Loeblein and Perkins (1999) [104] presented an economic analysis of unconstrained MPC scheme operating under constrained systems. The authors determined the most cost-effective model predictive regulatory control structure by using the back-off approach to satisfy the constraints. Zanin et al. (2002) [105] addressed the discrepancy between the real-time optimization (RTO) and control layers by incorporating the economic optimization problem in the controller and feeding the same piece of information in both layers. The proposed formulation diminishes the discrepancy between the decision layers to yield more economical operations, but the resulting control scheme does not guarantee the stability of the process for the entirety of operations. Rawlings and Amrit (2009) [106] developed asymptotic stability criteria by formulating the so-called "economic MPC" (or EMPC), where the objective function of the MPC is designed to minimize the operational costs instead of maintaining the steady state of the process. This approach aims to replace the conventional two-layer structure with RTO and dynamic regulatory control by a single control layer, where the economic optimization and process regulation are conducted simultaneously. Amrit et al. (2011) [107] further extended the stability criteria by (i) imposing a region constraint on the terminal state instead of a point constraint, and (ii) adding a penalty on the terminal state to the regulator cost.

Similar to the monolithic "top-down" scheduling and control approach, EMPC has been shown to be too complex to be solved in the control time steps. This limitation has led the researchers to develop decomposition algorithms for faster computational times. Würth et al. (2011) [108] proposed a decomposition framework for the single layer dynamic RTO formulation, where the slow trends and process uncertainty is handled in the upper layer, while the lower layer accounts for the fast disturbances acting on the process. Ellis and Christofides (2014) [109] focused on selecting a suitable input configuration for such two-layered dynamic RTO structures such that the asymptotic stability is guaranteed. Jamaludin and Swartz (2017) [110] and Li and Swartz (2019) [111] employed a convex MPC problem in the lower level regulatory control, which enabled its exact substitution with KKT optimality conditions. Simkoff and Baldea (2019) [112] used the same substitution strategy on a production scheduling problem.

Design optimization accounting for the scheduling and control decisions with closed-loop implementation is relatively recent in the literature. Patil et al. (2015) [3] modeled the product transitions in design optimization, while maintaining the stability of the closed-loop system governed by a PI control scheme. The authors formulated an MINLP similar to Equation (6) with the contribution of the criterion, $eig(A_i^z(x_{lin})) < 0$, which enforces the stability of the linearized states for all products i in a multiproduct unit under all critical scenarios z. Due to the linearization of the controllers around the operating point, this approach requires repetitive identification of the states at every optimization iteration. Koller and Ricardez-Sandoval (2017) [4] improved this approach by applying orthogonal collocation on finite elements on the integrated problem, and Koller et al. (2018) [5] employed the back-off method to satisfy the constraints under uncertainty by using Monte Carlo sampling techniques to determine the back-off terms.

Recently, Burnak et al. (2019) [6] introduced a multiparametric programming-based theory and framework for the integration of process design, scheduling, and control. The authors derived offline design dependent control and scheduling schemes that can be incorporated in a MIDO formulation in a multi-level fashion, as presented by Equation (10).

$$
\begin{aligned}
\min_{u,s,des} \quad & \int_0^T C(x(t),y(t),u(t),s(t),des,d(t))dt \\
s.t. \quad & \dot{x}(t) = f(x(t),y(t),u(t),s(t),des,d(t),t) \\
& \underline{y} \leq y(t) = g(x(t),y(t),u(t),s(t),des,d(t),t) \leq \overline{y} \\
& \underline{x} \leq x(t) \leq \overline{x}, \quad \underline{des} \leq des \leq \overline{des}, \quad \underline{d} \leq d(t) \leq \overline{d} \\
& s_t(\theta_s) = \arg\min_s \quad \sum_{t_s \in N_s} C_s(x_{t_s}, y_{t_s}, s_{t_s}, des, d_{t_s}) \\
& \qquad s.t. \quad \underline{x}_{t_s} \leq x_{t_s+1} = A_{t_s}x_{t_s} + B_{t_s}s_{t_s} + C_{t_s}d_{t_s} \leq \overline{x}_{t_s} \\
& \qquad\qquad \underline{y}_{t_s} \leq y_{t_s} = D_{t_s}x_{t_s} + E_{t_s}s_{t_s} + F_{t_s}d_{t_s} \leq \overline{y}_{t_s} \\
& \qquad\qquad \underline{s}_{t_s} \leq s_{t_s} \leq \overline{s}_{t_s}, \quad \underline{d}_{t_s} \leq d_{t_s} \leq \overline{d}_{t_s} \\
& \qquad\qquad \underline{\theta}_s \leq \theta_s = [x_{t_s=0}^T, y_{t_s=0}^T, d_{t_s}, des]^T \leq \overline{\theta}_s \\
& u_t(\theta_c) = \arg\min_c \quad \sum_{t_c \in N_c} C_c(x_{t_c}, y_{t_c}, u_{t_c}, des, d_{t_c}) \\
& \qquad s.t. \quad \underline{x}_{t_c} \leq x_{t_c+1} = A_{t_c}x_{t_c} + B_{t_c}u_{t_c} + C_{t_c}d_{t_c} \leq \overline{x}_{t_c} \\
& \qquad\qquad \underline{y}_{t_c} \leq y_{t_c} = D_{t_c}x_{t_c} + E_{t_c}u_{t_c} + F_{t_c}d_{t_c} \leq \overline{y}_{t_c} \\
& \qquad\qquad \underline{u}_{t_c} \leq u_{t_c} \leq \overline{u}_{t_c}, \quad \underline{d}_{t_c} \leq d_{t_c} \leq \overline{d}_{t_c} \\
& \qquad\qquad \underline{\theta}_c \leq \theta_c = [x_{t_c=0}^T, y_{t_c=0}^T, d_{t_c}, des]^T \leq \overline{\theta}_c
\end{aligned}
\tag{10}
$$

where s and u denote the scheduling and control decisions, respectively. Note that the proposed formulation postulates explicit expressions for the scheduling and control strategies as functions of a set of parameters, θ, which includes the design of the process. The design dependence of the operational strategies allows for their direct integration in the MIDO formulation. The postulated formulation has two main benefits, (i) due to the explicit form of the follower problems, the multi-level MIDO problem is reduced to a single level, and (ii) only the design variables are left as the degrees of freedom of the problem, since the remaining are determined as a function of the design.

5. Current Challenges and Future Directions

The PSE community has achieved unequivocally remarkable progress in realizing and advancing the set goals of Professor Sargent on systematic design optimization in five decades. Today, using design optimization tools to at least some extent has long become the standard practice in many industries. Commercial modeling and simulation software tools such as gPROMS (https://www.psenterprise.com/products/gproms) and Aspen Plus Dynamics (https://www.aspentech.com/en/products/pages/aspen-plus-dynamics) have been featuring robust and efficient solvers for dynamic optimization problems for a few years. Despite these milestones in PSE, we still must make significant assumptions and simplifications regarding the operational decisions in the process design phase, even though the impact of their interdependence on process economics and operability has been articulated in numerous studies. Hence, the academia still needs to mature the theoretical foundations and the applicability of unified design optimization approaches before it gains wide industrial recognition. Here, we discuss some of the bottlenecks and potential directions to improve the state-of-the-art for industrial practice.

5.1. The Need for an Industrial Benchmark Problem

As we have presented in this paper, there is a plethora of proposed modeling techniques and solution approaches for the next generation unified design optimization problems. Therefore, it is clear that we need a generally accepted benchmark problem, preferably in industrial scale, to validate the effectiveness of proposed methodologies. The PSE community has benefited greatly from such standardized problems, such as the famous Tennessee Eastman Process detailed by Downs and Vogel (1993) [113] for process control studies. We believe that a well-defined problem will clarify the objectives in unified design optimization and accelerate the research towards industrial expectations. The problem should describe at least the following.

1. *A high-fidelity model that describes the dynamics of the process.* The model should feature appropriate design variables to exhibit the dynamic consequences of scaling up/down the process. Furthermore, considering the reduction in capital investment that the multipurpose and multiproduct operating units provide, the process should comprise such units to examine the scheduling/design and scheduling/control trade-offs. Recent research that consider process design, scheduling, and closed-loop control problems simultaneously [3,5,6] have studied only a single processing unit, which reflects a limited fraction of the overall benefit that the grand unification can provide.

2. *Cost relations for investment, utility, and raw materials.* A functional form of the investment cost with respect to the capacity of the process is required to have standardized comparable results. Also, utility costs and raw materials may vary significantly, which inevitably impacts the optimal scheduling decisions. For instance, grid electricity costs are known to exhibit considerable differences during the day and night times. Thus, operational loads in energy intensive processes may fluctuate heavily. The impact of such changes in operating levels on design and control decisions were discussed in Section 3.

3. *Product demand and availability of the utility, raw materials, and operating units over a time horizon.* Production allocation and timing is a key aspect of scheduling problem, which are heavily dictated by the product demand and availability of resources. However, it is not a trivial practice to estimate the future of these quantities. Therefore, probability distributions of these components will be beneficial to determine their expected values, while being able to take into account their worst-case scenarios.

5.2. Robust Advanced Control and Scheduling Strategies

Incorporation of advanced control schemes seamlessly in the design optimization problem requires the controller to capture the dynamics of the process for the entire range of design variables. Burnak et al. (2019) [6] attempted to approximately model the design configuration as a right-hand uncertainty in the constraint set, validated by closed-loop simulations and closed-loop MIDO problems. However, the design variables impose uncertainty in the left-hand side of the constraints, as well as the nonlinear and bilinear terms in the objective function. Therefore, robust control strategies need to be developed for accurate predictions of future states in the control level prior to the realization of the design, and to guarantee the stability of the closed-loop operations in simultaneous approaches.

Analogously, scheduling schemes should be robustified in the design optimization to minimize the rescheduling due to unexpected disruptive events, such as unit failure, drastic changes in product demand rate and raw material availability. Excluding these events in the scheduling scheme may result in steep changes in the target operation, and thus unattainable set points for the controller.

5.3. Considering Flowsheet Optimization, Process Intensification, and Modular Design Opportunities

Optimization-based plant design techniques have been used and developed for more than four decades [114,115]. These techniques postulate "superstructures" that systematically simulate and compare every combination of flowsheet possibility to determine the optimal process. More recently, superstructures have been formulated at the phenomena level to capture the fundamental relations between the mass and energy, which in turn yields intensified processes [116–122]. Such intensified processes are expected to deliver significantly increased operational efficiency and decreased unit volumes, making them very attractive options both in academia and industry [123]. This rapidly growing interest in intensified processes is one of the most pronounced directions that the PSE community has been taking. Therefore, studying these intensified processes in the context of unified design optimization will attract a wider audience from the industry. Clearly, modeling the spatial (synthesis/intensification) and temporal (scheduling/control) decisions simultaneously in a single problem formulation will capture even more synergistic interactions, which will increase the process profitability.

Furthermore, the researchers studying process intensification can benefit from the tools and methodologies on unification of design, scheduling, and control. Baldea (2015) [124] reported a theoretical justification for the loss of control degrees of freedom due to process intensification, which poses a significant limitation on intensification activities. Tian and Pistikopoulos (2019) [125] and Dias and Ierapetritou (2019) [126] discuss the limitations on the operability of such intensified systems and potential directions to overcome these limitations in their excellent review papers. The researchers on process intensification technologies can adopt the techniques, ranging from steady-state and dynamic flexibility to integration of scheduling and control decisions, in order to address the operability issues.

5.4. Theoretical and Algorithmic Developments in MIDO

The most limiting bottleneck of the simultaneous approaches is the size of the integrated MIDO problems. The time component of the problem significantly increases the computational complexity, yielding infinitely many NP-hard problems to acquire an optimal solution profile. However, tailored algorithms can be developed by using the special structure of such integrated problems. For instance, the open loop design optimization problem is relatively simpler than the integrated MIDO, and constitutes a lower bound on the optimal solution of the overall problem. Such properties can be exploited in decomposing the MIDO into subproblems to significantly reduce the search space for faster algorithms.

5.5. Software Development

Despite the theoretical and practical advances in the unified design problem among the academia, there is no commercially available platform or a software prototype. Such a tool will make the integrated approaches more accessible to the process designers in industry who are not necessarily experts on process control and scheduling, and it will attract more researchers from different disciplines and backgrounds. Pistikopoulos et al. (2015) [127] introduced the PARametric Optimization & Control (PAROC) framework to design explicit controllers based on high-fidelity models, which can be a viable option to address the grand unification challenge [6,84,102,128,129]. However, it is clear that more progress is needed to engage a wider audience.

Author Contributions: Conceptualization, B.B. and E.N.P.; writing, B.B.; review, N.A.D. and E.N.P.; resources E.N.P.; supervision, E.N.P.

Funding: Financial support from the National Science Foundation (Grant No. 1705423) and Texas A&M Energy Institute, Shell Oil Company, Rapid Advancement in Process Intensification Deployment (RAPID) Institute, and Clean Energy Smart Manufacturing Innovation Institute (CESMII) is greatly acknowledged.

Conflicts of Interest: The authors declare no conflicts of interest.

References

1. Sargent, R.W.H. Integrated design and optimization of processes. *Chem. Eng. Prog.* **1967**, *63*, 71–78.
2. Terrazas-Moreno, S.; Flores-Tlacuahuac, A.; Grossmann, I.E. Simultaneous design, scheduling, and optimal control of a methyl-methacrylate continuous polymerization reactor. *AIChE J.* **2008**, *54*, 3160–3170. [CrossRef]
3. Patil, B.P.; Maia, E.; Ricardez-sandoval, L.A. Integration of Scheduling, Design, and Control of Multiproduct Chemical Processes Under Uncertainty. *AIChE J.* **2015**, *61*. [CrossRef]
4. Koller, R.W.; Ricardez-Sandoval, L.A. A dynamic optimization framework for integration of design, control and scheduling of multi-product chemical processes under disturbance and uncertainty. *Comput. Chem. Eng.* **2017**, *106*, 147–159, doi:10.1016/j.compchemeng.2017.05.007. [CrossRef]
5. Koller, R.W.; Ricardez-Sandoval, L.A.; Biegler, L.T. Stochastic back-off algorithm for simultaneous design, control, and scheduling of multiproduct systems under uncertainty. *AIChE J.* **2018**, *64*, 2379–2389. [CrossRef]
6. Burnak, B.; Diangelakis, N.A.; Katz, J.; Pistikopoulos, E.N. Integrated process design, scheduling, and control using multiparametric programming. *Comput. Chem. Eng.* **2019**, *125*, 164–184. [CrossRef]
7. Takamatsu, T.; Hashimoto, I.; Ohno, H. Optimal Design of a Large Complex System from the Viewpoint of Sensitivity Analysis. *Ind. Eng. Chem. Process Des. Dev.* **1970**, *9*, 368–379. [CrossRef]
8. Nishida, N.; Ichikawa, A.; Tazaki, E. Synthesis of Optimal Process Systems with Uncertainty. *Ind. Eng. Chem. Process Des. Dev.* **1974**, *13*, 209–214. [CrossRef]
9. Grossmann, I.E.; Sargent, R.W.H. Optimum design of chemical plants with uncertain parameters. *AIChE J.* **1978**, *24*, 1021–1028. [CrossRef]
10. Kwak, B.M.; Haug, E.J. Optimum design in the presence of parametric uncertainty. *J. Optim. Theory Appl.* **1976**, *19*, 527–546. [CrossRef]
11. Halemane, K.P.; Grossmann, I.E. Optimal process design under uncertainty. *AIChE J.* **1983**, *29*, 425–433. [CrossRef]
12. Swaney, R.E.; Grossmann, I.E. An index for operational flexibility in chemical process design. Part I: Formulation and theory. *AIChE J.* **1985**, *31*, 621–630. [CrossRef]

13. Swaney, R.E.; Grossmann, I.E. An index for operational flexibility in chemical process design. Part II: Computational algorithms. *AIChE J.* **1985**, *31*, 631–641. [CrossRef]

14. Grossmann, I.E.; Floudas, C.A. Active constraint strategy for flexibility analysis in chemical processes. *Comput. Chem. Eng.* **1987**, *11*, 675–693. [CrossRef]

15. Floudas, C.A.; Grossmann, I.E. Synthesis of flexible heat exchanger networks with uncertain flowrates and temperatures. *Comput. Chem. Eng.* **1987**, *11*, 319–336. [CrossRef]

16. Shimizu, Y. A plain approach for dealing with flexibility problems in linear systems. *Comput. Chem. Eng.* **1989**, *13*, 1189–1191. [CrossRef]

17. Shimizu, Y. Application of flexibility analysis for compromise solution in large-scale linear systems. *J. Chem. Eng. Jpn.* **1989**, *22*, 189–194. [CrossRef]

18. Bansal, V.; Perkins, J.D.; Pistikopoulos, E.N. Flexibility analysis and design of linear systems by parametric programming. *AIChE J.* **2000**, *46*, 335–354. [CrossRef]

19. Pistikopoulos, E.N.; Grossmann, I.E. Optimal retrofit design for improving process flexibility in linear systems. *Comput. Chem. Eng.* **1988**, *12*, 719–731, doi:10.1016/0098-1354(88)80010-3. [CrossRef]

20. Pistikopoulos, E.N.; Grossmann, I.E. Stochastic optimization of flexibility in retrofit design of linear systems. *Comput. Chem. Eng.* **1988**, *12*, 1215–1227. [CrossRef]

21. Pistikopoulos, E.N.; Grossmann, I.E. Optimal retrofit design for improving process flexibility in nonlinear systems—I. Fixed degree of flexibility. *Comput. Chem. Eng.* **1989**, *13*, 1003–1016. [CrossRef]

22. Pistikopoulos, E.N.; Grossmann, I.E. Optimal retrofit design for improving process flexibility in nonlinear systems—II. Optimal level of flexibility. *Comput. Chem. Eng.* **1989**, *13*, 1087–1096. [CrossRef]

23. Pistikopoulos, E.N.; Grossmann, I.E. Evaluation and redesign for improving flexibility in linear systems with infeasible nominal conditions. *Comput. Chem. Eng.* **1988**, *12*, 841–843. [CrossRef]

24. Raspanti, C.; Bandoni, J.; Biegler, L. New strategies for flexibility analysis and design under uncertainty. *Comput. Chem. Eng.* **2000**, *24*, 2193–2209. [CrossRef]

25. Kreisselmeier, G.; Steinhauser, R. Systematic Control Design by Optimizing a Vector Performance Index. In *Computer Aided Design of Control Systems*; Cuenod, M., Ed.; Pergamon: Oxford, UK, 1980; pp. 113–117. [CrossRef]

26. Chen, C.; Mangasarian, O.L. A class of smoothing functions for nonlinear and mixed complementarity problems. *Comput. Optim. Appl.* **1996**, *5*, 97–138. [CrossRef]

27. Pistikopoulos, E.N.; Mazzuchi, T.A. A novel flexibility analysis approach for processes with stochastic parameters. *Comput. Chem. Eng.* **1990**, *14*, 991–1000. [CrossRef]

28. Straub, D.A.; Grossmann, I.E. Integrated stochastic metric of flexibility for systems with discrete state and continuous parameter uncertainties. *Comput. Chem. Eng.* **1990**, *14*, 967–985. [CrossRef]

29. Straub, D.A.; Grossmann, I.E. Design optimization of stochastic flexibility. *Comput. Chem. Eng.* **1993**, *17*, 339–354, doi:10.1016/0098-1354(93)80025-I. [CrossRef]

30. Dimitriadis, V.D.; Pistikopoulos, E.N. Flexibility Analysis of Dynamic Systems. *Ind. Eng. Chem. Res.* **1995**, *34*, 4451–4462. [CrossRef]

31. Zhou, H.; Li, X.; Qian, Y.; Chen, Y.; Kraslawski, A. Optimizing the Initial Conditions to Improve the Dynamic Flexibility of Batch Processes. *Ind. Eng. Chem. Res.* **2009**, *48*, 6321–6326. [CrossRef]

32. Mohideen, M.J.; Perkins, J.D.; Pistikopoulos, E.N. Optimal design of dynamic systems under uncertainty. *AIChE J.* **1996**, *42*, 2251–2272. [CrossRef]

33. Mohideen, M.J.; Perkins, J.D.; Pistikopoulos, E.N. Optimal synthesis and design of dynamic systems under uncertainty. *Comput. Chem. Eng.* **1996**, *20*, S895–S900, doi:10.1016/0098-1354(96)00157-3. [CrossRef]

34. Mohideen, M.J.; Perkins, J.D.; Pistikopoulos, E.N. Robust stability considerations in optimal design of dynamic systems under uncertainty. *J. Process Control* **1997**, *7*, 371–385, doi:10.1016/S0959-1524(97)00014-0. [CrossRef]

35. Pretoro, A.D.; Montastruc, L.; Manenti, F.; Joulia, X. Flexibility analysis of a distillation column: Indexes comparison and economic assessment. *Comput. Chem. Eng.* **2019**, *124*, 93–108. [CrossRef]

36. Zhu, Y.; Legg, S.; Laird, C.D. Optimal design of cryogenic air separation columns under uncertainty. *Comput. Chem. Eng.* **2010**, *34*, 1377–1384.10.1016/j.compchemeng.2010.02.007. [CrossRef]

37. Huang, W.; Li, X.; Yang, S.; Qian, Y. Dynamic flexibility analysis of chemical reaction systems with time delay: Using a modified finite element collocation method. *Chem. Eng. Res. Des.* **2011**, *89*, 1938–1946. [CrossRef]

38. Konukman, A.E.S.; Çamurdan, M.C.; Akman, U. Simultaneous flexibility targeting and synthesis of minimum-utility heat-exchanger networks with superstructure-based MILP formulation. *Chem. Eng. Process. Process Intensif.* **2002**, *41*, 501–518. [CrossRef]

39. Konukman, A.E.S.; Akman, U. Flexibility and operability analysis of a HEN-integrated natural gas expander plant. *Chem. Eng. Sci.* **2005**, *60*, 7057–7074. [CrossRef]

40. Escobar, M.; Trierweiler, J.O.; Grossmann, I.E. Simultaneous synthesis of heat exchanger networks with operability considerations: Flexibility and controllability. *Comput. Chem. Eng.* **2013**, *55*, 158–180. [CrossRef]

41. Varvarezos, D.K.; Grossmann, I.E.; Biegler, L.T. An outer-approximation method for multiperiod design optimization. *Ind. Eng. Chem. Res.* **1992**, *31*, 1466–1477. [CrossRef]

42. Pistikopoulos, E.; Ierapetritou, M. Novel approach for optimal process design under uncertainty. *Comput. Chem. Eng.* **1995**, *19*, 1089–1110. [CrossRef]

43. Morari, M. Design of resilient processing plants—III: A general framework for the assessment of dynamic resilience. *Chem. Eng. Sci.* **1983**, *38*, 1881–1891. [CrossRef]

44. Morari, M. Flexibility and resiliency of process systems. *Comput. Chem. Eng.* **1983**, *7*, 423–437. [CrossRef]

45. Grossmann, I.E.; Morari, M. *Operability, Resiliency, and Flexibility: Process Design Objectives for a Changing World*; Carnegie-Mellon University: Pittsburgh, PA, USA, 1983, doi:10.1184/R1/6467234.v1.

46. Morari, M.; Grimm, W.; Oglesby, M.J.; Prosser, I.D. Design of resilient processing plants—VII. Design of energy management system for unstable reactors—New insights. *Chem. Eng. Sci.* **1985**, *40*, 187–198. [CrossRef]

47. Palazoglu, A.; Manousiouthakis, B.; Arkun, Y. Design of chemical plants with improved dynamic operability in an environment of uncertainty. *Ind. Eng. Chem. Process Des. Dev.* **1985**, *24*, 802–813. [CrossRef]

48. Palazoglu, A.; Arkun, Y. A multiobjective approach to design chemical plants with robust dynamic operability characteristics. *Comput. Chem. Eng.* **1986**, *10*, 567–575. [CrossRef]

49. Skogestad, S.; Morari, M. Design of resilient processing plants-IX. Effect of model uncertainty on dynamic resilience. *Chem. Eng. Sci.* **1987**, *42*, 1765–1780. [CrossRef]

50. Colberg, R.D.; Morari, M.; Townsend, D.W. A Resilience target for heat exchanger network synthesis. *Comput. Chem. Eng.* **1989**, *13*, 821–837. [CrossRef]

51. Perkins, J.D.; Wong, M.P.F. Assessing controllability of chemical plants. *Chem. Eng. Res. Des.* **1985**, *63*, 358–362.

52. Rosenbrock, H.H. *State-Space and Multivariable Theory*; Studies in Dynamical Systems Series; Wiley Interscience Division: Hoboken, NJ, USA, 1970.

53. Psarris, P.; Floudas, C.A. Improving dynamic operability in mimo systems with time delays. *Chem. Eng. Sci.* **1990**, *45*, 3505–3524. [CrossRef]

54. Psarris, P.; Floudas, C.A. Dynamic operability of mimo systems with time delays and transmission zeroes—I. Assessment. *Chem. Eng. Sci.* **1991**, *46*, 2691–2707. [CrossRef]

55. Psarris, P.; Floudas, C.A. Dynamic operability of mimo systems with time delays and transmission zeroes—II. Enhancement. *Chem. Eng. Sci.* **1991**, *46*, 2709–2728. [CrossRef]

56. Barton, G.W.; Chan, W.K.; Perkins, J.D. Interaction between process design and process control: The role of open-loop indicators. *J. Process Control* **1991**, *1*, 161–170. [CrossRef]

57. Narraway, L.T.; Perkins, J.D.; Barton, G.W. Interaction between process design and process control: Economic analysis of process dynamics. *J. Process Control* **1991**, *1*, 243–250. [CrossRef]

58. Narraway, L.; Perkins, J. Selection of process control structure based on economics. *Comput. Chem. Eng.* **1994**, *18*, S511–S515, doi:10.1016/0098-1354(94)80083-9. [CrossRef]

59. Bahri, P.; Bandoni, J.; Barton, G.; Romagnoli, J. Back-off calculations in optimising control: A dynamic approach. *Comput. Chem. Eng.* **1995**, *19*, 699–708. [CrossRef]

60. Walsh, S.; Perkins, J. Integrated Design of Effluent Treatment Systems. *IFAC Proc. Vol.* **1992**, *25*, 107–112, doi:10.1016/S1474-6670(17)54018-5. [CrossRef]

61. Luyben, M.L.; Floudas, C.A. A Multiobjective Optimization Approach for Analyzing the Interaction of Design and Control. *IFAC Proc. Vol.* **1992**, *25*, 101–106, doi:10.1016/S1474-6670(17)54017-3. [CrossRef]

62. Shah, N.; Pantelides, C.C.; Sargent, R.W.H. The Design and Scheduling of Multipurpose Batch Plants. *IFAC Proc. Vol.* **1992**, *25*, 203–208, doi:10.1016/S1474-6670(17)54032-X. [CrossRef]

63. Kondili, E.; Pantelides, C.C.; Sargent, R.W.H. A general algorithm for short-term scheduling of batch operations—I. MILP formulation. *Comput. Chem. Eng.* **1993**, *17*, 211–227. [CrossRef]

64. Thomaidis, T.V.; Pistikopoulos, E.N. Design of Flexible and Reliable Process Systems. *IFAC Proc. Vol.* **1992**, *25*, 235–240, doi:10.1016/S1474-6670(17)54037-9. [CrossRef]

65. Walsh, S.; Perkins, J. Application of integrated process and control system design to waste water neutralisation. *Comput. Chem. Eng.* **1994**, *18*, S183–S187, doi:10.1016/0098-1354(94)80031-6. [CrossRef]

66. Narraway, L.T.; Perkins, J.D. Selection of process control structure based on linear dynamic economics. *Ind. Eng. Chem. Res.* **1993**, *32*, 2681–2692. [CrossRef]

67. Luyben, M.L.; Floudas, C.A. Analyzing the interaction of design and control—1. A multiobjective framework and application to binary distillation synthesis. *Comput. Chem. Eng.* **1994**, *18*, 933–969. [CrossRef]

68. Luyben, M.L.; Fluodas, C.A. Analyzing the interaction of design and control—2. reactor-separator-recycle system. *Comput. Chem. Eng.* **1994**, *18*, 971–993, doi:10.1016/0098-1354(94)85006-2. [CrossRef]

69. Bansal, V.; Perkins, J.D.; Pistikopoulos, E.; Ross, R.; van Schijndel, J.M.G. Simultaneous design and control optimisation under uncertainty. *Comput. Chem. Eng.* **2000**, *24*, 261–266. [CrossRef]

70. Sandoval, L.A.R.; Budman, H.M.; Douglas, P.L. Simultaneous design and control of processes under uncertainty: A robust modelling approach. *J. Process Control* **2008**, *18*, 735–752. [CrossRef]

71. Ricardez-Sandoval, L.A.; Budman, H.M.; Douglas, P.L. Application of Robust Control Tools to the Simultaneous Design and Control of Dynamic Systems. *Ind. Eng. Chem. Res.* **2009**, *48*, 801–813. [CrossRef]

72. Ricardez-Sandoval, L.A.; Budman, H.M.; Douglas, P.L. Simultaneous design and control of chemical processes with application to the Tennessee Eastman process. *J. Process Control* **2009**, *19*, 1377–1391, doi:10.1016/j.jprocont.2009.04.009. [CrossRef]

73. Kookos, I.K.; Perkins, J.D. Control structure selection based on economics: Generalization of the back-off methodology. *AIChE J.* **2016**, *62*, 3056–3064. [CrossRef]

74. Mehta, S.; Ricardez-Sandoval, L.A. Integration of Design and Control of Dynamic Systems under Uncertainty: A New Back-Off Approach. *Ind. Eng. Chem. Res.* **2016**, *55*, 485–498. [CrossRef]

75. Rafiei-Shishavan, M.; Mehta, S.; Ricardez-Sandoval, L.A. Simultaneous design and control under uncertainty: A back-off approach using power series expansions. *Comput. Chem. Eng.* **2017**, *99*, 66–81. [CrossRef]

76. Rafiei, M.; Ricardez-Sandoval, L.A. Stochastic Back-Off Approach for Integration of Design and Control under Uncertainty. *Ind. Eng. Chem. Res.* **2018**, *57*, 4351–4365. [CrossRef]

77. Kookos, I.K.; Perkins, J.D. An Algorithm for Simultaneous Process Design and Control. *Ind. Eng. Chem. Res.* **2001**, *40*, 4079–4088. [CrossRef]

78. Malcolm, A.; Polan, J.; Zhang, L.; Ogunnaike, B.A.; Linninger, A.A. Integrating systems design and control using dynamic flexibility analysis. *AIChE J.* **2007**, *53*, 2048–2061. [CrossRef]

79. Moon, J.; Kim, S.; Linninger, A.A. Integrated design and control under uncertainty: Embedded control optimization for plantwide processes. *Comput. Chem. Eng.* **2011**, *35*, 1718–1724. [CrossRef]

80. Qin, S.; Badgwell, T.A. A survey of industrial model predictive control technology. *Control Eng. Pract.* **2003**, *11*, 733–764. [CrossRef]

81. Brengel, D.D.; Seider, W.D. Coordinated design and control optimization of nonlinear processes. *Comput. Chem. Eng.* **1992**, *16*, 861–886, doi:10.1016/0098-1354(92)80038-B. [CrossRef]

82. Bemporad, A.; Morari, M.; Dua, V.; Pistikopoulos, E.N. The explicit linear quadratic regulator for constrained systems. *Automatica* **2002**, *38*, 3–20. [CrossRef]

83. Sakizlis, V.; Perkins, J.D.; Pistikopoulos, E.N. Parametric Controllers in Simultaneous Process and Control Design Optimization. *Ind. Eng. Chem. Res.* **2003**, *42*, 4545–4563. [CrossRef]

84. Diangelakis, N.A.; Burnak, B.; Katz, J.; Pistikopoulos, E.N. Process design and control optimization: A simultaneous approach by multi-parametric programming. *AIChE J.* **2017**, *63*, 4827–4846. [CrossRef]

85. Diangelakis, N.A.; Pistikopoulos, E.N. A multi-scale energy systems engineering approach to residential combined heat and power systems. *Comput. Chem. Eng.* **2017**, *102*, 128–138. [CrossRef]

86. Sanchez-Sanchez, K.B.; Ricardez-Sandoval, L.A. Simultaneous Design and Control under Uncertainty Using Model Predictive Control. *Ind. Eng. Chem. Res.* **2013**, *52*, 4815–4833. [CrossRef]

87. Birewar, D.B.; Grossmann, I.E. Incorporating scheduling in the optimal design of multiproduct batch plants. *Comput. Chem. Eng.* **1989**, *13*, 141–161, doi:10.1016/0098-1354(89)89014-3. [CrossRef]

88. White, V.; Perkins, J.; Espie, D. Switchability analysis. *Comput. Chem. Eng.* **1996**, *20*, 469–474. [CrossRef]

89. Bhatia, T.; Biegler, L.T. Dynamic Optimization in the Design and Scheduling of Multiproduct Batch Plants. *Ind. Eng. Chem. Res.* **1996**, *35*, 2234–2246. [CrossRef]

90. Bhatia, T.K.; Biegler, L.T. Dynamic Optimization for Batch Design and Scheduling with Process Model Uncertainty. *Ind. Eng. Chem. Res.* **1997**, *36*, 3708–3717. [CrossRef]

91. Baldea, M.; Harjunkoski, I. Integrated production scheduling and process control: A systematic review. *Comput. Chem. Eng.* **2014**, *71*, 377–390. [CrossRef]

92. Mahadevan, R.; Doyle, F.J.; Allcock, A.C. Control-relevant scheduling of polymer grade transitions. *AIChE J.* **2002**, *48*, 1754–1764. [CrossRef]

93. Chatzidoukas, C.; Perkins, J.; Pistikopoulos, E.; Kiparissides, C. Optimal grade transition and selection of closed-loop controllers in a gas-phase olefin polymerization fluidized bed reactor. *Chem. Eng. Sci.* **2003**, *58*, 3643–3658. [CrossRef]

94. Flores-Tlacuahuac, A.; Grossmann, I.E. Simultaneous Cyclic Scheduling and Control of a Multiproduct CSTR. *Ind. Eng. Chem. Res.* **2006**, *45*, 6698–6712. [CrossRef]

95. Terrazas-Moreno, S.; Flores-Tlacuahuac, A.; Grossmann, I.E. Simultaneous cyclic scheduling and optimal control of polymerization reactors. *AIChE J.* **2007**, *53*, 2301–2315. [CrossRef]

96. Flores-Tlacuahuac, A.; Grossmann, I.E. Simultaneous Cyclic Scheduling and Control of Tubular Reactors: Single Production Lines. *Ind. Eng. Chem. Res.* **2010**, *49*, 11453–11463. [CrossRef]

97. Flores-Tlacuahuac, A.; Grossmann, I.E. Simultaneous Cyclic Scheduling and Control of Tubular Reactors: Parallel Production Lines. *Ind. Eng. Chem. Res.* **2011**, *50*, 8086–8096. [CrossRef]

98. Zhuge, J.; Ierapetritou, M.G. Integration of Scheduling and Control with Closed Loop Implementation. *Ind. Eng. Chem. Res.* **2012**, *51*, 8550–8565. [CrossRef]

99. Gutiérrez-Limón, M.A.; Flores-Tlacuahuac, A.; Grossmann, I.E. MINLP Formulation for Simultaneous Planning, Scheduling, and Control of Short-Period Single-Unit Processing Systems. *Ind. Eng. Chem. Res.* **2014**, *53*, 14679–14694. [CrossRef]

100. Du, J.; Park, J.; Harjunkoski, I.; Baldea, M. A time scale-bridging approach for integrating production scheduling and process control. *Comput. Chem. Eng.* **2015**, *79*, 59–69. [CrossRef]

101. Baldea, M.; Du, J.; Park, J.; Harjunkoski, I. Integrated production scheduling and model predictive control of continuous processes. *AIChE J.* **2015**, *61*, 4179–4190. [CrossRef]

102. Burnak, B.; Katz, J.; Diangelakis, N.A.; Pistikopoulos, E.N. Simultaneous Process Scheduling and Control: A Multiparametric Programming-Based Approach. *Ind. Eng. Chem. Res.* **2018**, *57*, 3963–3976. [CrossRef]

103. Charitopoulos, V.M.; Papageorgiou, L.G.; Dua, V. Closed-loop integration of planning, scheduling and multi-parametric nonlinear control. *Comput. Chem. Eng.* **2019**, *122*, 172–192. [CrossRef]

104. Loeblein, C.; Perkins, J.D. Structural design for on-line process optimization: I. Dynamic economics of MPC. *AIChE J.* **1999**, *45*, 1018–1029. [CrossRef]

105. Zanin, A.C.; de Gouvêa, M.T.; Odloak, D. Integrating real-time optimization into the model predictive controller of the FCC system. *Control Eng. Pract.* **2002**, *10*, 819–831. [CrossRef]

106. Rawlings, J.B.; Amrit, R. Optimizing Process Economic Performance Using Model Predictive Control. In *Nonlinear Model Predictive Control: Towards New Challenging Applications*; Magni, L., Raimondo, D.M., Allgöwer, F., Eds.; Springer: Berlin/Heidelberg, Germany, 2009; pp. 119–138, doi:10.1007/978-3-642-01094-1_10.

107. Amrit, R.; Rawlings, J.B.; Angeli, D. Economic optimization using model predictive control with a terminal cost. *Annu. Rev. Control* **2011**, *35*, 178–186. [CrossRef]

108. Würth, L.; Hannemann, R.; Marquardt, W. A two-layer architecture for economically optimal process control and operation. *J. Process Control* **2011**, *21*, 311–321, doi:10.1016/j.jprocont.2010.12.008. [CrossRef]

109. Ellis, M.; Christofides, P.D. Selection of control configurations for economic model predictive control systems. *AIChE J.* **2014**, *60*, 3230–3242. [CrossRef]

110. Jamaludin, M.Z.; Swartz, C.L.E. Dynamic real-time optimization with closed-loop prediction. *AIChE J.* **2017**, *63*, 3896–3911. [CrossRef]

111. Li, H.; Swartz, C.L.E. Dynamic real-time optimization of distributed MPC systems using rigorous closed-loop prediction. *Comput. Chem. Eng.* **2019**, *122*, 356–371, doi:10.1016/j.compchemeng.2018.08.028. [CrossRef]

112. Simkoff, J.M.; Baldea, M. Production scheduling and linear MPC: Complete integration via complementarity conditions. *Comput. Chem. Eng.* **2019**, *125*, 287–305. [CrossRef]

113. Downs, J.; Vogel, E. A plant-wide industrial process control problem. *Comput. Chem. Eng.* **1993**, *17*, 245–255, doi:10.1016/0098-1354(93)80018-I. [CrossRef]

114. Grossmann, I.E.; Sargent, R.W.H. Optimum design of heat exchanger networks. *Comput. Chem. Eng.* **1978**, *2*, 1–7. [CrossRef]

115. Nishida, N.; Stephanopoulos, G.; Westerberg, A.W. A review of process synthesis. *AIChE J.* **1981**, *27*, 321–351. [CrossRef]

116. Papalexandri, K.P.; Pistikopoulos, E.N. Generalized modular representation framework for process synthesis. *AIChE J.* **1996**, *42*, 1010–1032. [CrossRef]

117. Demirel, S.E.; Li, J.; Hasan, M.M.F. Systematic process intensification using building blocks. *Comput. Chem. Eng.* **2017**, *105*, 2–38, doi:10.1016/j.compchemeng.2017.01.044. [CrossRef]

118. Tula, A.K.; Babi, D.K.; Bottlaender, J.; Eden, M.R.; Gani, R. A computer-aided software-tool for sustainable process synthesis-intensification. *Comput. Chem. Eng.* **2017**, *105*, 74–95, doi:10.1016/j.compchemeng.2017.01.001. [CrossRef]

119. da Cruz, F.E.; Manousiouthakis, V.I. Process intensification of reactive separator networks through the IDEAS conceptual framework. *Comput. Chem. Eng.* **2017**, *105*, 39–55, doi:10.1016/j.compchemeng.2016.12.006. [CrossRef]

120. Tian, Y.; Pistikopoulos, E.N. Synthesis of Operable Process Intensification Systems—Steady-State Design with Safety and Operability Considerations. *Ind. Eng. Chem. Res.* **2019**, *58*, 6049–6068. [CrossRef]

121. Demirel, S.E.; Li, J.; Hasan, M.M.F. Systematic process intensification. *Curr. Opin. Chem. Eng.* **2019**. [CrossRef]

122. Demirel, S.E.; Li, J.; Hasan, M.M.F. A General Framework for Process Synthesis, Integration, and Intensification. *Ind. Eng. Chem. Res.* **2019**, *58*, 5950–5967. [CrossRef]

123. Tian, Y.; Demirel, S.E.; Hasan, M.M.F.; Pistikopoulos, E.N. An overview of process systems engineering approaches for process intensification: State of the art. *Chem. Eng. Process. Process Intensif.* **2018**, *133*, 160–210. [CrossRef]

124. Baldea, M. From process integration to process intensification. *Comput. Chem. Eng.* **2015**, *81*, 104–114, doi:10.1016/j.compchemeng.2015.03.011. [CrossRef]

125. Tian, Y.; Pistikopoulos, E.N. Synthesis of operable process intensification systems: Advances and challenges. *Curr. Opin. Chem. Eng.* **2019**. [CrossRef]

126. Dias, L.S.; Ierapetritou, M.G. Optimal operation and control of intensified processes—Challenges and opportunities. *Curr. Opin. Chem. Eng.* **2019**. [CrossRef]

127. Pistikopoulos, E.N.; Diangelakis, N.A.; Oberdieck, R.; Papathanasiou, M.M.; Nascu, I.; Sun, M. PAROC—An integrated framework and software platform for the optimisation and advanced model-based control of process systems. *Chem. Eng. Sci.* **2015**, *136*, 115–138, doi:10.1016/j.ces.2015.02.030. [CrossRef]

128. Diangelakis, N.; Burnak, B.; Pistikopoulos, E. A multi-parametric programming approach for the simultaneous process scheduling and control—Application to a domestic cogeneration unit. In Proceedings of the Chemical Process Control 2017, Tucson, AZ, USA, 8–12 January 2017; pp. 8–12.

129. Pistikopoulos, E.N.; Diangelakis, N.A. Towards the integration of process design, control and scheduling: Are we getting closer? *Comput. Chem. Eng.* **2016**, *91*, 85–92, doi:10.1016/j.compchemeng.2015.11.002. [CrossRef]

 *processes*

Article

Optimization-Based Scheduling for the Process Industries: From Theory to Real-Life Industrial Applications

Georgios P. Georgiadis [1,2], **Apostolos P. Elekidis** [1,2] and **Michael C. Georgiadis** [1,2,*]

1   Department of Chemical Engineering, Aristotle University of Thessaloniki, 54124 Thessaloniki, Greece
2   Chemical Process and Energy Resources Institute (CPERI), Centre for Research and Technology Hellas (CERTH), P.O. Box 60361, 57001 Thessaloniki, Greece
*   Correspondence: mgeorg@auth.gr; Tel.: +30-2310-994184

Received: 27 May 2019; Accepted: 4 July 2019; Published: 10 July 2019

Abstract: Scheduling is a major component for the efficient operation of the process industries. Especially in the current competitive globalized market, scheduling is of vital importance to most industries, since profit margins are miniscule. Prof. Sargent was one of the first to acknowledge this. His breakthrough contributions paved the way to other researchers to develop optimization-based methods that can address a plethora of process scheduling problems. Despite the plethora of works published by the scientific community, the practical implementation of optimization-based scheduling in industrial real-life applications is limited. In most industries, the optimization of production scheduling is seen as an extremely complex task and most schedulers prefer the use of a simulation-based software or manual decision, which result to suboptimal solutions. This work presents a comprehensive review of the theoretical concepts that emerged in the last 30 years. Moreover, an overview of the contributions that address real-life industrial case studies of process scheduling is illustrated. Finally, the major reasons that impede the application of optimization-based scheduling are critically analyzed and possible remedies are discussed.

Keywords: process scheduling; optimization; process system engineering; mixed-integer programming

---

1. Introduction

Scheduling is concerned with the allocation of scarce resources among competing activities over time. It is a decision-making process aiming to optimize one or more objectives by taking into account the processes taking place and their interactions with the environment. Scheduling problems exist in many manufacturing and production systems, in transportation and distribution of people and goods, and in other types of industries. The three elements which need to be mapped out are time, tasks and resources: The time at which the tasks have to be performed needs to be optimized considering the availability and restrictions on the required resources. The resources may include processing, material storage and transportation equipment, manpower, utilities (e.g., steam, electricity), any supplementary equipment and so on. The tasks typically include processing operations (e.g., reaction, separation, blending, packaging) as well as other activities like transportation, cleaning in place, changeovers, etc. Both external and internal elements of the production need to be considered. The external element originates from the need to co-ordinate manufacturing and inventory levels based on a given demand, as well as arrival time of raw materials and even maintenance activities. The internal element considers the execution of tasks in an appropriate sequence and time, while taking into account all external considerations and resource availabilities. Overall, the sequencing and timing of tasks over time and the assignment of appropriate resources to the tasks must be performed in an efficient manner, that will, as far as possible, optimize an objective. Typical objectives include the minimization of cost or maximization of profit, the maximization of throughput, the minimization of tardy jobs, etc.

Flexible multipurpose plants are able to produce a wide range of different products using a variety of production routes. This characteristic makes such plants particularly effective for the manufacture of classes of products that exhibit a large degree of diversity and which are subject to fast-varying demands. Due to their inherent flexibility, the scheduling of such plants is a problem of high complexity. Compared to other parts of the supply chain management (e.g., distribution management and inventory control), the production scheduling is often by far the most computationally demanding part. The most general "multipurpose" plants can be viewed as collections of production resources (e.g., raw materials, processing and storage equipment, utilities, manpower) shared by a number of processing operations manufacturing a number of products over a given time horizon. The process may involve intermediates shared among two or more products, recycles of unconverted material, and multiple routes to the same end product. Single or multiple stage multi-product plants are thus special cases of multipurpose plants.

Roger Sargent was one of the first researchers in chemical engineering who foresaw the value of, and need for optimization in the design, control, and operation of process systems. One of the major steps in his research was in the area of multipurpose batch and continuous process scheduling, where the introduction of the state-task network (STN) concept was a major breakthrough. This generic representation allowed researchers to efficiently address arbitrary configurations of recipe-based batch operations. The major novelty was that equipment was not preassigned, like in previous contributions [1]. The utilization of the STN representations for the production scheduling problem resulted to a discrete time mixed-integer linear programming (MILP) model. Despite its novelty, that paper was rejected twice, since at that time MILP algorithms were considered inefficient and incapable of solving large-scale problems, which of course has changed drastically over the last decades [2]. Furthermore, progress with the STN modelling approach was also due to the improved formulation proposed by Nilay Shah in which big-M constraints were replaced by fewer and tighter sets of constraints [3].

An indicative example of the impact of Sargent's research work in the process scheduling area is his paper, "A General Algorithm for Short-Term Scheduling of Batch-Operations. 1. MILP Formulation" by E. Kondili, C. C. Pantelides, and R. W. H. Sargent, Computers & Chemical Engineering, 17, 211–227 (1993). This is one of the most widely cited contributions in the PSE (Process Systems Engineering) community (over 800 citations in SCOPUS as of April 2019), and has been recognized by researchers all over the world as the major framework for mathematically modeling batch and continuous operations through the state-task-network representation.

The formulation of Kondili et al. [4] relies on binary variables that specify whether a task starts in an equipment at the start of each time period. Other key variables denote the amount of material in each state, and the amount of utility required for processing tasks over each time interval. Equipment and utility usage constraints as well as material balances and capacity constraints are considered in the formulation. A common, discrete time grid is employed to capture all plant resource utilizations in a straightforward manner. This approach was hindered in its ability to handle large problems mainly due to the limitations of discrete-time approaches that require relatively large numbers of grid points, thus resulting to large-sized models.

Inspired by the work of Kondili et al. [1] a number of contributions appeared in the literature utilizing the STN representation and the overall mathematical programming framework to address general classes of batch and continuous process scheduling problems. Shah et al. [3] was able to generate the smallest possible integrality gap for this type of formulation by efficiently modifying the allocation constraints. They additionally proposed a tailored branch-and-bound solution procedure that uses a significantly smaller LP (Linear Programming) relaxation in order to further improve integrality at each node. In the same research, authors addressed the cyclic scheduling problem, where they simultaneously derived optimal schedules as well as the frequency at which they should be repeated [5]. Papageorgiou and Pantelides [6,7] further expanded this work to cover the case of multiple campaigns. Yee and Shah [8,9] also considered variable elimination to improve the performance of general discrete-time scheduling models. More specifically, they recognized that only about 5–15% of the variables are active at the integer solution, and it would be beneficial to identify and eliminate as far as possible inactive variables prior to solving the scheduling problem. To achieve that, they introduce an LP-based heuristic, alongside a flexibility and sequence reduction technique and a formal branch-and-price method.

Pantelides et al. [10] presented an STN-based approach for the scheduling problem of pipeless plants, where material is transferred between processing stations in vessels, thus requiring the simultaneous scheduling of the movement and processing operations. Pantelides [11] criticized the STN, arguing that despite its advantages, it inherently suffers from a number of drawbacks. For example, the fact that each equipment if treated as a distinct entity that results to solution degeneracy in case of multiple equivalent items exist. Therefore, he proposed a differentiated representation, the resource-task network (RTN), which is based on the equable description of all resources [11]. Contrary to the STN representation, where different states are consumed or produced by a task utilizing the equipment and the utilities, in this approach even the items of equipment or the plants' utilities are considered as resources. Production units are assumed to be consumed at the start and produced at the end of a task. Furthermore, different equipment conditions (e.g., "clean" or "dirty") can be treated as separate resources, with different activities (e.g., "processing" or "cleaning") consuming and generating them—this allows for a simple representation of changeovers. Pantelides illustrated that the integrality gap of RTN formulations is never worse than the most efficient form of STN formulation, and the ability to adapt additional problem features in a straightforward way, made it a favorable framework for future research.

The review above has mainly focused on the development of discrete-time models. As pointed out by Schilling [12], while discrete-time models have been capable to handle numerous industrially-relevant problems (see, e.g., [13]), they are characterized by a number of inherent drawbacks:

1. A large number of time periods is required to capture all significant events and extract a high quality solution—this usually results to extremely large models;
2. Operations in which the processing time is dependent on the batch size are difficult to be modelled;
3. The modelling of continuous and semi-continuous operations must be approximately modelled.

In order to address these issues a number of researchers have attempted to develop scheduling models that employ a continuous representation of time. As a result, fewer grid points are required leading to fewer variables and smaller model sizes.

Dimitriadis et al. [14] describe two rolling horizon procedures for medium-term planning and scheduling, based on the more general RTN formulation. They take advantage of the unique properties of Wilkinson et al. [15] and aggregation in this context. In the forwards rolling horizon algorithm, the horizon is divided into two-time blocks. The first is relatively short and modelled in detail, while the second is relatively long and modelled using the aggregate scheduling formulation. The solution of this MILP gives rise to a detailed solution for the first period and an aggregate one for the second. Dimitriadis et al. [14] recognized that, rather than fix all the variables in the first period at the next iteration of the procedure, it makes sense only to fix the complicating integer variables and leave the continuous ones free for further optimization. At the next iteration, there are three time blocks, the first one with fixed integer variables, the second one modelled in detail and the third (the remainder of the horizon) modelled at an aggregate level. The algorithm proceeds until a detailed solution is obtained for the entire horizon.

As noted in the excellent review by Shah [16] a common conclusion in most PSE contributions is that one of the most important advances in the area of process scheduling over the past 25 years has been the increasing usage of rigorous mathematical programming approaches. In addition, the importance of the establishment of frameworks for process scheduling which can be used for the description of a wide variety of processes and for the development of general solution algorithms has been emphasized.

The contributions described above, inspired many researchers from the PSE community to further investigate the production scheduling problem. Numerous novel approaches have been proposed by different research teams, providing novel efficient models and solution techniques. Network representations [17], event-based formulations [18] and precedence-based models [19] have been developed. Furthermore, a high interest has been expressed for real-life industrial study cases and problem specific solutions have been generated for real industrial facilities. Moreover, the ever-increasing computational power, allowed the handling of larger problem instances. However, there is still a significant gap between the academic research and the industrial practice, as only a few contributions have been successfully applied in real-life scheduling problems.

The rest of this paper is organized as follows. In Section 2, a detailed analysis of the theoretical concepts of optimization-based process scheduling is presented, including a classification of the different mathematical models, as well as a characterization of the problems they are able to address. Section 3 illustrates a systematic review on the application of optimization methods in real-life industrial scheduling problems in the process industries. The main modelling features and the industrial case study characteristics are summarized. In Section 4, we highlight the major challenges in applying optimization methods in real industrial problems and discuss potential remedies to close the existing gap between theoretical advancements and practical implementations. Finally, Section 5, draws up the main concluding remarks of this work

2. Theoretical Aspects of Optimization-Based Process Scheduling

As noted by Gabow [20], all scheduling problems are NP-hard, meaning that no known solution algorithms exist that are of polynomial complexity in the problem size. Therefore, the development of efficient optimization-based solution strategies for production scheduling has been a great challenge to the research community. As a result, a significant contribution emerged in the last decades aiming to develop either tailored algorithms for specific problem instances or efficient generic methods.

2.1. Classification of Scheduling Problems

Main goal of all scheduling problems is to propose a schedule that reaches the production targets, while respecting all operational, logistical and technical constraints, and achieves a certain objective, such as the maximization of profit, the minimization of the total cost, earliness and/or tardiness, and production makespan.

The general scheduling problem seeks to optimally answer the following questions (Figure 1):

- What tasks must be executed to satisfy the given demand (batching/lot-sizing)?
- How should the given resources be utilized (task-resource assignment)?
- In what order are batches/lots processed (sequencing and/or timing)?

Note that depending on the specifics of the problem in hand, some of these decisions are not considered in the scheduling level. When developing a model for the optimal scheduling all characteristics of the production must be considered to ensure the feasibility of the proposed schedules. However, the production needs to be portrayed in an abstract way to reduce the computational complexity of the problem. This is even more crucial when dealing with real-life industrial applications, which are typically characterized by complex structures, ever-expanding product portfolios and a huge number of constraints that must be considered.

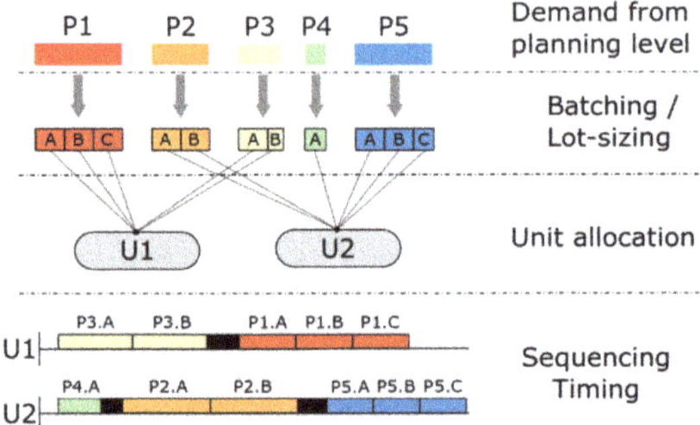

Figure 1. Decisions of production scheduling in the process industries.

Traditionally, scheduling problems are defined in terms of a triplet $\alpha/\beta/\gamma$ [21]. The α field describes the production environment, while the β field denotes the special characteristics and production constraints. Finally, field γ describes the problem's objective, e.g., minimization of cost. The entries of this triplet can be extremely diverse between process industries, since a great variety of aspects needs to be considered when developing optimization models for process scheduling. As a result, many classes of scheduling problems exist. However, the general production scheduling problem can be summarized as follows:

- Facility data; e.g., processing stages and units, storage vessels, processing rates, unit to task compatibility.
- Production targets that need to be satisfied.
- Availability of raw materials and resource limitations; e.g., maintenance of units, availability of utilities.

The first term denotes the characteristics of the facility and can be considered static input to the scheduling problem, since it remains the same for all problem instances of a facility, unless any redesign studies are considered. The remaining terms are inputs from other decision-making processes in the manufacturing environment. Scheduling is not a standalone problem; it is part of the manufacturing supply chain and has strong connections to other planning functions. Production targets and materials availability come from the planning level, while resource availability is an output of the control level, thus there is a significant flow of information from other planning functions to scheduling (Figure 2).

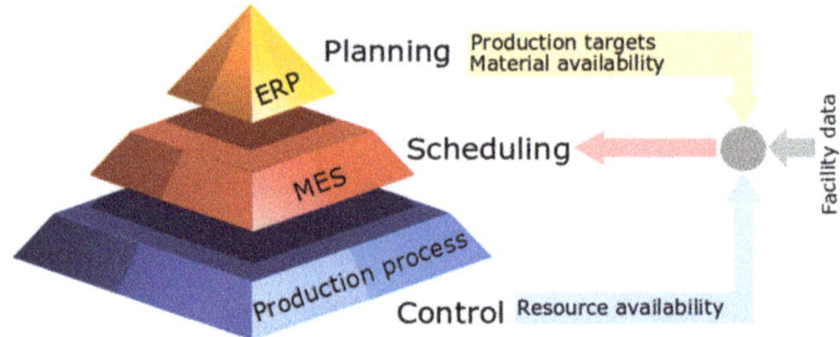

Figure 2. Information flow towards scheduling level.

Scheduling is a critical decision-making process in all process industries, from the chemical and pharmaceutical to the food and beverage and the petrochemical sector. Besides the aforementioned general description of scheduling, industrial applications display strong differences to each other, due to the facility itself, the production policy or market and business considerations. First step when approaching an industrial scheduling problem is to identify its problem specifics, in order to accurately portray the problem in hand. Moreover, a strong correlation between different classes of scheduling problems and the available mathematical modelling frameworks exist. The scheduling problems found in process industries are classified in terms of: (a) The production facility, (b) the interaction with the rest of the production supply chain, and (c) the specific processing characteristics and constraints. A short description of these terms follows, the interested reader can find details in the excellent reviews of Maravelias [22] and Harjunkoski et al. [23].

2.1.1. The Production Facility

At this point we should note that the following analysis focuses on production scheduling. However, many scheduling problems in the process industries target to the optimization of material transfer operations rather than production operations. Characteristic examples are crude oil and pipeline scheduling. With this in mind, the production facility is classified based on the type of process (batch/continuous) and the production environment (sequential or network).

Process Type

The type of production processes found in the process industries can be defined as continuous or batch. In the continuous mode, units are continuously fed and yield constant flow. Continuous processes are appropriate for mass production of similar products, since they can achieve consistency of product quality, while manufacturing costs are reduced, due to economies of scale. The main characteristic of batch processes is that all components are completed at a unit before they continue to the next one. Batch production is advantageous for production of low-volume high-added value products, or for production of seasonal demands which are difficult to forecast. One of the main advantages of batch production is the reduced initial capital investment, therefore, it is especially profitable for small business or trial runs of new facilities. From a scheduling point-of-view, both batch and continuous processes require the same type of decisions. Tasks can be characterized as batches or lots. Assignment (batches/lots to units), sequencing (between batches/lots) and timing (of batches/lots) decisions are identical, while selection and sizing of tasks (batching/lot-sizing) display more degrees of freedom in continuous processes. Capacity restrictions in continuous processes refer to processing rates and processing times and are usually unrestricted, thus a given order can be satisfied in a single lot (campaign) or multiple shorter ones. On the other hand, batch production is capacitated by the amount of processed material that a unit can process, thus affecting the number and size of batches to be scheduled. Another difference lies in the way inventory levels are affected. At this point, it is worth mentioning that many facilities are characterized by more than one type of processes. A characteristic example is the "make-and-pack" type of production, where several batch or continuous processing stages are followed by a packing (continuous) stage. This production flow is very common in the food and beverage and the consumer goods industries and requires the consideration of both the characteristics of batch and continuous production processes [24,25].

Production Environment

Production facilities can be classified as sequential or network based on the material handling restrictions. In sequential processing, each batch/lot follows a sequence of stages based on a specific recipe. Throughout its recipe a batch retains its identity, since it cannot be mixed with other batches or split into multiple downstream batches. Network facilities are characterized as more general and complex and have usually an arbitrary topology. Moreover, no restrictions exist for the handling of input and output materials, thus mixing and splitting operations are included. Based on their topological characteristics, sequential facilities can be further categorized into the following:

- Single stage: Production facility that consists of just one processing stage, which may consist of a single unit or multiple parallel units. The product to unit compatibility may be fixed (batch can be processed in a single unit) or flexible (batch can be processed in multiple units), but in all cases each batch must be processed in a single unit.

- Multistage: Each batch must be processed in more than one processing stages, each consisting of a single unit or multiple parallel units. The multistage environment can be further categorized into multiproduct and multipurpose, depending on the imposed routing restrictions. Multiproduct facilities are equivalent to flowshop environments in discrete manufacturing, where all products go through the same sequence of processing stages. In contrast, a facility is characterized as multipurpose when the routings are product-specific, or when a processing unit belongs to different processing stages depending on the product, thus being equivalent to jobshop environments in discrete manufacturing.

Early studies mainly focused on scheduling problems that are characterized as sequential [26,27]. Process industries with a sequential environment are very similar to discrete manufacturing, from a scheduling point-of-view. Sequential facilities can be easily modelled in terms of batches and production stages, like jobs and operations in discrete manufacturing. However, this does not hold true for network facilities, thus they cannot be modelled in a similar straightforward manner. Kondili et al. [1] followed by Pantelides [11] were the first to propose general representations of network facilities (STN, RTN), allowing the development of optimization models for scheduling problems of such complex structures. A classification of the production environments for process industries is illustrated in Figure 3.

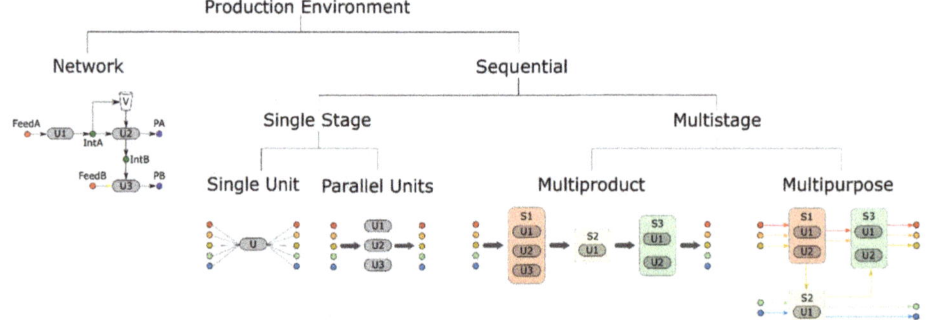

Figure 3. Categorization of scheduling problems based on the production environment.

2.1.2. Interaction with Other Planning Functions

Scheduling is strongly interconnected to the rest of the planning functions of the manufacturing supply chain; therefore, it cannot be approached as a standalone problem. The interactions between scheduling and the other decision making processes in a manufacturing environment must be accounted for, since they determine significant aspects of the scheduling problem; in particular: (a) the input parameters of the scheduling problem, (b) the decisions to be optimized by the scheduler, (c) the type of scheduling problem to be solved and (d) the problem's objective.

Planning and scheduling are two interdependent, however, distinct decision-making processes. Their differences lie in the level of detail of the used models, the time horizon and the problem's objective. In contrast to production scheduling, aggregate models are usually employed in planning, in order to specify the required produced amounts and storage levels that are able to satisfy a given demand at the minimum cost. Moreover, the planning horizon is much larger as it spans from weeks to months. The solution of planning determines the input of the scheduling problem in terms of production targets like order sizes, due dates and release dates. Additionally, batching/lot-sizing decisions can be made in the planning level, thus affecting the type of decisions that needs to be made in the scheduling level. In that case batching/lot-sizing decisions are pre-fixed and the scheduling decisions are narrowed down to just unit to task assignment, sequencing and timing of tasks. There is also an important flow of information between scheduling and control; more specifically, the optimized schedule provides the reference points to the control level while resource availability is in turn provided to the scheduling level. Most studies until the early 2000s, approach production scheduling as a standalone problem. However, the scientific community acknowledged the importance of integrating the decision-making process of the various functions (planning, scheduling and control) that comprise the supply chain of a process industry [28]. The integrated planning and scheduling problem has been studied in multiple works in the last decades [29,30] and also implemented in industrial case studies with great success [31]. In contrast the integrated scheduling and control and integrated planning, scheduling and control problems have been only recently examined [32,33].

The demand volume and variability defined by the market environment in which an enterprise operates plays a pivotal role, since it specifies the type of the scheduling problem to be solved. On the one hand, high-volume production with relative constant demand based on forecasting favors a "make-to-stock" production policy, while the low-volume production with irregular demand follows a "make-to-order" policy. In the former the generated schedule is repeated periodically ("cyclic scheduling"), while in the latter a short-term schedule must be frequently generated. The choice of a meaningful objective for any production scheduling problem is a challenging task due to the numerous competing goals. The production characteristic that usually imposes the objective function is the relation between the capacity of the plant and the demand to be satisfied. In particular, when the demand overcomes the capacity of the plant, then objectives such as, the minimization of backlogs or the maximization of throughput are favored. On the contrary, if the capacity is enough to satisfy the demand, then the minimization of total cost is usually preferred as the overarching production goal. However, the definition of the production scheduling objective also strongly depends on market considerations and goals originating from other planning functions. For example, the maximization of throughput cannot be a valid objective in a production that must be fixed to the amounts defined in the planning level. It must be also noted that production scheduling is an inherently dynamic process, so the objective can be adjusted at any time due to market-related reasons, e.g., new or changed contracts or fluctuations in the demand.

2.1.3. Processing Characteristics and Constraints

Scheduling problems may refer to facilities that exhibit various special processing characteristics and constraints. These aspects complicate the problem but must be considered, in order to ensure the feasibility of the generated production schedules. In the next section we will shortly review some of them and further details can be found in [34].

Resource considerations, aside from task-unit assignments and task-task sequences, are of great importance. These may involve auxiliary units (e.g., storage vessels), utilities (e.g., steam and water) and manpower. Resources are mainly classified into renewable (recover their capacity after being used in a task, e.g., labor) and non-renewable (their capacity is not recovered after being consumed by a task, e.g., raw materials). Renewable resources can be further classified into discrete (e.g., manpower) and continuous (e.g., electricity, cooling water). Another important characteristic in process industries is the handling of storage, which is usually referred to as the storage policy. Depending on the duration a material can be stored, the storage policies are described as (i) unlimited intermediate storage (UIS), (ii) non-intermediate storage (NIS), (iii) finite intermediate storage (FIS) and (iv) zero wait (ZW). Setups are a critical factor in most processing facilities as they represent operations like re-tooling of equipment, cleaning or transitions between steady states. They are associated with a specific downtime that can be sequence-independent or sequence-dependent (changeovers) and a cost is induced to the production process. To reduce the complexity associated with the consideration of setups, products are categorized into families. In that case setups exist only between products of different families.

This classification illustrates the complexity of scheduling problems and the tremendous diversity of aspects that must be accounted for when dealing with real industrial applications (Figure 4). The inherent diversification of scheduling problems in the process industries hindered the initial efforts of the academic community to propose a unified general mathematical framework. Therefore, research turned into the development of less general methods that can address industrial cases that share similar characteristics. As a result, a multitude of efficient specialized methods for the optimization of scheduling in the process industries have been proposed in the last 30 years.

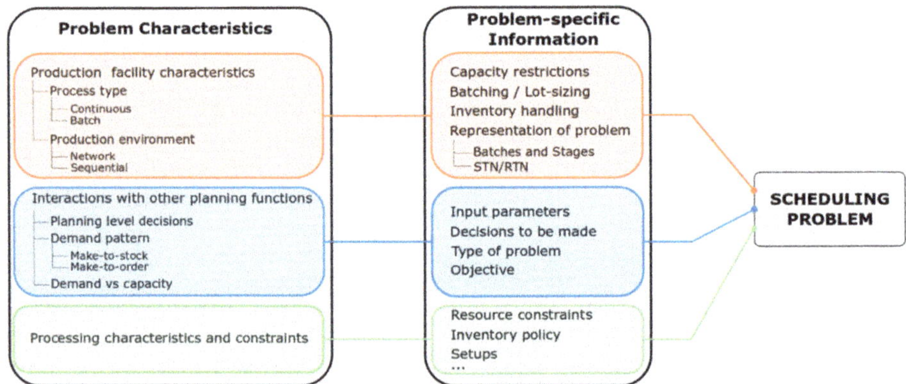

Figure 4. Information extracted from problem characteristics.

2.2. Classification of Modelling Approaches

As mentioned in the previous subsection, scheduling problems in the process industries are defined by extremely diverse features (e.g., production environment, processing characteristics etc.), while different aspects need to be taken into account based on external parameters, like the market environment in which the industry under study operates. Therefore, the initial attempts of proposing a mathematical framework that would constitute a panacea to all scheduling problems, were unsuccessful and soon solutions that take advantage of the problem-specific characteristics emerged. The struggle to overcome the computational complexity associated with scheduling problems, gave rise to numerous scheduling models. It should be noted that in this work we focus on optimization-based approaches, more specifically, the models presented are mixed-integer programming (MIP) models. Nevertheless, we should mention that an abundance of alternative solution approaches, e.g., constraint programming models [35,36], heuristics [37] and metaheuristics [38], exist in the literature. These methods can provide fast and feasible solutions, thus being a very attractive option for industrial case studies. However, their superiority in terms of computational complexity comes with a cost, since optimality of the generated schedules is not ensured. To combine the advantages of both optimization and non-optimization approaches, hybrid methods have emerged that are able to provide near-optimal solutions in low computational time [39].

The three main aspects that describe all optimization models for scheduling are: (i) The optimization decisions to be made, (ii) the modelling elements and (iii) the representation of time (Figure 5).

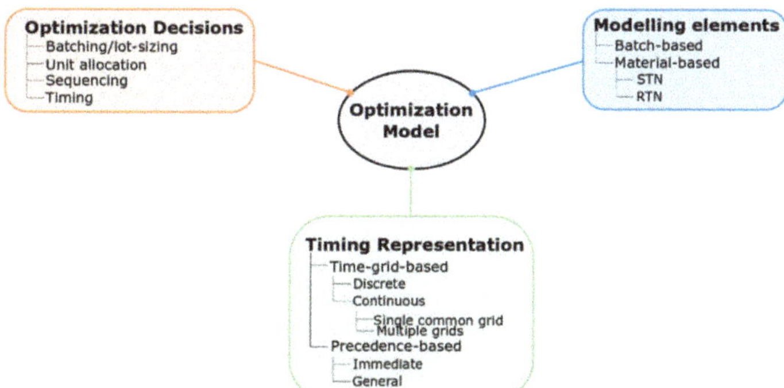

Figure 5. Main aspects of models for optimal production scheduling.

2.2.1. Optimization Decisions

The optimization decisions are affected by the handling of batches/lots. As we underlined in Section 2.1.2, batching decisions may be optimized in the planning level, thus be prefixed and be an input to the scheduling problem. Even if this is not the case, the scheduler has the flexibility to decide whether the batching decisions will be part of the optimization model. For example, the decision-maker can heuristically specify the number and size of batches and then utilize an optimization approach for the unit allocation, sequencing and timing decisions. Usually models for sequential environments favor this two-step approach. In contrast, a monolithic approach, consisting of batching/lot-sizing, unit assignment, sequencing and timing decisions, is used for network environments. Few recent works have proposed a monolithic approach to deal with scheduling problems in sequential environments [40–42]. In some special cases, like in the single machine problems, only sequencing and timing decisions are optimized, thus reducing the scheduling problem to a traditional travelling salesman problem.

2.2.2. Modelling Elements

According to the entity used to ensure the resource constraints on processing units, modelling approaches are classified into two categories: Batch-based and material-based. In sequential environments, where the identity of each batch remains the same throughout the processing stages, batch-based approaches are used. On the contrary a material-based approach is favoured, when dealing with network environments, where batches are mixed or split. It is important to mention that the modelling elements used are tied to the optimization decisions. More specifically, in monolithic approaches the scheduling problems are modelled using a material-based approach, while a batch-based approach is followed, whenever the batching decisions are known a priori.

The modelling elements are strongly tied with the representation of the manufacturing process, which is the core of every scheduling model. The goal of a successful representation is to translate the real problem (orders, units, stages) into mathematical entities (variables, constraints) in an abstract way, that will allow for the fast generation of optimal and feasible schedules. Even a simple manufacturing process may consist of multiple operations, therefore, the use of a simplified representation is essential. The oldest type of manufacturing process representation is utilized to model scheduling problems of sequential production environment and is based on (i) processing stages, (ii) processing units in each stage and (iii) batches or products (depending on whether batching decisions are prefixed or not). The second type of representation emerged in the early 1990s from the novel works of Kondili [1] and Pantelides [11], who introduced the STN and RTN, both based on the modelling of materials, tasks, units and utilities. The STN represents manufacturing processes as a collection of material states (feeds, intermediate final products) that are consumed or produced by tasks. The main difference between STN and RTN is that in the latter states, units and utilities are represented uniformly as resources that are produced and consumed by tasks. While originally introduced for scheduling problems in network environments, recent works have addressed problems in sequential environments [43,44] using the RTN representation.

2.2.3. Time Representations

The most studied topic and the one that mostly differentiates optimization models for scheduling is the representation of time. Depending on the way sequencing and timing of tasks are considered, modelling approaches are categorized in two broad approaches, in particular precedence-based and time-grid-based. Based on their type, precedence-based models are classified into general, immediate and unit-specific general precedence models and time-grid-based into discrete and continuous. Continuous-time formulation may employ single or multiple-time grids. Figure 6 illustrates the various time representation approaches in optimization models for scheduling.

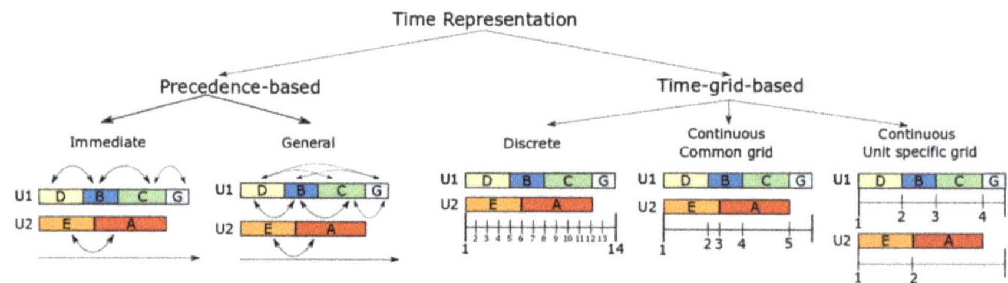

Figure 6. Categorization of modelling approaches based on time representation.

All precedence-based models consist of unit-task allocation and task-task sequencing constraints [45]. The latter are expressed as precedence relationships between tasks processed in the same unit, while the former ensure that each batch/lot is processed by exactly one unit in each stage. Binary sequencing variables are introduced to enforce the precedence relationships and ensure the generation of a feasible schedule (no processing of multiple tasks simultaneously in the same unit). Another main characteristic of any precedence model is that the timing variables are not mapped onto an external time reference, rather their exact values are specified within the scheduling horizon based on the interactions (timing constraints) between pairs of batches/lots or between processing stages of the same batch. Two types of precedence variables exist: (i) General, where precedence relationships are established between all pairs of batches/lots and (ii) immediate, where they are established only between consecutive pairs. General precedence models require fewer variables, so they are more computationally efficient. However, these models do not identify subsequent tasks, making it difficult to consider changeover costs and heuristics, such as pre-fixing or forbidding certain processing sequences. To overcome this limitation, Kopanos et al. [39] proposed the unit-specific general precedence approach that combines both general and immediate sequencing variables. In all cases precedence-based models can provide high quality solutions with low computational cost, thus being an attractive alternative when dealing with real-life industrial problems. One of the main disadvantages of this approach is the quadratic increase of the size of the model with the number of batches/products considered. The use of information such as product families or pre-fixing of sequences mitigates this phenomenon and vastly improves the efficiency of the models [46].

Time-grid-based models enforce timing and sequencing constraints through the utilization of a single or multiple time grids, onto which events (e.g., starting or completion of task) are mapped. A great variety of time-grid-based approaches exist depending on the representation of events (time slots, global periods, time points or events), which are classified into discrete and continuous. In discrete-time models the time-grid is portioned into a pre-fixed number of global time periods of a known duration, both of which need to be specified by the modeler. Most discrete formulations use a common time frame for all shared resources. However, Velez and Maravelias [47] proposed a discrete model that employs multiple time frames. One of the main challenges when setting up discrete models is the proper selection of the number of time periods that needs to be employed. A fine grid results to solutions of higher quality but in cost of larger less computationally efficient models. An advantage of discrete-time models is their capability of monitoring inventory and backlog levels, material balances, as well as the availability and consumption of utilities without introducing nonlinearities. Moreover, time-dependent utility-pricing and holding and backlog costs can be linearly modelled, while integration with higher planning levels is straightforward [48]. Additionally, discrete-time formulations are superior to their continuous counterparts in terms of solution quality [49]. Nevertheless, discrete formulations result in very large, however tight, models, especially when small discretization of time is mandatory. In continuous models, the horizon is subdivided into a fixed number of periods of variable length, which is defined as part of the optimization procedure. Single, common and multiple, unit-specific time frames have all been successfully employed to continuous-time models. Continuous formulations can alleviate some of the computational issues associated with discrete-time models, since fewer time periods, thus variables, are required for the representation of the same scheduling problem. However, they are not necessarily more computationally efficient compared to their discrete counterparts. Finally, it should be mentioned, that few models that utilize multiple ways of representing time have been proposed, thus combining both the advantages of discrete- and continuous-time formulations [29,50].

2.3. Alternative MILP Models for Process Scheduling

We already illustrated a classification of the various scheduling problems as well as the main modelling approaches that have been suggested in the last 30 years. A scheduling model is determined by both externally specified (problem class) and user selected (modelling approach) factors. On the one hand, the model should be suitable for the examined problem environment and the processing specifics of the facility under study, and on the other, it should be developed in terms of the chosen modelling approach's characteristics. A given problem can be represented in multiple ways, however there is a significant relationship between these two aspects. In this subsection we will demonstrate the basic aspects of the mathematical models that have been proposed by the scientific community. More specifically, we present an overview of the models based on the problems they are used for and we analyse the basic constraints and variables of representative models. Further details on the different mathematical models for production scheduling can be found in the excellent review of Méndez et al. [34].

2.3.1. Models for Network Production Environments

In network environments batches do not maintain their identity, since mixing and splitting of batches is allowed. Therefore, the problem is presented utilizing either the STN or the RTN process representation (batch-based approaches). Moreover, the complexity of the production arrangement, with tasks consuming or producing multiple materials and materials being processed in different tasks and units, requires the proper monitoring of material balances, status of units and utility and inventory levels. This necessitates the utilization of a time-grid based approach.

A plethora of modelling formulation emerged after the introduction of the discrete STN and RTN models. Reklaitis and Mockus [51] were the first to propose a continuous-time STN formulation. A single common grid is used, in which the timing of the grid points ("event orders") was determined by the optimization. The model is an MINLP, which can be further simplified to a mixed integer bilinear problem that is solved using an outer-approximation algorithm. Zhang and Sargent [52,53] developed an RTN-based continuous time formulation that can address both batch and continuous operations. The ensued MINLP model is solved using a local linearization procedure in combination with a column generation algorithm.

One of the major drawbacks of the first models developed according to the continuous STN and RTN mathematical frameworks was the large integrality gap. This deficiency was addressed by Schilling and Pantelides [12,54]. They modified the formulation of Zhang and Sargent [53], simplifying it and improving its general solution characteristics, while they developed a hybrid branch-and-bound solution method which branches in the space of the interval durations as well as in the space of the integer variables.

Castro et al. [17] proposed a relaxation of Schilling's formulation [12], allowing tasks to last longer than the actual processing time. Consequently, their model is less degenerate and less CPU time is required. Some of the co-authors further improved this formulation in [55], allowing the optimization of continuous processes. A novel common-grid STN-continuous formulation was introduced by Giannelos and Georgiadis [56]. They utilized a non-uniform time grid, that eliminates any unnecessary time events, thus leading to small MILP models. Maravelias and Grossmann [57] suggested a general continuous STN-model that accounts for various processing characteristics such as different storage policies, shared storage, changeover times and variable batch sizes. The contribution of Sundaramoorthy and Karimi [58] is another well-known continuous MILP model that introduced the idea of several balances (resource, time, masses etc.).

The concept of multiple unit-specific time grids was first proposed by Ierapetritou and Floudas [18]. This approach decouples the task events from the unit events, thus less slots are required. As a result, smaller MILP models are generated, leading to a significant decrease in computational effort. Multiple works have been proposed ever since, improving the computational characteristics and expanding the scope of the initial formulation [59–61].

Velez and Maravelias [47] were the first to introduce the concept of multiple, non-uniform discrete time grids. The multiple grids can be unit-, task- and material-specific. The same authors extended this work in [62] with the consideration of general resources and characteristics like changeovers and intermediate storages. It should be noted that while these formulations were initially proposed for network facilities, they can be also used for the scheduling of sequential environments.

We will now focus our attention on two representative scheduling models for network environments. First, we will consider the continuous common-grid model by Castro et al. [55]. Here an RTN representation is employed, while the model utilizes a common grid to express the timing constraints. More specifically, a set of global time points T is predefined throughout the scheduling horizon. The major decisions are expressed through the binary allocation variable $\overline{N}_{i,t,t'}$ that is enabled whenever a task starts at time point t and is completed at or before point t'. The rest of the decision variables are continuous and express the exact time that corresponds to each time point T_t, the size of a batch/lot of a task $\overline{\xi}_{i,t}$ and the amount of resource being consumed at each time point $R_{r,t}$. The major constraints of the model can be summarized as follows:

$$T_{t'} - T_t \geq \sum_{i \in I_r} (\alpha_i \cdot \overline{N}_{i,t,t'} + \beta_i \cdot \overline{\xi}_{i,t,t'}), \qquad \forall r \in R^J, t \in T, t' \in T, t \leq t' \tag{1}$$

$$T_{t'} - T_t \leq H \cdot \left(1 - \sum_{i \in I_r} \overline{N}_{i,t,t'}\right) + \sum_{i \in I_r} (\alpha_i \cdot \overline{N}_{i,t,t'} + \beta_i \cdot \overline{\xi}_{i,t,t'}), \qquad \forall r \in R^J, t \in T, t' \in T, t \leq t' \tag{2}$$

$$V_i^{\min} \cdot \overline{N}_{i,t,t'} \leq \overline{\xi}_{i,t,t'} \leq V_i^{\max} \cdot \overline{N}_{i,t,t'}, \qquad \forall i \in I, t \in T, t' \in T, t \leq t' \tag{3}$$

$$V_i^{\min} \cdot \overline{N}_{i,t,t+1} \leq \sum_{r \in R_i^{ST}} R_{r,t} \leq V_i^{\max} \cdot \overline{N}_{i,t,t+1}, \qquad \forall i \in I, t \in T \tag{4}$$

$$V_i^{\min} \cdot \overline{N}_{i,t-1,t} \leq \sum_{r \in R_i^{ST}} R_{r,t} \leq V_i^{\max} \cdot \overline{N}_{i,t-1,t}, \qquad \forall i \in I, t \in T \tag{5}$$

$$R_r^{\min} \leq R_{r,t} \leq R_r^{\max} \qquad \forall r \in R, t \in T \tag{6}$$

$$R_{r,t} = R_{r,t-1} + \sum_{i \in I_r} \left[\sum_{t'<t} \left(\mu^p_{r,i} \cdot \overline{N}_{i,t,t'} + v^p_{r,i} \cdot \xi_{i,t,t'} \right) + \sum_{t'>t} \mu^c_{r,i} \cdot \overline{N}_{i,t,t'} + v^c_{r,i} \cdot \overline{\xi}_{i,t,t'} \right] +$$
$$\sum_{i \in I_{ST}} \left(\mu^p_{r,i} \cdot \overline{N}_{i,t-1,t} + \mu^c_{r,i} \cdot \overline{N}_{i,t,t+1} \right), \qquad \forall r \in R,\, t \in T,\, t > 1 \tag{7}$$

Assuming that no more than one task can be executed in each unit at a certain time (unary resource), constraint sets (1) and (2) guarantee that the time difference between any pair of time points t and t' must be at least equal to the processing time of all tasks starting and finishing at those points. Furthermore, the batch/lot size is bounded by the unit capacity (3), while constraints (4) and (5) impose the storage constraints. They ensure that in case of a resource excess at time t, the corresponding storage task has to take place for both $t - 1$ and t. Finally, constraint (7) guarantees that the resource balance considerations are not violated.

Next, we present the model of Janak et al. [60]. In contrast to the previous model, an STN-representation is chosen. Tasks are mapped onto multiple time grids through the concept of event points. These are time instances located along the time axis of each unit that represent the starting of a task. Due to the incorporation of a unit-specific grid, fewer time points are required compared to common-grid formulations, thus the number of binary variables is significantly reduced. The main variables of the model are $W_{i,t}$, $W^s_{i,t}$ and $W^f_{i,t}$ denoting that a task i is active, started or finished at event point t, accordingly. This formulation is one of the most general of the ones that employ a unit-specific grid, since it has the ability to account for various storage policies, batch splitting and mixing, changeovers and variable batch sizes. As a result, it involves a huge number of constraints, hence only the major ones will be presented here.

$$\sum_{i \in I_j} W_{i,t} \leq 1, \qquad \forall j \in J,\, t \in T \tag{8}$$

$$W_{i,t} = \sum_{t' \leq t} W^s_{i,t'} - \sum_{t' \leq t} W^f_{i,t''}, \qquad \forall i \in I,\, t \in T \tag{9}$$

$$\sum_{t \in T} W^s_{i,t} = \sum_{t \in T} W^f_{i,t'}, \qquad \forall i \in I \tag{10}$$

$$B^s_{i,j,t} \leq B_{i,j,t}, \qquad \forall i \in I,\, j \in J_i,\, t \in T \tag{11}$$

$$B^s_{i,j,t} \leq V^{\max}_i \cdot W^s_{i,t}, \qquad \forall i \in I,\, j \in J_i,\, t \in T \tag{12}$$

$$B^s_{i,j,t} \geq B_{i,j,t} - V^{\max}_i \cdot \left(1 - W^s_{i,t}\right), \qquad \forall i \in I,\, j \in J_i,\, t \in T \tag{13}$$

$$S_{s,t} = S_{s,t-1} + \sum_{i \in I^P_s} \rho_{i,s} \cdot B^f_{i,j,t-1} + \sum_{i^{ST} \in I^{ST}_s} B^{ST}_{i^{ST},t-1} - \sum_{i \in I^C_s} \rho_{i,s} \cdot B^f_{i,j,t} + \sum_{i^{ST} \in I^{ST}_s} B^{ST}_{i^{ST},t'}, \qquad \forall s \in S,\, t \in T,\, t > 1 \tag{14}$$

$$T^f_{i,j,t} \geq T^s_{i,j,t}, \qquad \forall i \in I,\, j \in J_i,\, t \in T \tag{15}$$

$$T^f_{i,j,t} \leq T^s_{i,j,t} + H \cdot W_{i,t}, \qquad \forall i \in I,\, j \in J_i,\, t \in T \tag{16}$$

$$T^s_{i,j,t} \geq T^f_{i',j,t-1} + H \cdot (1 - W_{i',t-1}), \qquad \forall j \in J_i,\, i \in I_j,\, i' \in I_j,\, i \neq i',\, t \in T,\, t > 1 \tag{17}$$

$$T^s_{i,j,t} \geq T^f_{i',j',t-1} + H \cdot \left(1 - W^f_{i',t-1}\right), \qquad \forall s \in S,\, i \in I^c_s,\, i' \in I^p_s,\, j \in J,$$
$$j \in J_i,\, j' \in J_{i'},\, j \neq j',\, t \in T,\, t > 1 \tag{18}$$

The major assignment constraints (8)–(10) impose that: (i) At most, one task can be executed by unit j at time t (unary resource), (ii) the assignment variable $W_{i,t}$ will be active only if the task i has started but not finished at or before time t and (iii) each task i must start and finish within the given scheduling horizon. Batch-size considerations are employed by constraints (11)–(13). In particular, they bound the amount of material starting processing at time t, $B^s_{i,j,t}$ according to the unit capacity and relate it to the amount undertaking task i in unit j at time t, $B_{i,j,t}$. Constraint set (14) enforces the material balances, stating that the amount of state s at time t, $S_{s,t}$, is increased by the amount produced and stored at time $t-1$ and decreased by the amount consumed or stored at time t. The timing constraints (15) and (16) relate the starting $T^s_{i,j,t}$ and completion $T^f_{i,j,t}$ times of a task i in unit j and at time t. More specifically, they impose that the completion time must be larger or equal to the starting time, and that if the task i is not processed in unit j at time t, the completion time is set equal to the starting time. Constraint (17) ensures that if task i finishes at time $t-1$ and task i starts at time t in the same unit, then task i must start after the completion of task i'. Finally, let us consider a task i' that produces a state s at time point $t-1$ that is used by task i in time t. To respect the production recipe task i must start after the completion of task i'. This sequencing consideration is enforced by constraint (18).

2.3.2. Models for Sequential Production Environments

Scheduling problems of sequential environments do not share the same complexity, in terms of problem representation, with the ones encountered in network environments. Therefore, both precedence-based and time-grid based approaches can be employed. Each of these approaches display specific advantages and drawbacks. On the one hand, precedence-based models generate smaller, more intuitive models that provide high quality solutions, on the other hand time-grid based models are usually tighter and computationally superior. As a result, a great variety of models have been proposed to address sequential production environments.

One of the most impactful time-grid based models is [63] from Pinto and Grossmann. They described an MILP model for the minimization of earliness of orders for a multiproduct plant with multiple equipment items at each stage. The interesting feature of the model is the representation of time, where two types of individual time grids are used: One for units and one for orders. Castro and Grossmann [64] proposed a non-uniform time grid representation for the scheduling problem of multistage multiproduct plants. They tested their formulation for various objectives, e.g., minimization of makespan, total cost and total earliness and compared it with other known formulations, concluding that the efficiency of a model highly depends on the objective and the problem characteristics. The same authors extended their work in [43] with the consideration of sequence-dependent setup times.

Unlike to most of the other contributions, which propose continuous-time models, the work of Maravelias and co-workers thoroughly investigated the employment of discrete-time models in sequential environments. Sundaramoorthy et al. [65] suggested a discrete time model to incorporate utility constraints for the scheduling problem of multistage batch processes. Merchan et al. [66] developed four novel formulations, two of them based on the STN and RTN representation and two more inspired by the resource-constrained project scheduling problem (RCPSP). Moreover, the authors introduced tightening constraints and reformulations that allowed for significant computational enhancements. Recently, Lee and Maravelias [67] presented two new MIP models for scheduling in multipurpose environments using network representations. Interestingly, states and tasks were defined based on batches instead of materials, making possible the consideration of material handling constraints in sequential production environments. The authors displayed the potential of the proposed models by incorporating important process features, such as time-varying data and limited shared resources, and by solving medium-size problem instances to optimality.

The concept of precedence has been extensively studied by the PSE community [68–70]. Numerous unit-specific immediate [71], immediate [72] and general precedence [19,73] models have been proposed for scheduling problems in sequential environments. In initial studies the batches to be scheduled was a problem data, however later contributions suggested models for the simultaneous batching and scheduling problem [74].

Let us consider the general scheduling problem of a multistage multiproduct facility with multiple units operating in parallel in each stage. Moreover, we assume that the batching decisions are fixed and provided to the scheduler from the planning decision level. This problem can be efficiently tackled by numerous precedence-based models. Here we use the formulation proposed by Méndez et al. [19] and present its core constraints. The main decision variable of all precedence-based models is a Boolean indicating the sequential relation between any pair of orders. More specifically, in the presented formulation $X_{o,o',l}$ defines whether an order o is processed prior to order o' at stage s. Other characteristic decision variables are the binary allocation variable $Y_{o,j}$, defining whether an order o is executed by unit j or not, and $C_{o,l}$ that denotes the completion of order o in each stage. The main constraints of the model are illustrated below:

$$\sum_{j \in J_{ol}} Y_{o,j} = 1, \qquad \forall o \in O, l \in L \tag{19}$$

$$C_{o,l} \geq \sum_{j \in J_{o,l}} \left[Y_{o,j} \cdot \left[pt_{o,j} + su_{o,j} \right] \right], \qquad \forall o \in O, l \in L \tag{20}$$

$$C_{o',l} - pt_{o'j} \geq C_{o,l} + su_{o',j} + \tau_{o,o',j} - M \cdot \left(1 - X_{o,o',l}\right) - M \cdot \left(2 - Y_{o,j} - Y_{o',j}\right),$$
$$\forall o, o' \in O, o' > o, l \in L_{o,o'}, j \in J_{o,o',l} \tag{21}$$

$$C_{o,l} - pt_{o'j} \geq C_{o',l} + su_{o,j} + \tau_{o',i,j} - M \cdot X_{o,o',l} - M \cdot \left(2 - Y_{o,j} - Y_{o',j}\right),$$
$$\forall o, o' \in O, o' > o, l \in L_{o,o'}, j \in J_{o,o',l} \tag{22}$$

$$C_{o,l} \leq C_{o,l+1} - \sum_{j \in J_{o,l+1}} \left(pt_{o,j} \cdot Y_{o,j} \right) \tag{23}$$

Constraint (19) ensures that each order o is processed by exactly one unit j in each stage l. The main timing considerations are specified by constraint (20), which enforces the completion time of an order o executed by unit j to be at least equal to the required processing and setup time. Big-M parameters are employed to express the sequencing constraints (21) and (22), between any pair of orders o in each stage l. A major difference to immediate precedence models, is that here only one sequencing variable is defined for every pair of orders, as a result the size of the model is significantly reduced. Moreover, both constraints become active only when both orders are processed by the same unit, i.e., $Y_{o,j} = Y_{o',j} = 1$, therefore the unit index is omitted from the precedence variables. If order o is processed before order o' in the same unit, constraint (21) becomes active, ensuring that order o' will be completed after the completion of order o plus the required processing time of o' and any sequence-dependent or -independent setup times, while constraint (22) becomes redundant. In the opposite case where order o is processed earlier than o', constraint (22) is activated and (21) becomes redundant. Finally, constraint (23) guarantees the correct sequence between processing stages for the same order.

At this point we should emphasize that no modelling approach exists that is computationally superior to the others in every type of scheduling problem. While discrete-time approaches generate tighter models, their continuous-time counterparts (precedence-based, continuous time-grid-based) require less variables, thus generating smaller-sized models. Extensive comparative studies on scheduling problems in sequential environments conclude that time-grid-based models tend to be generally superior to precedence-based ones [43,64]. However, we must note that the computational efficiency of a model can drastically change even with small alterations in the facility characteristics and the final objective. This will be more evident in our analysis in Section 3, which accentuates the case-specific nature of the problem. Finally, consider that most modelling developments have been tested in small or medium sized study cases, that usually do not represent real-life industrial scheduling problems. Consequently, the computational efficient of any optimization-based model itself is not sufficient enough to address large-sized industrial problems. Thus, as we present in the following section, the introduction of techniques, such as heuristics and decomposition algorithms, is inevitable.

3. Real-Life Process Systems Industrial Applications

As described in the previous section, a plethora of different mathematical models has been proposed for tackling the production scheduling problem. Except from solving literature problem examples, several researchers, mainly from the PSE community, expressed a high interest for handling real-life industrial case studies. Numerous modelling approaches and methods can be found in the open literature, addressing a great variety of industrial process scheduling problems. A categorization based on the industrial sectors, such as chemical, pharmaceutical, petrochemical, steel, food and consumer goods industries, is presented below, along with the proposed modeling approaches. We focus our attention on MILP-based approaches for the offline scheduling problem, excluding other solution methods (e.g., heuristic rules, metaheuristic algorithms etc.).

3.1. Chemical Industries

One of the main industrial sectors widely studied, considers chemical plants, where a variety of new products is produced via the chemical transformation of multiple raw materials. The use of mixed batch and continuous processes, the special equipment technologies and the necessity to achieve a specific quality of products are the main challenges in these problems. In chemical plants, various types of products can be manufactured via the same or a similar sequence of operations by sharing the several plant's production units, intermediate materials, and other production resources. Lin and Floudas [75] proposed a continuous time, event-based MILP scheduling model and a decomposition methodology, to solve large-scale industrial cases of multiproduct batch plants. A real-life study case of a chemical plant, including 3 stages, 35 final products and 10 pieces of equipment is considered. To systematically apply the proposed approach, a graphical user interface is developed. Depending on each problem instance, the computational time of the proposed approach ranges from 15 min to 7 h. Janak et al. [76] extended the previous approach, by adapting intermediate due dates and other technical constraints. A unit specific, event-based formulation is applied in parallel with a decomposition-based approach, utilizing the rolling horizon technique. Problem instances with up to 67 product orders have been considered and a termination criterion of 3 h CPU time has been used. Westerlund et al. [77] introduced a mixed discrete-continuous time formulation to tackle short-term and periodic scheduling problems of multi-product plants, including intermediate storage constraints. As the suggested approach is focused on industrial applications, good quality solutions are targeted in reasonable computational times instead of global optimal solutions. The mixed discrete-continuous model provides better solutions in smaller computational times, in comparison with the discrete-time approach. A strategic planning tool was developed based on the proposed model and applied to an industrial plant, importing demand data from the plant's ERP (Enterprise Resource Planning) system. Additionally, four scheduling approaches have been developed by Velez et al. [78]. Here the idea of multiple discrete-time grids is utilized, as each material, task and unit has its own time grid. Upper bounds on the total production of each material are defined using the concept of the effective time window for the executed tasks. Further extensions are adapted in order to solve a variety of different problems. The introduced methods have been applied to benchmark problem instances that can be found in literature [79] and to a real case study from Dow company [80], including five main product lines. Near optimal solutions are achieved in 1 h CPU time on average. A comparison of the proposed approaches and four other continuous time formulations has been also presented. The results indicate that the discrete time models generate better solutions in less computational time.

3.2. Pharmaceutical Industries

A special subsector of the chemical plants is the pharmaceutical industry. The majority of the operations taking place in these facilities are batch, as there is a high necessity to ensure the quality of the final products. Moniz et al. [81] motivated by a real-world scheduling problem of a chemical-pharmaceutical industry, developed a case-specific discrete-time MILP scheduling model, for batch plants. All the data used by the mathematical formulation, is taken automatically from a decision-making tool and a process representation, developed as a prototype in Microsoft Visio. A representative industrial case, including four products, nine shared processing units and 40 tasks, has been studied. The solutions can be generated in acceptable computational times according to the plant operators and even for larger problem instances, suboptimal but good quality solutions are provided in 1 h CPU time. Stefansson et al. [82] studied a large-scale industrial case study from a pharmaceutical company, including even 73 products and 35 product families. Mathematical frameworks based both on discrete and continuous time representations have been proposed and a comparison of them is illustrated. The initial problem is decomposed into two subproblems and the stage which constitutes the main production bottleneck is scheduled first. The continuous-time formulation can provide better solutions even for larger problem instances. Case studies with up to 400 orders can be solved by utilizing the continuous time formulation and schedules with 9.8% integrality gap are generated in 1408 min. Optimal schedules for smaller case studies, involving up to 150 product orders, can be generated in less than 1 h. On the other hand, only up to 75 products can be scheduled to optimality by utilizing the discrete time formulation, as suboptimal solutions with 10% integrality gap are generated for instances with up to 300 products. Castro et al. [83], presented a decomposition-based algorithm for tackling the high complexity of large-scale problems of multiproduct facilities. The production orders are inserted iteratively into the generated schedule, allowing some flexibility to provide better solutions. A case study comprising of 50 production orders, 17 units and six stages is efficiently solved in less than 1 min. The same pharmaceutical study case has been also considered by Kopanos et al. [39]. They proposed a decomposition-based solution strategy relying on two precedence-based MILP models in order to optimize different objectives, such as makespan, changeover-time and cost minimization. A feasible schedule is rapidly generated, and it is enhanced by applying an improvement algorithm. High quality solutions are provided for industrial cases with up to 60 products allocated to 17 units. Liu et al. [84] focused on the production and maintenance planning problem of biopharmaceutical process, consist of a fermentation and a purification stage. Maintenance activities related to the regeneration of the column resin, taken place in the purification stage, are considered. Two industrial indicative problem instances are illustrated to assess the applicability of the proposed MILP model and global optimal solutions are found, without exceeding the time limitation of 1 h CPU time. An event-based continuous time mathematical framework based on the STN representation has also been proposed for a general multiperiod biopharmaceutical scheduling problem [85]. Optimal solutions can be found in computational time in the range of 1–2 min.

3.3. Petrochemical Industries

A special interest is expressed for the scheduling problem of oil refineries or petroleum industries. A variety of products are produced by this specific industrial sector, such as gasoline, diesel jet fuel and others. Many different and complex processes are taken place in the oil refineries; therefore, their efficient scheduling constitutes a great challenge. Shah et al. [86] motivated by a study case provided by Honeywell Process Solutions (HPS), considered an MILP based heuristic algorithm. The initial oil refinery problem is spatially decomposed into two subproblems, one considering the production and blending and the other the delivery of the finished products. Feasible solutions are generated by solving the two subproblems iteratively, via a six-step heuristic algorithm as the resolution of the direct proposed MILP model is characterized by a high computational cost. Ten different problem instances were presented, for the production of diesel and jet fuel and nearly optimal solutions were generated within less computational time. In particular, the computational time ranged from 2 s to 1 h depending on the cases' complexity. Zhang and Hua [87] deployed a plant-wide multi-period planning model, aiming to the integration of the plant processes and the utility system, in order to reduce the energy consumption. The plant-wide model is extended by considering the utility system model and constraints referring to the utilities' balances such as steam, fuel oil and gas are adapted. The maximization of the total profit of the whole refinery plant is considered as the objective.

The optimal operation modes of units and stream flow are defined by the model. Product blending and maintenance activities are taken into account. As the process system and the utility system are optimized separately by the suggested hierarchical method and the subproblems' complexity decreased, good quality, but not global optimal solutions are generated in acceptable computational times. The applicability of the approach was illustrated in a real study case that considers a refinery industry, located in South China. The refinery, except from importing of electricity to cover its power needs, was also able to export the surplus power back to the network or other power companies. The integrated problem of investment planning and operation scheduling of offshore oil facilities was also addressed, by utilizing a multiperiod MILP model in order to maximize the general profit [88]. Various operational nonlinear constraints related to the reservoir performance and to other resources are efficiently adapted into the proposed model which was solved by utilizing a decomposition algorithm in order to handle the high complexity. A real-life, large-scale illustrative example was considered. Although a feasible solution can't be returned by solving the exact MILP model, a feasible solution within 6600 s was obtained, by utilizing the proposed decomposition algorithm. The operation scheduling of a crude oil terminal has been considered by Assis et al. [89]. A real-life case study, oriented by the national refinery of Uruguay was considered and near optimal solutions were obtained by using an iterative two step MILP-NLP algorithm within a time limit of 3600 s. A domain reduction relaxation was also adapted for handling the emerging bilinear terms.

Other than the processes taking place in the refinery industries, a special interest has been also expressed from the PSE community, in the scheduling of liquid transportation via pipeline systems in petroleum supply chain. The crude oil is gathered and transported to the refineries, as the final refined products are sent to the retail market and distributed to customers. In order to reduce the transportation time, pipelines are preferred instead of using trucks or other means of transport, providing also more safety and lower CO_2 emissions. Castro and Mostafaei [90] motivated by the scheduling problem of liquid transportation, proposed an event-point MILP formulation for treelike transportation systems, where a single input node leads to multiple outputs. A continuous time representation was utilized and novel constraints for ensuring the avoidance of forbidden product sequences were adapted. A real-life study case from the Iranian Oil Pipelines and Telecommunication Company network was considered and the optimal schedules could lead to even a 6.2% capacity increase, as the given demand can be efficiently covered fourteen hours earlier. A comparison with previous methods, proposed from one of the co-authors [91], indicates the efficiency of the approach. A time termination criterion of 5 h has been used for the proposed formulation. The number of the event points has been identified as key parameter with high impact on the computational time and solution quality. Nearly optimal solutions can be generated in less than 1 h by reducing the available event points. A similar problem, referring to the scheduling of a transportation system of petroleum products, produced from a single oil refinery industry was tackled by Cafaro and Cerdá [92]. They proposed an MILP continuous time model in order to define the optimal lot size, the batch sequence, as well as the delivery time of batch order. A variety of constraints were taken into account, such as tank availability and quality control operations. A real-life study case consisting of six different oil derivatives produced by a unique oil refinery to a single distribution center was scheduled, and the results indicated that better solutions were produced in comparison with other approaches for the same problem, in less than 60 s CPU time. The same problem was also addressed by Cafaro et al. [93], but now allowing simultaneous product deliveries, thus providing more realistic solutions. The proposed two-level MILP-based solution technique aimed for the minimization of the total number of operations in order to reduce the number of restarts and stoppages of the pipeline. On the upper level, the feasibility of the problem was ensured, as more detailed decisions such as lot sizing, lot sequencing and timing decisions were defined on the lower level. A study case related to REPLAN refinery industry, consisting of five distribution centers at Brazil, is used to illustrate the applicability of the model. Significant savings were noticed in CPU time using the multiple delivery policy, as the illustrative examples under consideration can be solved in less than 125 s CPU time. Rejowski and Pinto [94], inspired also from the REPLAN refinery, proposed two discrete-time, MILP-based models, to solve a real-life problem, including the distribution of various petroleum products to five depots. An indicative instance of a 75-h time horizon that is discretized in 5-h intervals was presented. A good quality solution with integrality gap of 5.8% was returned within the time limit of 10,000 CPU s. Boschetto et al. [95], proposed an MILP-based solution algorithm for solving a large scale real-life pipeline network problem, by determining the delivery and the pumping times of 14 different oil products and ethanol, to a number of distribution centers. Efficient heuristic rules were utilized in parallel with a continuous time representation, for tackling the daily scheduling problem, in reasonable computational times within 3–5 min. The generated solution for various studied cases, have been also validated by the planners.

3.4. Food Industries

The PSE community has also shown significant interest for the scheduling of food industries. Common characteristics of food processing industrial facilities, such as intermediate due dates, shelf life considerations and multiple mixed batch and continuous processing stages, substantially complicate the optimization of scheduling decisions. The above combined with market trends that enforce the gradual increase of the product portfolio, the demand profile (high variability-low volumes), and the multiple identical machines and shared resources, make the consideration of real-life industrial cases extremely challenging.

As the food industry focuses mainly on the production of perishable final products a make-to-stock production policy is not efficient, as the generation of high inventory levels should be avoided. A plethora of industrial study cases have been considered from various subsectors of the food industry. Baldo et al. [96] motivated by a real study case from a Portuguese brewery industry, proposed a novel MILP-based relax and fix heuristic algorithm, for the integrated fermentation and packing problem. The time horizon is discretized in two subperiods. The first subperiod is scheduled in detail, as for the second subperiod only the main planning decisions, such as the inventory levels, are optimized. Small and big sized problem instances have been considered, with five filling lines and up to 40 products. Although a direct comparison with the company plan was not possible, good quality schedules were generated, using a termination criterion of runtime limit equal to 3600 s or 7200 s. An immediate precedence-based MILP formulation for the packing stage of a brewery company was developed using a mixed discrete-continuous time representation in [29]. The scheduling decisions were defined in a continuous manner, while material balances were expressed at each discrete time period to ensure the generation of feasible schedules. The idea of grouping the products into product families leads to significant reduction of the computational cost. Changeover times among sequential time periods were also taken into account. The industrial study case under consideration consists of eight processing units and 162 products are produced in total, which are grouped into 22 product families. The generated solutions were better than the ones extracted by commercial tools. An upper bound of 300 s CPU time was utilized, for all cases under consideration. Abakarov and Simpson [97] investigated the scheduling problem of food canneries focusing on the sterilization stage and allowing the possibility of the simultaneous sterilization of different products in the same retort. A graphic user interface, able to identify the nondominated simultaneous sterilization vectors, was connected to the proposed MILP model. Different cases were solved, including 16 products with randomly generated product demand values, depicting a reduction of up to 25% in total plant operation time. The usage of COIN-OR as software tool can decrease the model's computational time to 7.38 s. Georgiadis et al. [98] studied the integrated sterilization and packing stage scheduling problem in a large-scale canned fish Spanish industry. An MILP based decomposition algorithm was utilized to tackle the high computational cost, as the products are inserted in an iterative way until the final schedule was generated. A general precedence model efficiently describes the batch (sterilization) and the continuous (packing) processes of the plant. Nearly optimal schedules of a large-scale problem instance, with 100 final products and 362 product batches, have been generated for both stages, in less than 20 min. A study case of a real-world edible-oil deodorized industry was studied by Liu et al. [99]. The plant was described as a single-stage multiproduct batch process. The final products were grouped into product families having the same due date. The proposed approaches relied on mixed discrete and continuous MILP mathematical formulations and classic TSP (travelling salesman problem) constraints. A real study case of 128 hours' time horizon of interest was studied. 70 product orders of 30 different final products of seven groups of different release time were scheduled. The new formulations are shown to be more efficient than previously proposed methods found in the literature. Solutions with approximately 2% integrality gap can be generated in 20 CPU s without allowing the backlog generation and 1075 CPU s by allowing the possibility of backlogs. Polon et al. [100] studied a sausage production industry aiming to the profit maximization by solving an MILP scheduling model for batch processes. The packaging stage, which often constitutes the main production bottleneck has not been considered. The plant operates in a single campaign mode and eight products are produced in total.

A special subsector of food industries is dairy manufacturing. Numerous products are produced, such as yoghurt, cheese and butter and distributed to customers worldwide. Doganis and Sarimveis [101] solved the scheduling problem of a single yoghurt production line taking into account inventory, manpower and capacity restrictions. The model was tested using data from a yoghurt production line of a Greek dairy industry, where 18 products were produced and global optimum schedules have been generated in less than 15 s. The integrated planning and scheduling problem of a small size Balkan type semi-continuous yoghurt facility, with eight final product types, produced by three intermediates has also been investigated [102]. The evaluation of the proposed MILP approach has been utilized via a simulation model. Thirty-two different scenarios were considered and a significant decrease in the total waste and makespan was achieved in approximately 1 h of CPU time. Touil et al. [103] deployed an MILP model for a small multiproduct milk industry, located in Morocco, aiming at the minimization of makespan. The stages of homogenization, pasteurization and packaging were scheduled for four final products, seven packing lines, two pasteurization units and one homogenizer. Efficient solutions were illustrated for the cases under consideration, as optimal schedules can be found in 2 min CPU time. A novel mixed discrete-continuous MILP formulation was deployed by Kopanos et al. [104] for the scheduling problem of a Greek yoghurt production facility. The idea of "product families" was adapted similarly to the other aforementioned works from the same authors. The packing stage was scheduled in detail, but mass balance constraints related to the production stage were also adapted, using a discrete time representation. Ninety-three final products (grouped into 23 product families) were allocated in four packing lines. Novel resource constraints can adapt realistic limitations to various types of resources (e.g., manpower) and ensure the generation of feasible optimal solutions in less than 10 min, depending on the case complexity. Based on a similar approach, the scheduling problem of another large-scale Greek dairy industry has been studied [105]. A rolling horizon technique was embedded to reactively adjust the schedule in case of disturbances, like the cancellation or modification of orders, or the sudden arrival of new orders or any digressions from the planned production. One hundred and fifty-eight final products (grouped into 44 product families) were allocated to six parallel packing lines, while the time horizon of interest was five days. A total cost decrease of 20% was achieved in comparison with the schedules generated by the company. An integrated software tool with a user-friendly graphical interface has been developed to connect the proposed MILP model to the input data, located in excel files (parameter values such as changeover times etc.) and the ERP system (providing the demand values). As a result, optimal solutions can be generated automatically in less than 10 min.

3.5. Consumer Goods Industries

Consumer goods, or final goods, are described as products consumed by the average customer. Depending on the shelf-life duration, they can be further categorized to durable goods (such as detergents) and nondurable (e.g., beverages). One of the main consumer goods group is the fast-moving consumer goods (FMCG), which are characterized by frequent purchases, rapid consumption and low prices. Elzakker et al. [106], presented a problem-specific model for the short-term scheduling problem, considering a fast-moving consumer goods (FMCG) industry. An algorithm based on a unit-specific, continuous time interval MILP model is proposed. Dedicated time intervals to specific product types are adapted to decrease the computational time. In order to assess the efficiency and the applicability of the proposed formulation ten industrial case studies are considered, as provided by Unilever, related to an ice cream production process. Optimal schedules have been generated for problem instances of up to 73 batches of eight products allocated to six storage tanks and two packing lines within 170 s. The time-horizon under consideration was 120 h. The production scheduling problem of an ice cream facility has also been tackled by Kopanos et al. [107]. A real-life study case of eight final ice cream products, two packing lines and six aging vessels is addressed. The simultaneous optimization of all processing stages is achieved, and 50 problem instances are optimally solved. An MILP-based decomposition strategy is proposed to handle scheduling problems of large-scale food process industries. High quality solutions were generated for larger cases of up to 24 final products utilizing the proposed decomposition technique. Industry related needs imposed the adaptation of a 600 CPU s as a time limit, but global optimal solutions can be found in less than 10 s for smaller problem instances.

Giannelos and Georgiadis [108] developed an MILP model to address the scheduling problem in fast consumer goods manufacturing processes. The proposed MILP model relied on the STN process representation and a continuous time formulation was used to reduce the computational complexity of the problem. The formulation was tested on a medium-sized industrial consumer goods manufacturing process, considering cases with up to 35 final products and 5 packing lines. Feasible schedules were generated within a 5–10% integrality gap in computational times, smaller than 5 min. Méndez and Cerdá [109] proposed a general precedence MILP formulation based on a continuous time representation, while they introduced constraints related to sequence dependent setup times and products' due dates. Furthermore, efficient preordering rules were considered in order to provide solutions for industrial study cases of up to 18 final products, produced from five intermediates over a scheduling period of five days. The CPU time needed was gradually depleted to less than 10 s for small problem instances and even to 10 min to larger ones, by applying the proposed approaches and the suggested preordering rules. Baumann and Trautmann [110] proposed a hybrid method for large-scale, short-term scheduling problems of packed consumer goods products. A subset of the operations were scheduled iteratively by solving a general precedence MILP model [24]. Also, an iterative improvement step is applied to the initial schedule, by following a reinsertion policy identifying some critical operations. Ten large-sized instances provided by The Procter and Gamble Company that consisted of up to 1391 operations have been solved within reasonable CPU times of less than 1 h, as a 5 s time limit has been set for each iteration [111]. Smaller-scale problem instances with known optimal solutions, have also been considered, and optimal or near-optimal schedules were generated by applying the aforementioned hybrid method. Elekidis et al. [112] investigated the short-term scheduling problem of a large-scale consumer's goods industry. An immediate-general precedence-based model was illustrated, focusing mainly on the packing stage. Constraints related to the previous stages were also taken into account. The production orders are inserted iteratively, utilizing a decomposition algorithm. Various real-life study cases have been considered that include up to six packing lines and 130 final products. Near optimal schedules are generated and significant savings in the changeover time are noticed within a CPU time of 10 min.

Georgiadis et al. [113] presented two different scheduling approaches, based on the RTN and the STN representations respectively. The work was focused on the scheduling problem of large-scale manufacturing industries of electrical appliances. A case study provided from a large manufacturing company located in Greece was used to assess the applicability of the proposed approaches. The generated schedules can be visualized via the Microsoft Excel application. A significant decrease in the operational cost was reported in a variety of problem instances. Although, the necessary computational time was in the range of some seconds, this could differ as the considered problem instances are described as data-driven.

3.6. Steel Plants

Another important field of interest is the steel-making process industries. Various challenges arise, due to the large variety of final products, the complex process that take place and the volatile electricity prices. The steel production is often divided into three stages: Molten steel is produced first (melt shop) and then the produced slabs are transformed (the hot rolling) into intermediate or final-products, (e.g., coils, billets etc.). In the last stage (cold casting), the dimensions and the desired mechanical properties are achieved. Biondi et al. [114], studied the scheduling problem of a hot rolling mill in a steel plant. Strict production constraints related to metallurgic production were taken into account. Decisions regarding the production planning on a hot rolling mill were taken, by applying intelligent heuristics, resulting to efficient production programs. A slot-based MILP formulation was proposed and the sequence of the aforementioned programs was defined. The technique described above, has also been implemented as a web-service, that gathers information from both the ERP system and the DCS (Distributed Control System). Rolling programs including up to 3000–5000 coil orders can be generated within some seconds. Yang et al. [115], proposed an MILP mathematical formulation to tackle the scheduling problem by optimizing the byproduct gas systems in steel plants. Optimal solutions can be found in approximately 10 s. A representative case study from a steel plant in China has been considered to illustrate the proposed approach. A significant reduction of even 7.8% in the operation cost was noticed. Li et al. [116] considered the scheduling problem of steel making industries, focusing mainly on the steelmaking continuous casting process, as it constitutes the main production bottleneck. A novel unit-specific event-based continuous-time MILP model was proposed, relied on material continuity and other technological requirements constraints in order to ensure the generation of feasible schedules. An extension of previous rolling horizon approaches [75,76] is also applied due to the high complexity of the large-scale problems under consideration. Four representative industrial problems have been considered, to assess the efficiency of the proposed approach. Although, not even a feasible solution could be returned by solving the MILP model for the two larger problem instances within the time limit of 80,000 s, good quality solutions were generated by using the new proposed approach in 3 s and 12,287 s respectively.

Gajic et al. [117] studied the integrated scheduling and electricity optimization problem of a hot rolling mill, taking also into account electricity costs and prices. An MILP-based model was proposed in parallel with intelligent heuristics, aiming to group the individual heats into casting sequences and decompose the large problem into several sub-problems with lower complexity. The proposed approach has been successfully implemented via offline tests and it has been deployed in the melt shop at Acciai Speciali Terni S.p.A., a member of ThyssenKrupp AG and one of the world's leading producers of stainless steel based in Italy. The scheduling solutions could be generated within a few minutes and it has been shown that the electricity costs can be reduced by 3% as the coordination among the different production stages was significantly improved. Hadera et al. [118] proposed a new general precedence MILP scheduling model adapting energy awareness. Optimal production schedules were generated, simultaneously optimizing the electricity purchase and solving the load commitment problem. The case of selling electricity back to the grid was also taken into account. In order to handle large-scale industrial problem instances of a melt shop section of a stainless-steel plant, a bi-level heuristic algorithm was used. Solutions in the range of 9% and 25% integrality gap were obtained within the time limit of 600 s.

The scheduling problem of multiproduct plants with parallel units, implementing energy intensive tasks has also been considered. Continuous and discrete time RTN-based mathematical formulations have been proposed and tested to a few study cases from a real industrial problem [119]. Efficient solutions were generated and optimality gaps of 1% were achieved within 5 min of computational time. Significant savings of electricity costs, even up to 20%, were reported [120]. The same problem was also considered by Kong et al. [121]. The authors proposed an MILP model, targeting the minimization of the operational cost by optimizing the by-product gas distribution. The proposed MILP model has been successfully tested in a real-life case study, provided by an iron and steel plant located in

China. The results show that the operational cost was reduced by up to 6.2% by applying the proposed approach. Wang et al. [122] investigated the bi-objective single machine batch scheduling problem of a real-world scheduling problem in a glass company located in Shanghai, China. An exact ε-constraint method was adapted to the MILP model in order to minimize the makespan and the total energy costs. Two heuristic methods were proposed to tackle the high complexity of larger scale problems. A representative real-life case study, including 13 batches has been studied and a pareto curve was generated, illustrating the tradeoffs between the two objectives. Not even a feasible solution can be obtained after 31,823.64 CPU s by utilizing the direct ε-constraint method. However, approximate Pareto fronts, including 11 different solution points, can be generated in 5648.05 and 6359.25 CPU s, by using the two proposed heuristic methods respectively.

3.7. Paper Industries

A special interest has been expressed for the problem of trim loss minimization, mainly in the paper industry. In case the final products are to be divided in sizes of specific dimensions, significant trim losses are unavoidably generated, leading to important increase in the operational cost. Westerlund et al. [123] studied the trim-loss problem of a Finnish paper-converting mill. A two-step optimization procedure based on an MILP model was solved by CPLEX solver in fractions of a second, resulting to waste savings of 2% of the turnover. A sequential updating procedure [124] has been also presented. The researchers proposed that by resolving an MILP model iteratively, feasible schedules can be promptly extracted. A case with up to 10 products and three machines was tackled and the provided solutions are better than the ones generated manually. According to Roslöf et al. [125] various sophisticated heuristics can be utilized in large scale industrial problems to provide feasible suboptimal solutions in reasonable computational times. Extending the previous approach, an extra improvement reordering step was introduced that can lead to nearly optimal solutions. Various problem instances have been considered and the proposed approach has been compared with a heuristic based policy, and manually generated schedules. Better solutions can be provided by targeting at tardiness and makespan minimization utilizing the proposed method. A real-life case study provided by a Finnish paper mill included 61 scheduling jobs and a single processing unit was solved in 3755.1 s of CPU time. Giannelos and Georgiadis [126] proposed a slot-based MILP scheduling model, which relied on a continuous time representation, to examine the problem of cutting operations on parallel slitting machines. A CPU time limit of 1500 s has been adapted, and good quality solutions have been obtained within 6% to 9% integrality gap. The proposed approach has been applied to an industrial case study, provided by a paper mill company (Macedonian Paper Mills, S.A., Greece), including eight final products and consisting of three parallel cutting machines. Castro et al. [127] proposed an MILP and an MINLP mathematical model, which were based on a continuous and a discrete time RTN representation. The non-linearity of the second formulation can be eliminated assuming a constant throughput. The aforementioned frameworks were applied to an industrial case study from a pulp mill plant, located in Portugal. According to the detailed comparisons, the discrete time formulation seems to be more efficient, as optimal schedules can be generated in less than 600 s of CPU time, depending on the different level of discretization and the differentiation of the problem instances. On the other hand, the computational cost of the continuous time formulation seems prohibitively high, as by increasing the number of the event points by one, an increase of one order of magnitude appeared also to the computational time. Castro et al. [128] proposed an RTN-based formulation and showcased a detailed comparison between continuous and discrete time models by applying them in an industrial problem consisting of three raw materials, five intermediates and five product qualities. Novel recycling policies were also adapted and as a result, significant reduction of raw materials can be achieved, providing higher profits and lower waste. On the contrary, with the previous research, the proposed continuous time formulation led to a better quality and faster solution than the discrete one. A similar pattern also appeared in the computational time, as an increase of one order of magnitude appeared by increasing the number of the event points by one.

Table 1 presents the contributions reviewed in this section, illustrating the industrial sector for which the optimization method has been developed, as well as their main features.

Table 1. Summary of the industrial applications using optimization-based scheduling approaches.

Author	Industrial Sector	Main Research Features
Lin and Floudas [75]	Chemical industry	• Continuous time event-based mixed-integer linear programming (MILP) • Decomposition methodology
Janak et al. [129]	Chemical industry	• Graphical user interface development • Rolling horizon approach
Westerlund et al. [77]	Chemical industry	• Planning tool connected with the plant's ERP system
Velez, Merchan and Maravelias [78]	Chemical industry	• Multiple discrete-time grids • A real case study from Dow company
Moniz et al. [81]	Pharmaceutical industry	• A Visio-based decision-making tool development
Stefansson et al. [82]	Pharmaceutical industry	• Discrete and continuous time representations • Stage decomposition
Castro, Harjunkoski and Grossmann [83]	Pharmaceutical industry	• Decomposition-based algorithm
Kopanos, Méndez and Puigjaner [39]	Pharmaceutical industry	• Precedence-based MILP models • Decomposition-based solution strategy
Liu et al. [84]	Pharmaceutical industry	• Maintenance planning • 1 h CPU time
Kabra et al. [85]	Pharmaceutical industry	• State-task network (STN) representation • Computational time in the range of 1–2 min.
Shah, Sahay and Ierapetritou [86]	Oil refineries	• Six-step MILP based heuristic algorithm • A case study provided by Honeywell Process Solutions (HPS).
Zhang and Hua [87]	Oil refineries	• Integration of the plant processes and the utility system
Iyer et al. [88]	Oil refineries	• Decomposition algorithm • Feasible solutions within 6600 s are obtained
Assis et al. [89]	Oil refineries	• Scheduling of a crude oil terminal • A case study by the national refinery of Uruguay
Casrto and Mostafei [90]	Pipeline systems	• A case study from the Iranian Oil Pipelines and Telecommunication Company • 6.2% capacity increase
Cafaro et al. [93]	Pipeline systems	• Simultaneous product deliveries are allowed • A case study, related to REPLAN refinery
Rejowski and Pinto [94]	Pipeline systems	• Integrality gap of 5.8% in 10,000 CPU s • A case study, related to REPLAN refinery
Boschetto et al. [95]	Pipeline systems	• Heuristic rules • Computational times within 3–5 min.
Baldo et al. [96]	Food industries	• A novel MILP-based relax and fix heuristic algorithm • A case study from a brewery industry
Kopanos, Puigjaner and Maravelias [29]	Food industries	• An immediate precedence-based MILP formulation • A case study from a brewery industry
Abakarov andSimpson [97]	Food industries	• A food cannery case study • Scheduling of the sterilization stage
Georgiadis et al. [98]	Food industries	• A case study from a large-scale canned fish industry case study • MILP based decomposition algorithm
Liu, Pinto and Papageorgiou [99]	Food industries	• An edible-oil deodorized industry case study • Mixed discrete and continuous MILP mathematical
Polon et al. [100]	Food industries	• A case study from a sausage production industry • Scheduling of the production stage
Doganis and Sarimveis [101]	Dairy industry	• A single yoghurt production line
Sel, Bilgen and Bloenhof-Ruwaard [102]	Dairy industry	• Integrated planning and scheduling of a yoghurt facility
Touil, Echchatbi and Charkaoui [103]	Dairy industry	• A case study from a milk industry
Kopanos, Puigjaner and Georgiadis [104]	Dairy industry	• A case study from a yoghurt industry • Novel resource constraints

Table 1. *Cont.*

Author	Industrial Sector	Main Research Features
Georgiadis et al. [105]	Dairy industry	• An integrated software tool connects the plant's ERP system with the proposed MILP model • A total cost decrease of 20% is achieved
Giannelos and Georgiadis [108]	Consumer goods industry	• STN continuous time formulation • Medium-size industrial consumer goods manufacturing process
Baumann and Trautmann [110]	Consumer goods industry	• General precedence MILP hybrid method • 10 large-scale problem instances provided by The Procter and Gamble Company
Elzakker et al. [106]	Fast-moving consumer goods (FMCG) industry	• A unit-specific, continuous time interval-based algorithm • Ice cream production process of Unilever
Kopanos, Puigjaner and Georgiadis [107]	Fast-moving consumer goods (FMCG) industry	• Ice cream production process
Elekidis, Corominas and Georgiadis [112]	Consumer goods industry	• An immediate-general precedence-based decomposition algorithm
Georgiadis et al. [113]	Manufacturing industries	• Resource task network (RTN) and STN based models • A comparison with a PSE scheduling tool and an MILP model • Development of a middleware interface for data transfer
Biondi, Saliba and Harjunkoski [114]	Steel industry	• Slot-based MILP formulation • communication with ERP and DCS
Li et al. [116]	Steel industry	• A unit-specific event-based continuous-time MILP model
Gajic et al. [117]	Steel industry	• Integrated scheduling and electricity optimization problem • A melt shop case study • 3% electricity cost reduction
Hadera et al. [118]	Steel industry	• A melt shop case study • Integrated scheduling and electricity optimization problem • A general precedence MILP scheduling model
Castro, Harjunkoski and Grossmann [120]	Steel industry	• Integrated scheduling and energy optimization problem • RTN-based MILP model • 20% electricity cost reduction
Wang et al. [122]	glass company	• Bi-objective optimization problem • Makespan and the total energy cost minimization
Westerlund, Isaksson and Harjunkoski [123]	Paper Industry	• Trim-loss problem of a paper converting mill
Roslöf et al. [124]	Paper Industry	• MILP based decomposition algorithm
Giannelos and Georgiadis [126]	Paper Industry	• A slot-based MILP model • A paper mill company case study
Castro, Barbosa-Povoa and Matos [127]	Paper Industry	• Continuous and discrete time RTN representation • A case study from a pulp mill plant
Castro, Westerlund and Forssell [128]	Paper Industry	• RTN-based formulation • Novel recycling policies

4. Industrial Applications of Optimization-Based Scheduling—Challenges

In the previous section, a wide range of real-life applications using various scheduling frameworks have been described. It can be noticed that most of the methods have efficiently handled small or medium sized problem instances, with just a few of them addressing large scale industrial problems.

A major issue, referred to the applicability of the scheduling approaches, is the accuracy of the input data. In many cases the generated schedules are manually modified by the planners in a time-consuming process, as some important information is missing [111]. If the data given by the plant managers are not accurate, the assessment of the model's efficiency is extremely difficult or even impossible. We can conclude that the connection of the proposed solution method with the plants' ERP and the other plant systems, plays a key role for a successful model implementation. Integrated software should be developed, to provide an easy way to transport the necessary input data to the model solver in an automated way. In addition, the direct communication between the mathematical model and the ERP systems, could also make possible the consideration of the several planning decisions of the plant (e.g., planned cleaning or maintenance activities).

The aforementioned suggestion could also give the chance to solve and analyze a plethora of different cases, in order to test the efficiency of the model. All possible operational scenarios should be provided from the plant managers to the model developers. Further important production parameters could also be identified by checking the feasibility of the schedules generated by the model during the test runs.

The proposed models could efficiently solve all problem cases. The company should identify the largest or the most complex cases of the plant. Different models could also be addressed for different problem instances. Also, as the scheduling and planning decisions are strongly connected, there is a high necessity to ensure that the planning decisions are feasible. For example, the customer's demand, given usually by the ERP systems, must be covered by the plant's capacity during the available time horizon. Otherwise, not even a feasible solution could be returned from the model. However, even in cases when the optimization fails, an output should be exported [130].

The model's output should also be visualized in an interactive way. The planners should be able to modify the schedules by hand, or adding late order deliveries. We have to keep in mind that the planners are the ones responsible for the final schedule and an optimization model can provide a suggested solution to them. Flexible Gannt charts, allowing the right or left swift movement of the production orders, could also be an efficient tool for the planners. Also, in that way, a number of modifications, based on the planners' experience can be easily adapted to the final schedule [23].

Moreover, as the industrial environment is highly dynamic and it is characterized by a high level of uncertainty, a lot of daily modifications should be applied. Therefore, efficient rescheduling methods should also be provided, utilizing fast execution models. The utilization of decomposition algorithms or heuristic rules in parallel with the MILP models could also be a good quality option, as nearly optimal solutions can be generated in reasonable computational times [96,107]. It should also be noticed, that the updated schedule has to consider the previous schedule as an initial condition. In this way more realistic solutions can be provided.

As in the majority of the plants, production scheduling is done on a daily basis, the computational time of the utilized models should not be extremely high. Even good quality suboptimal solutions could be acceptable in time limitations defined by the company. Moreover, in practice due to unexpected events, the initial schedule rarely is completely applied.

An analysis of the expected savings could also be accomplished. The consideration of case studies of past weeks could provide a good estimation of the optimization savings range. The company should analyze the investment that has to be made to install a new integrated software and the possible time or cost savings of it. The necessary investment has to lead to higher profits. The company personnel should also be trained to use and modify efficiently the new optimization tools.

The proposed model approach should also be able to adapt new information or different types of modifications. New products could be inserted, by identifying their basic production features, as well as new pieces of equipment. Parameter values, such as, units' production rates, or new product allocation policies should be easily changed by the planners. The new integrated software should be flexible enough to adapt to the plant's changes, as otherwise the solution approach could be considered obsolete. The optimization tools should be easily used, and modified by the planners in a daily basis, without the supervision of experts in model development [131].

5. Conclusions

Prof. Sargent is one of the greatest personalities in the area of chemical engineering. He is the pioneer and father of the field of process system engineering. He was one of the first researchers who foresaw the value of the need for the optimization of process scheduling. His contributions inspired numerous researches in the last three decades, resulting in a plethora of mathematical programming formulations for general classes of process scheduling problems. This work presents a review of the theoretical modelling aspects and solution approaches for the process scheduling problem. Although there is no single comprehensive approach for handling all possible industrial problems, and the majority of the proposed models are characterized as problem-specific, a categorization is possible by identifying basic common features. Various problem instances, as well as different optimization models are classified, presenting their most important characteristics and the main contributions of them. Furthermore, an overview on the applications of the modeling formulations in real-life industrial case studies is presented. The industrial studies under consideration are categorized according to the different process industries subsectors, focusing on chemical, pharmaceutical, food, consumer goods, steel and paper industries. It is concluded that only a limited number of the contributions are able to solve large-scale industrial problems. In most of the large-scale problem instances, a variety of decomposition techniques and heuristic rules are applied in parallel with the mathematical programming models, and as a result; good but suboptimal solutions are obtained. A very small number of the generated solutions were directly compared with the plant's schedules. The development of integrated software tools, aiming to the direct data transfer between the mathematical models and the plant's ERP systems, according also and to the general trend of digitalization, was identified as a crucial step for the successful industrial applications of scheduling methods. Another scheduling challenge to be confronted is the efficient and flexible visualization of the generated solutions, allowing also for manual modifications.

Author Contributions: All authors contributed to this work. Individual contributions of the authors can be specified as: "Conceptualization M.C.G.; Analysis and Investigation G.P.G. and A.P.E.; Writing – Original draft G.P.G. and A.P.E.; Writing – Review & Editing M.C.G., G.P.G. and A.P.E.; Supervision M.C.G.".

Funding: This research was funded by the European Union's Horizon 2020 research and innovation program (SPIRE PPP framework) grant number 723575 (Project CoPro).

Conflicts of Interest: The authors declare no conflict of interest.

Nomenclature

Indices/Sets

$i \in I$	processing tasks
$o \in O$	orders to be processed
$j \in J$	units
$t, t\prime \in T$	time events
$l \in L$	processing stages
$r \in R$	resources
$s \in S$	states
i^{ST}	storage tasks
I_r	tasks requiring resource r
I_j	tasks that can be executed in unit j
I_s^{ST}	storage tasks for state s
I_s^P	tasks that produce state s
I_s^C	tasks that consume state s
R^J	resources corresponding to unit j
R_i^{ST}	resources corresponding to storage that can be used for task i
J_i	units that can perform task i
$J_{o,o\prime,l}$	units that can execute both order i and order i' at stage l
$J_{o,l}$	units that can execute order i at stage l
$L_{o,o\prime}$	stage required for the production of order o and o'

Parameters

α_i	constant term for the processing time of task i
β_i	proportional term for the processing time of task i
H	time horizon
V_i^{min}	minimum batch size of task i

$V_i^{\max}$	maximum batch size of task i
$R_r^{\min}$	minimum available resource r
$R_r^{\max}$	maximum available resource r
$\mu_{r,i}^{p}$	fixed term for the production of resource r at the end of task i
$\mu_{r,i}^{c}$	fixed term for the consumption of resource r at the beginning of task i
$v_{r,i}^{p}$	variable term for the production of resource r at the end of task i
$v_{r,i}^{c}$	variable term for the consumption of resource r at the beginning of task i
$\rho_{i,s}$	proportion of state s consumed/produced by task i
$pt_{o,j}$	processing time of order o in unit j
$su_{o,j}$	setup time for order o in unit j
$\tau_{o,\prime,j}$	changeover time between orders o and o′ processed in unit j

Variables

T_t	exact time of event point t
$\overline{N}_{i,t,t\prime}$	defines a task I that starts at event point t and ends at time point t′
$\overline{\xi}_{i,t,t\prime}$	amount of material processed by task I, that starts at t and ends at t′
$R_{r,t}$	amount of resource r consumed at event point t
$W_{i,t}^{s}$	denotes that a task i starts at event point t
$W_{i,t}^{f}$	denotes that a task i ends at event point t
$W_{i,t}$	denotes that a task i is active at event point t
$B_{i,j,t}^{s}$	batch size of the task i in unit j started at event point t
$B_{i,j,t}$	batch size of the task i in unit j being processed at event point t
$B_{i,j,t}^{f}$	batch size of the task i in unit j finished at event point t
$B_{i^{ST},j,t}^{ST}$	batch size of storage task i^{st} at event point t
$T_{i,j,t}^{s}$	time at which the execution of task i by unit j at event point t starts
$T_{i,j,t}^{f}$	time at which the execution of task i by unit j at event point t ends
$Y_{o,j}$	binary variable denoting that order o is allocated to unit j
$C_{o,l}$	completion time of order o in unit j
$X_{o,o\prime,l}$	binary variable that is activated when order o is processed before order o′ at stage l

References

1. Kondili, E.; Pantelides, C.C.; Sargent, R.W.H. A General Algorithm for Short-term Scheduling of Batch Operations-I. MILP Formulation. *Comput. Chem. Eng.* **1993**, *17*, 211–227. [CrossRef]
2. Bixby, R.; Rothberg, E. Progress in Computational Mixed Integer Programming—A Look Back from the Other Side of the Tipping Point. *Ann. Oper. Res.* **2007**, *149*, 37–41. [CrossRef]
3. Shah, N.; Pantelides, C.C.; Sargent, R.W.H. A General Algorithm for Short-Term Scheduling of Batch-Operations 2. Computational Issues. *Comput. Chem. Eng.* **1993**, *17*, 229–244. [CrossRef]
4. Kondili, E.; Pantelides, C.C.; Sargent, R.W.H. A General Algorithm for Scheduling of Batch Operations. In Proceedings of the Third International Symposium on Process Systems Engineering (PSE'88), Sydney, Australia, 28 August–2 September 1998; pp. 62–75.
5. Shah, N.; Pantelides, C.C.; Sargent, R.W.H. Optimal Periodic Scheduling of Multipurpose Batch Plants. *Ann. Oper. Res.* **1993**, *42*, 193–228. [CrossRef]
6. Papageorgiou, L.G.; Pantelides, C.C. Optimal Campaign Planning/Scheduling of Multipurpose Batch/Semicontinuous Plants. 1. Mathematical Formulation. *Ind. Eng. Chem. Res.* **1996**, *35*, 488–509. [CrossRef]
7. Papageorgiou, L.G.; Pantelides, C.C. Optimal Campaign Planning/Scheduling of Multipurpose Batch/Semicontinuous Plants. 2. A Mathematical Decomposition Approach. *Ind. Eng. Chem. Res.* **1996**, *35*, 510–529. [CrossRef]
8. Yee, K.L.; Shah, N. Scheduling of Multistage Fast-Moving Consumer Goods Plants. *J. Oper. Res. Soc.* **1997**, *48*, 1201–1214. [CrossRef]
9. Yee, K.L.; Shah, N. Improving the Efficiency of Discrete Time Scheduling Formulation. *Comput. Chem. Eng.* **1998**, *22*, S403–S410. [CrossRef]

10. Pantelides, C.C.; Realff, M.J.; Shah, N. Short-term Scheduling of Pipeless Batch Plants. *Chem. Eng. Res. Des.* **1997**, *75*, S156–S169. [CrossRef]
11. Pantelides, C.C. Unified Frameworks for Optimal Process Planning and Scheduling. In Proceedings of the Second International Conference on Foundations of Computer-Aided Process Operations, Crested Butte, CO, USA, 18–23 July 1993; pp. 253–274.
12. Schilling, G.H. Algorithms for Short-Term and Periodic Process Scheduling and Rescheduling. Ph.D. Thesis, University of London, London, UK, 1997.
13. Tahmassebi, T. Industrial Experience with a Mathematical-programming Based System for Factory Systems Planning/Scheduling. *Comput. Chem. Eng.* **1996**, *20*, S1565–S1570. [CrossRef]
14. Dimitriadis, A.D.; Shah, N.; Pantelides, C.C. RTN-based Rolling Horizon Algorithms for Medium Term Scheduling of Multipurpose Plants. *Comput. Chem. Eng.* **1997**, *21*, S1061–S1066. [CrossRef]
15. Wilkinson, S.J. Aggregate Formulations for Large-Scale Process Scheduling. Ph.D. Thesis, University of London, London, UK, 1996.
16. Shah, N. Process Industry Supply Chains: Advances and Challenges. *Comput. Chem. Eng.* **2005**, *29*, 1225–1235. [CrossRef]
17. Castro, P.M.; Barbosa-Póvoa, A.P.; Matos, H. An Improved RTN Continuous-Time Formulation for the Short-term Scheduling of Multipurpose Batch Plants. *Ind. Eng. Chem. Res.* **2001**, *40*, 2059–2068. [CrossRef]
18. Ierapetritou, M.G.; Floudas, C.A. Effective Continuous-Time Formulation for Short-Term Scheduling. 1. Multipurpose Batch Processes. *Ind. Eng. Chem. Res.* **1998**, *37*, 4341–4359. [CrossRef]
19. Méndez, C.A.; Henning, G.P.; Cerdá, J. An MILP Continuous-Time Approach to Short-Term Scheduling of Resource-Constrained Multistage Flowshop Batch Facilities. *Comput. Chem. Eng.* **2001**, *25*, 701–711. [CrossRef]
20. Gabow, H.N. On the Design and Analysis of Efficient Algorithms for Deterministic Scheduling. In Proceedings of the 2nd International Conference Foundations of Computer-Aided Process Design, Snowmass, CO, USA, 19–24 June 1983; pp. 473–528.
21. Pinedo, M.L. *Scheduling: Theory, Algorithms and Systems*, 5th ed.; Springer: Berlin, Germany, 2016.
22. Maravelias, C.T. General Framework and Modeling Approach Classification for Chemical Production Scheduling. *AIChE J.* **2012**, *58*, 1812–1828. [CrossRef]
23. Harjunkoski, I.; Maravelias, C.T.; Bongers, P.; Castro, P.M.; Engell, S.; Grossmann, I.E.; Hooker, J.; Méndez, C.; Sand, G.; Wassick, J. Scope for Industrial Applications of Production Scheduling Models and Solution Methods. *Comput. Chem. Eng.* **2014**, *62*, 161–193. [CrossRef]
24. Baumann, P.; Trautmann, N.A. Continuous-Time MILP Model for Short-Term Scheduling of Make-and-Pack Production Processes. *Int. J. Prod. Res.* **2013**, *51*, 1707–1727. [CrossRef]
25. Georgiadis, G.P.; Ziogou, C.; Kopanos, G.M.; Pampin, B.M.; Cabo, D.; Lopez, M.; Georgiadis, M.C. On the Optimization of Production Scheduling in Industrial Food Processing Facilities. In Proceedings of the 29th European Symposium on Computer Aided Process Engineering, Eindhoven, The Netherlands, 16–19 June 2019. Accepted manuscript.
26. Egli, U.M.; Rippin, D.W.T. Short-term Scheduling for Multiproduct Batch Chemical Plants. *Comput. Chem. Eng.* **1986**, *10*, 303–325. [CrossRef]
27. Vaselenak, J.A.; Grossmann, I.E.; Westerberg, A.W. An Embedding Formulation for the Optimal Scheduling and Design of Multipurpose Batch Plants. *Ind. Eng. Chem. Res.* **1987**, *26*, 139–148. [CrossRef]
28. Grossmann, I. Enterprise-wide Optimization: A New Frontier in Process Systems Engineering. *AIChE J.* **2005**, *51*, 1846–1857. [CrossRef]
29. Kopanos, G.M.; Puigjaner, L.; Maravelias, C.T. Production Planning and Scheduling of Parallel Continuous Processes with Product Families. *Ind. Eng. Chem. Res.* **2011**, *50*, 1369–1378. [CrossRef]
30. Li, Z.; Ierapetritou, M.G. Rolling Horizon Based Planning and Scheduling Integration with Production Capacity Consideration. *Chem. Eng. Sci.* **2010**, *65*, 5887–5900. [CrossRef]
31. Sel, C.; Bilgen, B.; Bloemhof-Ruwaard, J.M.; van der Vorst, J.G.A.J. Multi-bucket Optimization for Integrated Planning and Scheduling in the Perishable Dairy Supply Chain. *Comput. Chem. Eng.* **2015**, *77*, 59–73. [CrossRef]
32. Du, J.; Park, J.; Harjunkoski, I.; Baldea, M. A Time Scale-bridging Approach for Integrating Production Scheduling and Process Control. *Comput. Chem. Eng.* **2015**, *79*, 59–69. [CrossRef]

33. Charitopoulos, V.M.; Dua, V.; Papageorgiou, L.G. Traveling salesman Problem-based Integration of Planning, Scheduling, and Optimal Control for Continuous Processes. *Ind. Eng. Chem. Res.* **2017**, *56*, 11186–11205. [CrossRef]

34. Méndez, C.A.; Cerdá, J.; Grossmann, I.E.; Harjunkoski, I.; Fahl, M. State-of-the-art Review of Optimization Methods for Short-term Scheduling of Batch Processess. *Comput. Chem. Eng.* **2006**, *30*, 913–946. [CrossRef]

35. Malapert, A.; Guéret, C.; Rousseau, L.M. A Constraint Programming Approach for a Batch Processing Problem with Non-identical Job Sizes. *Eur. J. Oper. Res.* **2012**, *221*, 533–545. [CrossRef]

36. Zeballos, L.J.; Novas, J.M.; Henning, G.P. A CP Formulation for Scheduling Multiproduct Multistage Batch Plants. *Comput. Chem. Eng.* **2011**, *35*, 2973–2989. [CrossRef]

37. Bassett, M.H.; Pekny, J.F.; Reklaitis, G.V. Decomposition Techniques for the Solution of Large-Scale Scheduling Problems. *AIChE J.* **1996**, *42*, 3373–3387. [CrossRef]

38. Panek, S.; Engell, S.; Subbiah, S.; Stursberg, O. Scheduling of Multi-Product Batch Plants Based Upon Timed Automata models. *Comput. Chem. Eng.* **2008**, *32*, 275–291. [CrossRef]

39. Kopanos, G.M.; Méndez, C.A.; Puigjaner, L. MIP-Based Decomposition Strategies for Large-Scale Scheduling Problems in Multiproduct Multistage Batch Plants: A Benchmark Scheduling Problem of the Pharmaceutical Industry. *Eur. J. Oper. Res.* **2010**, *207*, 644–655. [CrossRef]

40. Prasad, P.; Maravelias, C.T. Batch Selection, Assignment and Sequencing in Multi-Stage Multi-Product Processes. *Comput. Chem. Eng.* **2008**, *32*, 1106–1119. [CrossRef]

41. Sundaramoorthy, A.; Maravelias, C.T. Simultaneous Batching and Scheduling in Multistage Multiproduct Processes. *Ind. Eng. Chem. Res.* **2008**, *47*, 1546–1555. [CrossRef]

42. Lee, H.; Maravelias, C.T. Mixed-integer Programming Models for Simultaneous Batching and Scheduling in Multipurpose Batch Plants. *Comput. Chem. Eng.* **2017**, *106*, 621–644. [CrossRef]

43. Castro, P.M.; Grossmann, I.E.; Novais, A.Q. Two New Continuous-Time Models for the Scheduling of Multistage Batch Plants with Sequence Dependent Changeovers. *Ind. Eng. Chem. Res.* **2006**, *45*, 6210–6226. [CrossRef]

44. Velez, S.; Maravelias, C.T. Mixed-Integer Programming Model and Tightening Methods for Scheduling in General Chemical Production environments. *Ind. Eng. Chem. Res.* **2013**, *52*, 3407–3423. [CrossRef]

45. Pinto, J.M.; Grossmann, I.E. Assignment and Sequencing Models of the Scheduling of Process Systems. *Ann. Oper. Res.* **1998**, *1621*, 36–43. [CrossRef]

46. Kopanos, G.M.; Puigjaner, L.; Georgiadis, M.C. Optimal Production Scheduling and Lot-Sizing in Dairy Plants: The Yogurt Production Line. *Ind. Eng. Chem. Res.* **2009**, *49*, 701–718. [CrossRef]

47. Velez, S.; Maravelias, C.T. Multiple and Nonuniform Time Grids in Discrete-Time MIP Models for Chemical Production Scheduling. *Comput. Chem. Eng.* **2013**, *53*, 70–85. [CrossRef]

48. Maravelias, C.T.; Sung, C. Integration of Production Planning and Scheduling: Overview, Challenges and Opportunities. *Comput. Chem. Eng.* **2009**, *33*, 1919–1930. [CrossRef]

49. Sundaramoorthy, A.; Maravelias, C.T. Computational Study of Network-Based Mixed-Integer Programming Approaches for Chemical Production Scheduling. *Ind. Eng. Chem. Res.* **2011**, *50*, 5023–5040. [CrossRef]

50. Lee, H.; Maravelias, C.T. Combining the Advantages of Discrete- and Continuous-Time Scheduling Models: Part 1. Framework and Mathematical Formulations. *Comput. Chem. Eng.* **2018**, *116*, 176–190. [CrossRef]

51. Reklaitis, G.V.; Mockus, L. Mathematical Programming Formulation for Scheduling of Batch Operations based on Non-uniform Time Discretization. *Acta Chim. Slov.* **1995**, *42*, 81–86.

52. Xueja, Z.; Sargent, R.W.H. The Optimal Operation of Mixed Production Facilities—A General Formulation and some Approaches for the Solution. In Proceedings of the 5th International Symposium Process Systems Engineering, Kyongju, Korea, 30 May–3 June 1994; pp. 171–178.

53. Zhang, X.; Sargent, R.W.H. The Optimal Operation of Mixed Production Facilities—Extensions and Improvements. *Comput. Chem. Eng.* **1996**, *20*, S1287–S1293. [CrossRef]

54. Schilling, G.; Pantelides, C.C. A Simple Continuous-Time Process Scheduling Formulation and a Novel Solution Algorithm. *Comput. Chem. Eng.* **1996**, *20*, 1221–1226. [CrossRef]

55. Castro, P.M.; Barbosa-Póvoa, A.P.; Matos, H.A.; Novais, A.Q. Simple Continuous-Time Formulation for Short-Term Scheduling of Batch and Continuous Processes. *Ind. Eng. Chem. Res.* **2004**, *43*, 105–118. [CrossRef]

56. Giannelos, N.F.; Georgiadis, M.C. A Simple New Continuous-Time Formulation for Short-Term Scheduling of Multipurpose Batch Processes. *Ind. Eng. Chem. Res.* **2002**, *41*, 2178–2184. [CrossRef]

57. Maravelias, C.T.; Grossmann, I.E. New General Continuous-Time State–Task Network Formulation for Short-Term Scheduling of Multipurpose Batch Plants. *Ind. Eng. Chem. Res.* **2003**, *42*, 3056–3074. [CrossRef]
58. Sundaramoorthy, A.; Karimi, I.A. A Simpler Better Slot-Based Continuous-Time Formulation for Short-Term Scheduling in Multipurpose Batch Plants. *Chem. Eng. Sci.* **2005**, *60*, 2679–2702. [CrossRef]
59. Vin, J.P.; Ierapetritou, M.G. A New Approach for Efficient Rescheduling of Multiproduct Batch Plants. *Ind. Eng. Chem. Res.* **2000**, *39*, 4228–4238. [CrossRef]
60. Janak, S.L.; Lin, X.; Floudas, C.A. Enhanced Continuous-Time Unit-Specific Event-Based Formulation for Short-Term Scheduling of Multipurpose Batch Processes: Resource Constraints and Mixed Storage Policies. *Ind. Eng. Chem. Res.* **2004**, *43*, 2516–2533. [CrossRef]
61. Shaik, M.A.; Floudas, C.A. Novel Unified Modeling Approach for Short-Term Scheduling. *Ind. Eng. Chem. Res.* **2009**, *48*, 2947–2964. [CrossRef]
62. Velez, S.; Maravelias, C.T. Theoretical Framework for Formulating MIP Scheduling Models with Multiple and Non-Uniform Discrete-Time Grids. *Comput. Chem. Eng.* **2015**, *72*, 233–254. [CrossRef]
63. Pinto, J.M.; Grossmann, I.E. A Continuous Time Mixed Integer Linear Programming Model for Short Term Scheduling of Multistage Batch Plants. *Ind. Eng. Chem. Res.* **1995**, *34*, 3037–3051. [CrossRef]
64. Castro, P.M.; Grossmann, I. New Continuous-Time MILP Model for the Short-Temr Scheduling of Multistage Batch Plants. *Ind. Eng. Chem. Res.* **2005**, *44*, 9175–9190. [CrossRef]
65. Sundaramoorthy, A.; Maravelias, C.T.; Prasad, P. Scheduling of Multistage Batch Processes under Utility Constraints. *Ind. Eng. Chem. Res.* **2009**, *48*, 6050–6058. [CrossRef]
66. Merchan, A.F.; Lee, H.; Maravelias, C.T. Discrete-time Mixed-integer Programming Models and Solution Methods for Production Scheduling in Multistage Facilities. *Comput. Chem. Eng.* **2016**, *94*, 387–410. [CrossRef]
67. Lee, H.; Maravelias, C.T. Discrete-time mixed-integer programming models for short-term scheduling in multipurpose environments. *Comput. Chem. Eng.* **2017**, *107*, 171–183. [CrossRef]
68. Gupta, S.; Karimi, I.A. An Improved MILP Formulation for Scheduling Multiproduct, Multistage Batch Plants. *Ind. Eng. Chem. Res.* **2003**, *42*, 2365–2380. [CrossRef]
69. Kopanos, G.M.; Laınez, M.; Puigjaner, L. An Efficient Mixed-Integer Linear Programming Scheduling Framework for Addressing Sequence-Dependent Setup Issues in Batch Plants. *Ind. Eng. Chem. Res.* **2009**, *48*, 6346–6357. [CrossRef]
70. Harjunkoski, I.; Grossmann, I. Decomposition Techniques for Multistage Scheduing Problems Using Mixed-Integer and Constraint Programming Methods. *Comput. Chem. Eng.* **2002**, *26*, 1533–1552. [CrossRef]
71. Cerda, J.; Henning, G.P.; Grossmann, I.E. A Mixed-Integer Linear Programming Model for Short-Term Scheduling of Single-Stage Multiproduct Batch Plants with Parallel Lines. *Ind. Eng. Chem. Res.* **1997**, *36*, 1695–1707. [CrossRef]
72. Méndez, C.A.; Henning, G.P.; Cerdá, J. Optimal Scheduling of Batch Plants Satisfying Multiple Product Orders with Different Due-Dates. *Comput. Chem. Eng.* **2000**, *24*, 2223–2245. [CrossRef]
73. Mendez, C.A.; Cerdá, J. An MILP Framework for Batch Reactive Scheduling with Limited Discrete Resources. *Comput. Chem. Eng.* **2004**, *28*, 1059–1068. [CrossRef]
74. Castro, P.M.; Erdirik-Dogan, M.; Grossmann, I.E. Simultaneous Batching and Scheduling ofSingle Stage Batch Plants with Parallel Units. *AIChE J.* **2007**, *54*, 183–193. [CrossRef]
75. Lin, X.; Floudas, C.A. Continuous-Time Optimization Approach for Medium-Range Production Scheduling of a Multiproduct Batch Plant. *Ind. Eng. Chem. Res.* **2002**, *41*, 3884–3906. [CrossRef]
76. Janak, S.L.; Floudas, C.A.; Kallrath, J.; Vormbrock, N. Production Scheduling of a Large-Scale Industrial Batch Plant. I. Short-Term and Medium-Term Scheduling. *Ind. Eng. Chem. Res.* **2006**, *45*, 8234–8252. [CrossRef]
77. Westerlund, J.; Hästbacka, M.; Forssell, S.; Westerlund, T. Mixed-Time Mixed-Integer Linear Programming Scheduling Model. *Ind. Eng. Chem. Res.* **2007**, *46*, 2781–2796. [CrossRef]
78. Velez, S.; Merchan, A.F.; Maravelias, C.T. On the Solution of Large-Scale Mixed Integer Programming Scheduling Models. *Chem. Eng. Sci.* **2015**, *136*. [CrossRef]
79. Kallrath, J. Planning and Scheduling in the Process Industry. *OR Spectrum* **2002**, *24*, 219–250. [CrossRef]
80. Nie, Y.; Biegler, L.T.; Wassick, J.M.; Villa, C.M. Extended Discrete-Time Resource Task Network Formulation for the Reactive Scheduling of a Mixed Batch/Continuous Process. *Ind. Eng. Chem. Res.* **2014**, *53*, 17112–17123. [CrossRef]

81. Moniz, S.; Barbosa-Póvoa, A.P.; de Sousa, J.P.; Duarte, P. Solution Methodology for Scheduling Problems in Batch Plants. *Ind. Eng. Chem. Res.* **2014**, *53*, 19265–19281. [CrossRef]

82. Stefansson, H.; Sigmarsdottir, S.; Pall, J.; Shah, N. Discrete and Continuous Time Representations and Mathematical Models for Large Production Scheduling Problems: A Case Study from the Pharmaceutical Industry. *Eur. J. Oper. Res.* **2011**, *215*, 383–392. [CrossRef]

83. Castro, P.M.; Harjunkoski, I.; Grossmann, I.E. Optimal Short-Term Scheduling of Large-Scale Multistage Batch Plants. *Ind. Eng. Chem. Res.* **2009**, *48*, 11002–11016. [CrossRef]

84. Kabra, S.; Shaik, M.A.; Rathore, A.S. Multi-period scheduling of a multi-stage multi-product bio-pharmaceutical process. *Comp. Chem. Eng.* **2013**, *57*, 95–103. [CrossRef]

85. Liu, S.; Yahia, A.; Papageorgiou, L. Optimal Production and Maintenance Planning of biopharmaceutical Manufacturing under Performance Decay. *Ind. Eng. Chem. Res.* **2014**, *53*, 17075–17091. [CrossRef]

86. Shah, N.K.; Sahay, N.; Ierapetritou, M.G. Efficient Decomposition Approach for Large-Scale Refinery Scheduling. *Ind. Eng. Chem. Res.* **2015**, *54*, 9964–9991. [CrossRef]

87. Zhang, B.J.; Hua, B. Effective MILP Model for Oil Refinery-wide Production Planning and Better Energy Utilization. *J. Clean. Prod.* **2007**, *15*, 439–448. [CrossRef]

88. Iyer, R.; Grossmann, I.E.; Vasantharajan, S.; Cullick, A. Optimal Planning and Scheduling of Offshore Oil Field Infrastructure Investment and Operations. *Ind. Eng. Chem. Res.* **1998**, *37*, 1380–1397. [CrossRef]

89. Assis, L.; Camponogara, E.; Zimberg, B.; Ferreira, E.; Grossmann, I.E. A piecewise McCormick relaxation-based strategy for scheduling operations in a crude oil terminal. *Comp. Chem. Eng.* **2017**, *106*, 309–321. [CrossRef]

90. Castro, P.M.; Mostafaei, H. Batch-centric Scheduling Formulation for Treelike Pipeline Systems with Forbidden Product Sequences. *Comput. Chem. Eng.* **2018**, *122*, 2–18. [CrossRef]

91. Castro, P.M. Optimal Scheduling of Multiproduct Pipelines in Networks with Reversible Flow. *Ind. Eng. Chem. Res.* **2017**, *56*, 9638–9656. [CrossRef]

92. Cafaro, D.C.; Cerdá, J. Efficient Tool for the Scheduling of Multiproduct Pipelines and Terminal Operations. *Ind. Eng. Chem. Res.* **2008**, *47*, 9941–9956. [CrossRef]

93. Cafaro, V.G.; Cafaro, D.C.; Mendez, C.A.; Cerdá, J. Detailed Scheduling of Single-Source Pipelines with Simultaneous Deliveries to Multiple Offtake Stations. *Ind. Eng. Chem. Res.* **2012**, *51*, 6145–6165. [CrossRef]

94. Rejowski, R.; Pinto, J. Scheduling of a multiproduct pipeline system. *Comp. Chem. Res.* **2003**, *27*, 1229–1246. [CrossRef]

95. Boschetto, S.; Magatão, L.; Brondani, W.; Neves-Jr, F.; Arruda, L.; Barbosa-Povoa, A.; Relvas, S. An Operational Scheduling Model to Product Distribution through a Pipeline Network. *Ind. Eng. Chem. Res.* **2010**, *49*, 5661–5682. [CrossRef]

96. Baldo, T.A.; Santos, M.O.; Almada-Lobo, B.; Morabito, R. An Optimization Approach for the Lot Sizing and Scheduling Problem in the Brewery Industry. *Comput. Ind. Eng.* **2014**, *72*, 58–71. [CrossRef]

97. Simpson, R.; Abakarov, A. Optimal Scheduling of Canned Food Plants Including Simultaneous Sterilization. *J. Food Eng.* **2009**, *90*, 53–59. [CrossRef]

98. Georgiadis, G.P.; Ziogou, C.; Kopanos, G.M.; Garcia, M.; Cabo, D.; Lopez, M.; Georgiadis, M.C. Production Scheduling of Multi-Stage, Multi-product Food Process Industries. In Proceedings of the 28th European Symposium on Computer Aided Process Engineering, Graz, Austria, 10–13 June 2018; pp. 1075–1080.

99. Liu, S.; Pinto, J.M.; Papageorgiou, L.G. Single-Stage Scheduling of Multiproduct Batch Plants: An Edible-Oil Deodorizer Case Study. *Ind. Eng. Chem. Res.* **2010**, *49*, 8657–8669. [CrossRef]

100. Polon, P.E.; Alves, A.F.; Olivo, J.E.; Paraíso, P.R.; Andrade, C.M.G. Production Optimization in Sausage Industry Based on the Demand of the Products. *J. Food Process. Eng.* **2018**, *41*. [CrossRef]

101. Doganis, P.; Sarimveis, H. Optimal Scheduling in a Yogurt Production Line Based on Mixed Integer Linear Programming. *J. Food Eng.* **2007**, *80*, 445–453. [CrossRef]

102. Sel, C.; Bilgen, B.; Bloemhof-Ruwaard, J. Planning and Scheduling of the Make-and-Pack Dairy Production under Lifetime Uncertainty. *Appl. Math. Model.* **2017**, *51*, 129–144. [CrossRef]

103. Touil, A.; Echechatbi, A.; Charkaoui, A. An MILP Model for Scheduling Multistage, Multiproducts Milk Processing. In Proceedings of the IFAC-PapersOnLine, Troyes, France, 28–30 June 2016; pp. 869–874.

104. Kopanos, G.M.; Puigjaner, L.; Georgiadis, M.C. Resource-constrained production planning in semicontinuous food industries. *Comput. Chem. Eng.* **2011**, *35*, 2929–2944. [CrossRef]

105. Georgiadis, G.P.; Kopanos, G.M.; Karkaris, A.; Ksafopoulos, H.; Georgiadis, M.C. Optimal Production Scheduling in the Dairy Industries. *Ind. Eng. Chem. Res.* **2019**, *58*, 6537–6550. [CrossRef]
106. van Elzakker, M.A.H.; Zondervan, E.; Raikar, N.B.; Grossmann, I.E.; Bongers, P.M.M. Scheduling in the FMCG Industry: An Industrial Case Study. *Ind. Eng. Chem. Res.* **2012**, *51*, 7800–7815. [CrossRef]
107. Kopanos, G.M.; Puigjaner, L.; Georgiadis, M.C. Efficient mathematical frameworks for detailed production scheduling in food processing industries. *Comput. Chem. Eng.* **2012**, *42*, 206–216. [CrossRef]
108. Giannelos, N.F.; Georgiadis, M.C. Efficient scheduling of consumer goods manufacturing processes in the continuous time domain. *Comput. Oper. Res.* **2003**, *30*, 1367–1381. [CrossRef]
109. Méndez, C.A.; Cerdá, J. An MILP-based approach to the short-term scheduling of make-and-pack continuous production plants. *OR Spectr.* **2002**, *24*, 403–429. [CrossRef]
110. Baumann, P.; Trautmann, N. A Hybrid Method for Large-Scale Short-Term Scheduling of Make-and-Pack Production Processes. *Eur. J. Oper. Res.* **2014**, *236*, 718–735. [CrossRef]
111. Honkomp, S.J.; Lombardo, S.; Rosen, O.; Pekny, J.F. The curse of reality—Why process scheduling optimization problems are difficult in practice. *Comput. Chem. Eng.* **2000**, *24*, 323–328. [CrossRef]
112. Elekidis, A.; Corominas, F.; Georgiadis, M.C. Optimal short-term Scheduling of Industrial Packing Facilities. In Proceedings of the 29th European Symposium on Computer Aided Process Engineering, Eindhoven, The Netherlands, 16–19 June 2019. Accepted manuscript.
113. Georgiadis, M.C.; Levis, A.A.; Tsiakis, P.; Sanidiotis, I.; Pantelides, C.C.; Papageorgiou, L.G. Optimisation-based scheduling: A discrete manufacturing case study. *Comput. Ind. Eng.* **2005**, *49*, 118–145. [CrossRef]
114. Biondi, M.; Saliba, S.; Harjunkoski, I. Production Optimization and Scheduling in a Steel Plant: Hot Rolling Mill. In Proceedings of the IFAC, Milano, Italy, 28 August–2 September 2011; pp. 11750–11754.
115. Yang, J.; Cai, J.; Sun, W.; Liu, J. Optimization and Scheduling of Byproduct Gas System in Steel Plant. *J. Iron Steel Res. Int.* **2015**, *22*, 408–413. [CrossRef]
116. Li, J.; Xiao, X.; Tang, Q.; Floudas, C.A. Production Scheduling of a Large-Scale Steelmaking Continuous Casting Process via Unit-Specific Event-Based Continuous-Time Models: Short-Term and Medium-Term Scheduling. *Ind. Eng. Chem. Res.* **2012**, *51*, 7300–7319. [CrossRef]
117. Gajic, D.; Hadera, H.; Onofri, L.; Harjunkoski, I.; Di Gennaro, S. Implementation of an integrated production and electricity optimization system in melt shop. *J. Clean. Prod.* **2017**, *155*, 39–46. [CrossRef]
118. Hadera, H.; Harjunkoski, I.; Sand, G.; Grossmann, I.E.; Engell, S. Optimization of steel production scheduling with complex time-sensitive electricity cost. *Comput. Chem. Eng.* **2015**, *76*, 117–139. [CrossRef]
119. Castro, P.M.; Harjunkoski, I.; Grossmann, I.E. Optimal scheduling of continuous plants with energy constraints. *Comput. Chem. Eng.* **2011**, *35*, 372–387. [CrossRef]
120. Castro, P.M.; Harjunkoski, I.; Grossmann, I.E. New Continuous-Time Scheduling Formulation for Continuous Plants under Variable Electricity Cost. *Ind. Eng. Chem. Res.* **2009**, *48*, 6701–6714. [CrossRef]
121. Kong, H.; Qi, E.; He, S.; Gang, L. MILP Model for Plant-Wide Optimal By-Product Gas Scheduling in Iron and Steel Industry. *J. Iron Steel Res. Int.* **2010**, *17*, 34–37. [CrossRef]
122. Wang, S.; Liu, M.; Chu, F.; Chu, C. Bi-objective optimization of a single machine batch scheduling problem with energy cost consideration. *J. Clean. Prod.* **2016**, *137*, 1205–1215. [CrossRef]
123. Westerlund, J.; Isaksson, J.; Harjunkoski, I. Solving a two-dimensional trim-loss problem with MILP. *Eur. J. Oper. Res.* **1998**, *104*, 572–581. [CrossRef]
124. Roslöf, J.; Harjunkoski, I.; Westerlund, J.; Isaksson, J. A Short-Term Scheduling Problem in the Paper-Converting Industry. *Comput. Chem. Eng.* **1999**, *23*, S871–S874. [CrossRef]
125. Roslöf, J.; Harjunkoski, I.; Björkqvist, J.; Karlsson, S.; Westerlund, J. An MILP-based reordering algorithm for complex industrial scheduling and rescheduling. *Comput. Chem. Eng.* **2001**, *25*, 821–828. [CrossRef]
126. Giannelos, N.F.; Georgiadis, M.C. Scheduling of Cutting-Stock Processes on Multiple Parallel Machines. *Chem. Eng. Res. Des.* **2001**, *79*, 747–753. [CrossRef]
127. Castro, P.M.; Barbosa-Póvoa, A.P.; Matos, H.A. Optimal Periodic Scheduling of Batch Plants Using RTN-Based Discrete and Continuous-Time Formulations: A Case Study Approach. *Ind. Eng. Chem. Res.* **2003**, *42*, 3346–3360. [CrossRef]
128. Castro, P.M.; Westerlund, J.; Forssell, S. Scheduling of a continuous plant with recycling of byproducts: A case study from a tissue paper mill. *Comput. Chem. Eng.* **2009**, *33*, 347–358. [CrossRef]

129. Janak, S.L.; Floudas, C.A.; Kallrath, J.; Vormbrock, N. Production scheduling of a large-scale industrial batch plant. II. Reactive scheduling. *Ind. Eng. Chem. Res.* **2006**, *45*, 8253–8269. [CrossRef]
130. Harjunkoski, I. Deploying scheduling solutions in an industrial environment. *Comput. Chem. Eng.* **2016**, *91*, 127–135. [CrossRef]
131. Grossmann, I.E.; Harjunkoski, I. Process systems Engineering: Academic and industrial perspectives. *Comput. Chem. Eng.* **2019**, *126*, 474–484. [CrossRef]

 *processes*

Article

Adjustable Robust Optimization for Planning Logistics Operations in Downstream Oil Networks

Camilo Lima [1], Susana Relvas [1,*], Ana Barbosa-Póvoa [1] and Juan M. Morales [2]

[1] CEG—IST, Instituto Superior Técnico, Universidade de Lisboa, Av. Rovisco Pais, 1049-001 Lisboa, Portugal
[2] Department of Applied Mathematics, University of Málaga, 29071 Málaga, Spain
* Correspondence: susana.relvas@tecnico.ulisboa.pt; Tel.: +351-21-841-7729; Fax: +351-21-841-7979

Received: 27 April 2019; Accepted: 19 July 2019; Published: 2 August 2019

Abstract: The oil industry operates in a very uncertain marketplace, where uncertain conditions can engender oil production fluctuations, order cancellation, transportation delays, etc. Uncertainty may arise from several sources and inexorably affect its management by interfering in the associated decision-making, increasing costs and decreasing margins. In this context, companies often must make fast and precise decisions based on inaccurate information about their operations. The development of mathematical programming techniques in order to manage oil networks under uncertainty is thus a very relevant and timely issue. This paper proposes an adjustable robust optimization approach for the optimization of the refined products distribution in a downstream oil network under uncertainty in market demands. Alternative optimization techniques are studied and employed to tackle this planning problem under uncertainty, which is also cast as a non-adjustable robust optimization problem and a stochastic programing problem. The proposed models are then employed to solve a real case study based on the Portuguese oil industry. The results show minor discrepancies in terms of network profitability and material flows between the three approaches, while the major differences are related to problem sizes and computational effort. Also, the adjustable model shows to be the most adequate one to handle the uncertain distribution problem, because it balances more satisfactorily solution quality, feasibility and computational performance.

Keywords: distribution; planning; oil supply chain; robust optimization; uncertainty

---

1. Introduction

In the oil industry, companies develop a set of activities and processes so that crude oil can be duly transformed into oil products demanded by final consumers such as petrochemical industries, airports and individual users [1]. The associated decision-making process is framed within a deeply complex context and addresses the oil exploration, production and transportation at upstream, oil refining at midstream, and oil product distribution and marketing at the downstream segments [2]. Most of these activities are difficult tasks and need to be properly accomplished. Decision-support tools based on mathematical programming techniques are often applied to assist such activities and processes across the oil supply chain [3]. These optimization tools are tailored to cover the underlying reality under study, and hence they are often stochastic large-scale programs, which normally increases the problem complexity [4].

In management sciences and engineering, the most widely-used modeling framework to cope with uncertainty has been two-stage stochastic programming [2], where the first-stage variables must be decided before the uncertainty realizations and the second-stage variables are decided in accordance with the uncertainty outcomes [5]. When either the information is missing or even the probability distribution of the uncertainty is not available, fuzzy programming could be an alternative framework to handle uncertainty in optimization problems by considering random parameters as fuzzy numbers and treating constraints as fuzzy sets [6]. Robust optimization (RO) can be used, which also requires moderate information about uncertainty described by means of uncertainty sets, i.e., the key building blocks in this modeling framework [7]. Other frameworks have also been proposed to support decision making under uncertainty such as parametric programming, which relies on sensitivity analysis theory, and also chance-constrained programming, which uses probabilistic theory [8]. Currently, data-driven optimization methods are a new and timely research line to investigate the decision-making process under uncertainty, supported not only by the crescent access to uncertainty data, but also by the development in machine learning approaches [9]. Data-driven methods aims at building uncertainty sets directly from uncertainty [10]. In general, the choice of the mathematical modeling framework to solve a particular problem under uncertainty relies on the specific characteristics of this problem, as well as on the level of information about uncertainty.

The use of decision-support tools specifically for the DOSC problem under uncertainty has been a research challenge [11], and different approaches have been proposed in this direction. These include stochastic programming [1,2,4,8,12–21]; fuzzy programming [22,23]; and robust optimization [16,24]. In these optimization models, demand, supply, price, cost, process capacity, product yield, conversion rate and technology development are regarded as uncertain parameters [3].

Nowadays, RO has become attractive to solve optimization problems under uncertainty as it only requires the mean and range of the uncertain data, and other probabilistic information can be gradually incorporated when it becomes available [25]. In addition, it balances adequately solution quality, feasibility and computational performance [26], being formulated to hedge the optimization problem against any disturbances in the input data series [27]. As single-stage non-adjustable robust optimization (NARO) models may lead to over-conservative solutions [28–30], some authors have been emphasizing the need of exploring other methods such as the two-stage or multistage adjustable robust optimization (ARO) models, since less conservative solutions can be [28]. The first-stage and second-stage decisions are robust against all uncertain realizations of the random parameters, as well as the second-stage solution has full adaptability to the uncertainty [25].

Against this background, the purpose of this paper is to address the problem of refined products distribution under demand uncertainty in the downstream oil supply chain (DOSC). Extending previous work developed by the authors [1], where a stochastic programming approach was used considering uncertainty on price and demand for crude oil. Scenarios were built by using time series analysis. In the present work a robust modeling framework is developed that considers the problem as a two-stage adjustable robust optimization (ARO), aiming to conclude on the model conservativeness as well as on the associated solution performance. The framework includes a polyhedral uncertainty set, where the first-stage decision refers to purchase of oil volumes and the second-stage decisions include oil refining planning, oil product distribution planning, inventory management, international trade, and customer fulfillment. To evaluate the proposed model, a real-case study in the Portuguese oil industry is used. Results are compared with those obtained using the equivalent non-adjustable, stochastic and deterministic models. The robust model proposed in this paper is, to the best of the authors' knowledge, the first model to consider an adjustable robust framework to deal with uncertainty in DOSC and, therefore, an academic contribution for this area.

The rest of the paper is structured as follows: Section 2 describes the problem under study. In Section 3, the background on two-stage ARO modeling framework is explained. Section 4 proposes the ARO mathematical formulation. In Section 5, the case study based on the Portuguese oil industry is explored. In Section 6, the obtained numerical results are discussed. Finally, Section 7 draws some conclusions and presents some directions for future research.

2. Problem Description

The problem under study consists of determining the robust tactical and operational planning of a downstream oil network under uncertainty over a planning time horizon. It must be modeled satisfactorily in a robust optimization framework so as to maximize the worst-case DOSC profit. Profit is computed as the difference between the total cost of oil purchase associated with the first-stage decisions, and the worst-case operational profit associated with the second-stage recourse decisions, in order to satisfy the uncertain product demand, subject to material balance, refining capacity, product yield, supply availability, stock limitations and transportation capacity constraints.

The network under study is composed by three echelons: oil refineries, storage depots and local markets. Crude oil is acquired by oil refineries, which process it into oil products in order to satisfy the internal and external demands. The oil products are adequately stored across the network facilities, and thus conveyed by pipelines, tanker ships, tank wagons and tank trucks from oil refineries through the primary distribution, and by tank trucks from storage depots in the secondary distribution. Oil refineries are allowed to import and export oil products, while storage depots can only import. In summary, the problem can be generally defined as follows:

Given a downstream oil system, in which an oil company manages the refining processes, storage activities, logistics, distribution, and marketing;

Determine the robust tactical and operational planning under uncertainty so as to define the production levels, yield fractions, material flows, inventory levels, demand fulfillment, transport assignment and international trade throughout the network;

Subject to material balance, product yield, equipment capacities, supply limitations, and transportation capacity constraints;

So as to maximize the worst-case network profit that comprises the worst-case recourse profit minus the total cost of crude oil procurement.

To solve this problem, a two-stage adjustable robust modeling framework, considering uncertainty in demand for refined products, is developed.

3. Two-Stage ARO Formulation

In the two-stage ARO modeling framework, the first-stage variables comprise the set of decisions that shall be made before the uncertainty is revealed and the second-stage variables include the set of decisions that must be made after the uncertainty is disclosed. The second-stage decision variables have fully adaptability to any realization of the uncertainty [25]. Considering a linear formulation and let x and y be the corresponding first-stage and second-stage decision variables, while $\mathcal{U}$ is either a polyhedral or a discrete uncertainty set, according to Zeng and Zhao [28], the general form of a two-stage ARO formulation can be defined as follows:

$$\max_{x} c^T x + \min_{u \in \mathcal{U}} \max_{y \in F(x,u)} b^T y \tag{1}$$

$$\text{s.t.} Ax \leq d, x \in S_x \tag{2}$$

where $F(x,u) = \{y \in S_y : Gy \leq h - Ex - Mu\}$ with $S_x \in \mathbb{R}^n_+$ and $S_y \in \mathbb{R}^m_+$. If the uncertainty set $\mathcal{U}$ is continuous, the optimization problem (Equations (1) and (2)) will have an infinite number of constraints in the second-stage problem [27]. However, when $\mathcal{U}$ is a polyhedral uncertainty set, only its extreme points, i.e., its vertices, might belong to an optimal solution of that second-stage problem [25]. In accordance with Zeng and Zhao [28], when considering only a finite and discrete set of extreme points $\mathcal{U} = \{u_1, \ldots, u_r\}$, as well as the set of the respective recourse decision variables $\{y_1, \ldots, y_r\}$, the prior optimization problem can be recast as presented below:

$$\max_{x} c^T x + \eta \tag{3}$$

$$\text{s.t.} Ax \leq d \tag{4}$$

$$\eta \leq b^T y_v, v = 1, \ldots, \tag{5}$$

$$Ex + Gy_v \leq h - Mu_v, v = 1, \ldots, r \tag{6}$$

$$x \in S_x, y_v \in S_y, v = 1, \ldots, r. \tag{7}$$

consequently, we can find a solution for the two-stage ARO model (Equations (1) and (2)) through solving its mixed-integer linear program counterpart (Equations (3)–(7)). However, defining this equivalent formulation and determining its solution can often be difficult and dependent on the size of the finite and discrete uncertainty set $\mathcal{U} = \{u_1, \ldots, u_r\}$ [28].

4. Adjustable Robust Mathematical Programming Model

Having defined in a generic form the two-stage ARO approach, the tactical and operational planning for the DOSC is now considered, which aims at maximizing the worst-case network profit under refined product uncertainty. In the established two-stage ARO problem, the first-stage decisions, i.e., the *here-and-now* decisions, comprise the crude oil purchase, while the second-stage decisions, i.e., the recourse (*wait-and-see*) decisions, include production levels, product yields, material flows, inventory levels, demand fulfillment, transport allocation and international trade, which are made in accordance with the realization of the product demand uncertainty. The two-stage ARO model for the general distribution problem under uncertain demand for refined products is formulated hereinafter.

4.1. Robust Objective Function

The notation used to formulate the two-stage ARO model is outlined in Table A1 of Appendix A. As shown in Equation (8), the objective function aims at maximizing the worst-case network profit (WCP), which can be stated as follows:

$$\max_{XCO} WCP = - \sum_{i \in I, t \in T} XCO_{i,t} PO_t + Q_{\mathcal{U}}(XCO) \tag{8}$$

$$\text{s.t.} \quad XCO_{i,t} \leq DO_{i,t} \forall i \in I, t \in T \tag{9}$$

where $XCO_{i,t}$ represents the crude oil purchase at oil refinery i in time point t and refers to the first-stage decisions. $Q_{\mathcal{U}}(\cdot)$ is defined as a function of the first-stage (*here-and-now*) decisions and denotes the worst-case recourse profit, which must be determined for the worst-case scenario of the uncertain product demands in the second-stage problem. Here, uncertainty is modeled through a polyhedral uncertainty set $\mathcal{U}$, where the uncertain parameters can take values in, because such fact guides to define a robust counterpart that corresponds to a linear programming problem [27]. Equation (9) limits the amount of crude oil purchased by each oil refinery i at time point t by the corresponding oil demand.

The worst-case recourse profit $Q_{\mathcal{U}}(\cdot)$ is equal to the objective function value of the next robust formulation, Equation (10), which can be formulated as follows:

$$
\begin{aligned}
Q_{\mathcal{U}}(\cdot) = \min_{\Delta DP_{k,p,t} \in \mathcal{U}} \max_{\mathcal{R} \in \mathcal{F}(\mathcal{H}, \Delta DP_{k,p,t})} \Bigg(& \sum_{i \in I, t \in T} RR_{i,t} + \sum_{j \in J, p \in P, t \in T} MD_{j,p,t} + \sum_{k \in K, p \in P, t \in T} MC_{k,p,t} \\
& - \sum_{i \in I, p \in P, t \in T} CE_{i,p,t} - \sum_{h \in H, p \in P, t \in T} CI_{h,p,t} - \sum_{i \in I, p \in P, t \in T} CPD_{i,p,t} \\
& - \sum_{j \in J, p \in P, t \in T} CSD_{j,p,t} - \sum_{i \in I, t \in T} CIO_{i,t} - \sum_{i \in I, p \in P, t \in T} CIR_{i,p,t} \\
& - \sum_{j \in J, p \in P, t \in T} CID_{j,p,t} - \sum_{k \in K, p \in P, t \in T} CIM_{k,p,t} - \sum_{k \in K, p \in P, t \in T} CUD_{k,p,t} \Bigg)
\end{aligned}
\tag{10}
$$

where $\mathcal{U}$ represents the polyhedral uncertainty set that is used to describe the stochastic parameters, $\mathcal{R} \in \mathbb{R}_+$ denotes the set of recourse decisions, i.e., $\mathcal{R} = \{XR, XOP, XCO, XP, XS, XRP, XU, XE, XI, IO, IR, ID, IM\}$, $\mathcal{H} \in \mathbb{R}_+$ depicts the set of *here-and-now* decisions, $\mathcal{H} = \{XCO\}$, and $\mathcal{F}$ is a set defined as a function of the set of *here-and-now* decisions $\mathcal{H}$ and the worst-case realization of the deviations $\Delta DP_{k,p,t}$ from the nominal values $\widehat{DP}_{k,p,t}$. Notice that the set of recourse decisions $\mathcal{R}$ is made after the outer minimization problem selects the worst-case realization of the deviations of the product demands $\Delta DP_{k,p,t}$. In other words, the *min-max* problem picks the worst realization of the demand that minimizes the network profit. Therefore, the robust modeling framework concerns with the maximization of the recourse profit in the worst-case scenario of the random parameter within the uncertainty set $\mathcal{U}$. In summary, the two-stage RO model addresses the volumes of exported and imported products, refined oil, product yield fractions, primary and secondary product distributions, inventories of oil and products, and met and unmet demand as the second-stage decisions, which are adjusted to the volumes of purchased oil, i.e., the first-stage decisions, after uncertainty is disclosed.

The worst-case recourse profit $Q_{\mathcal{U}}(\cdot)$, i.e., Equation (10), includes the revenues from the sales of refined products by the oil refineries, margins from buying and selling refined products by the storage depots and local markets, and costs for exporting, importing, transporting, and storing refined products throughout the network. Before presenting the equations of the recourse optimization problem $Q_{\mathcal{U}}(\cdot)$, the associated revenues, margins and costs are now discussed, being outlined by the set of Equations (11)–(22) below:

$$RR_{i,t} = \sum_{p \in P} \left[XR_{i,p,t} \times \left(PP_{a,p,t} - TN_{r,p} \right) \right] \forall i \in I, t \in T, a = r == refinery \tag{11}$$

$$MD_{j,p,t} = \sum_{(k,m) \in Route_{i,j,m}} XS_{j,k,m,p,t} \times \left(PP_{a,p,t} - TN_{r,p} \right)$$
$$- \sum_{(i,m) \in Route_{i,j,m}} XP_{i,j,m,p,t} \times PP_{a_2,p,t}$$
$$\forall j \in J, p \in P, t \in T, a = r = depot, a_2 = refinery \tag{12}$$

$$MC_{k,p,t} = \left[\left(XRP_{k,p,t} - XU_{k,p,t} \right) \times \left(PP_{a_1,p,t} - TN_{r,p} \right) \right]$$
$$- \left[\left(\sum_{(i,m) \in Route_{i,k,m}} XP_{i,k,m,p,t} \right) \times PP_{a_2,p,t} + \left(\sum_{(j,m) \in Route_{j,k,m}} XS_{j,k,m,p,t} \right) \times PP_{a_3,p,t} \right]$$
$$\forall k \in K, p \in P, t \in T, a_1 = r = retail, a_2 = refinery, a_3 = depot \tag{13}$$

$$CE_{i,p,t} = XE_{i,p,t} \times TN_{r,p} \quad \forall i \in I, p \in P, t \in T, r = export \tag{14}$$

$$CI_{h,p,t} = XI_{h,p,t} \times \left(PP_{a,p} + TN_{r,p} \right) \quad \forall h \in H, p \in P, t \in T, a = r = import \tag{15}$$

$$CPD_{i,p,t} = \sum_{(l,m) \in Route_{i,l,m}} \left(XP_{i,l,m,p,t} \times CT_{m,p} \times Dist_{i,l,m} \right) \quad \forall i \in I, p \in P, t \in T \tag{16}$$

$$CSD_{j,p,t} = \sum_{(k,m) \in Route_{j,k,m}} \left(XS_{j,k,m,p,t} \times CT_{m,p} \times Dist_{j,k,m} \right) \quad \forall j \in J, p \in P, t \in T \tag{17}$$

$$CIO_{i,t} = CKI \times IO_{i,t} \times PO_t \quad \forall i \in I, t \in T, p = oil \tag{18}$$

$$CIR_{i,p,t} = CKI \times IR_{i,p,t} \times PP_{a,p,t} \quad \forall i \in I, p \in P, t \in T, a = refinery \tag{19}$$

$$CID_{j,p,t} = CKI \times ID_{j,p,t} \times PP_{a,p,t} \quad \forall j \in J, p \in P, t \in T, a = depot \tag{20}$$

$$CIM_{k,p,t} = CKI \times IM_{k,p,t} \times PP_{a,p,t} \quad \forall k \in K, p \in P, t \in T, a = retail \tag{21}$$

$$CUD_{k,p,t} = XU_{k,p,t} \times TN_{r,p} \quad \forall k \in K, p \in P, t \in T, r = unmet \tag{22}$$

Equation (11) defines the refinery revenues obtained from the sales of refined products. Equations (12) and (13) determine the storage depot margins and the local market margins, respectively, being the difference between the revenues from selling refined products and the costs for buying these ones. Equations (14) and (15) specify, respectively, the costs of exporting and importing refined products. Equations (16) and (17) determine the costs for dispatching refined products through the primary and secondary distributions, in that order. Equation (18) defines the crude oil inventory costs at oil refineries, whereas Equations (19)–(21) define, respectively, the refined product inventory costs at oil refineries, storage depots and local markets. Equation (22) represents the costs for not supplying the required product demand.

4.2. Equations of the Recourse Problem $Q_u(\cdot)$

Equations (23)–(45) of the second-stage problem $Q_u(\cdot)$ depend on the worst-case realization of the stochastic parameters and are used to model the refinery operation, network flow allocation, inventory policy, and so on:

$$XOP_{i,t} \leq PC_i \quad \forall i \in I, t \in T \tag{23}$$

$$XR_{i,p,t} = XOP_{i,t} \times YF_{i,p} \quad \forall i \in I, p \in P, t \in T \tag{24}$$

$$IO_{i,t} = ISO_i + XCO_{i,t} - XOP_{i,t} \quad \forall i \in I, t \in T, t = t_1 \tag{25}$$

$$IO_{i,t} = IO_{i,t-1} + XCO_{i,t} - XOP_{i,t} \quad \forall i \in I, t \in T, t > t_1 \tag{26}$$

$$IR_{i,p,t} = ISP_{i,p} + XR_{i,p,t} + XI_{i,p,t} - \sum_{(l,m)\in Route_{i,l,m}} XP_{i,l,m,p,t} - XE_{i,p,t}$$
$$\forall i \in I, p \in P, t \in T, t = t_1 \tag{27}$$

$$IR_{i,p,t} = IR_{i,p,t-1} + XR_{i,p,t} + XI_{i,p,t} - \sum_{(l,m)\in Route_{i,l,m}} XP_{i,l,m,p,t} - XE_{i,p,t}$$
$$\forall i \in I, p \in P, t \in T, t > t_1 \tag{28}$$

$$ID_{j,p,t} = ISP_{j,p} + XI_{j,p,t} + \sum_{(i,m)\in Route_{i,j,m}} XP_{i,j,m,p,t} - \sum_{(k,m)\in Route_{j,k,m}} XS_{j,k,m,p,t}$$
$$\forall j \in J, p \in P, t \in T, t = t_1 \tag{29}$$

$$ID_{j,p,t} = ID_{j,p,t-1} + XI_{j,p,t} + \sum_{(i,m)\in Route_{i,j,m}} XP_{i,j,m,p,t} - \sum_{(k,m)\in Route_{j,k,m}} XS_{j,k,m,p,t}$$
$$\forall j \in J, p \in P, t \in T, t > t_1 \tag{30}$$

$$IM_{k,p,t} = ISP_{k,p} + \sum_{(i,m)\in Route_{i,k,m}} XP_{i,k,m,p,t} + \sum_{(j,m)\in Route_{j,k,m}} XS_{j,k,m,p,t} + XU_{k,p,t} - XRP_{k,p,t} + \Delta DP_{k,p,t}$$
$$\forall k \in K, p \in P, t \in T, t = t_1 \tag{31}$$

$$IM_{k,p,t} = IM_{k,p,t-1} + \sum_{(i,m)\in Route_{i,k,m}} XP_{i,k,m,p,t} + \sum_{(j,m)\in Route_{j,k,m}} XS_{j,k,m,p,t} + XU_{k,p,t} - XRP_{k,p,t} + \Delta DP_{k,p,t}$$
$$\forall k \in K, p \in P, t \in T, t > t_1 \tag{32}$$

$$XRP_{k,p,t} \leq \left(\widehat{DP}_{k,p,t} + \Delta DP_{k,p,t}\right) \quad \forall k \in K, p \in P, t \in T \tag{33}$$

$$SSO_i \times SCO_i \leq IO_{i,t} \leq SCO_i \quad \forall i \in I, t \in T \tag{34}$$

$$SSP_i \times SC_{i,p} \leq IR_{i,p,t} \leq SC_{i,p} \quad \forall i \in I, p \in P, t \in T \tag{35}$$

$$SSP_j \times SC_{j,p} \leq ID_{j,p,t} \leq SC_{j,p} \quad \forall j \in J, p \in P, t \in T \tag{36}$$

$$SSP_k \times SC_{k,p} \leq IM_{k,p,t} \leq SC_{k,p} \quad \forall k \in K, p \in P, t \in T \tag{37}$$

$$\sum_{l\in Route_{i,l,m}} XP_{i,l,m,p,t} \leq ASPD_{i,m,p} \quad \forall i \in I, m \in M, p \in P, t \in T \tag{38}$$

$$\sum_{k\in Route_{j,k,m}} XS_{j,k,m,p,t} \leq ASSD_{j,m,p} \quad \forall j \in J, m \in M, p \in P, t \in T \tag{39}$$

$$\sum_{p\in P} XP_{i,l,m,p,t} \leq ACPD_{i,l,m} \quad \forall (i,l,m) \in Route_{i,l,m}, t \in T \tag{40}$$

$$\sum_{p\in P} XS_{j,k,m,p,t} \leq ACSD_{j,k,m} \quad \forall (j,k,m) \in Route_{j,k,m}, t \in T \tag{41}$$

$$\sum_{(i,m)\in Route_{i,j,m}} XP_{i,j,m,p,t} + XI_{j,p,t} \leq SC_{j,p} \times TCM_{j,p} \quad \forall j \in J, p \in P, t \in T \tag{42}$$

$$\sum_{(i,m)\in Route_{i,k,m}} XP_{i,k,m,p,t} + \sum_{(j,m)\in Route_{j,k,m}} XS_{j,k,m,p,t} \leq SC_{k,p} \times TCM_{k,p} \quad \forall k \in K, p \in P, t \in T \tag{43}$$

$$XP_{i,l,m,p,t} = 0 \quad \forall i \in I, l \in L, p \in P, t \in T, m = road, \mu_{i,j,m} > MTD \tag{44}$$

$$XS_{j,k,m,p,t} = 0 \quad \forall j \in J, k \in K, p \in P, t \in T, m = road, \mu_{j,k,m} > MTD \tag{45}$$

Equation (23) limits the oil processed at oil refineries. Equation (24) defines the fractions obtained through the crude oil distillation process. Equations (25) and (26) control the crude oil balance, as well as Equations (27) and (28) handle the refined product balance, at oil refineries. Likewise, Equations (29) and (30) control the refined product balance at storage depots, as well as Equations (31) and (32) do it at local markets. Equation (33) limits the product supply at local markets by the nominal product demand plus the associated deviation. Equations (34)–(37) determine the storage capacities for the facilities in the network. Equation (38) limits the capacity of supplying refined products in the primary distribution, as well as Equation (39) does so in the secondary distribution. Equation (40) defines the arc capacities for the primary distribution, whereas Equation (41) for the secondary distribution. Equations (42) and (43) determine the receiving capacities for the storage depots and local markets, respectively. Equation (44) and (45) suppress the primary and secondary distributions for long distances, respectively.

4.3. Definition of Uncertainty Set $\mathcal{U}$

The uncertainty set is the key building block of the robust modeling framework [7]. It is assumed that the deviation of the product demand $\Delta DP_{k,p,t}$ takes values within a convex and budget uncertainty set (Equations (46)–(48)) as follows—see [31]:

$$\Delta DP_{k,p,t} = \widetilde{DP}_{k,p,t} - \widehat{DP}_{k,p,t} \quad \forall k \in K, p \in P, t \in T \tag{46}$$

$$\left|\Delta DP_{k,p,t}\right| \leq \Delta DP_{k,p,t}^{max} \quad \forall k \in K, p \in P, t \in T \tag{47}$$

$$\sum_{t} \frac{\left|\Delta DP_{k,p,t}\right|}{\Delta DP_{k,p,t}^{max}} \leq \Gamma_{k,p} \quad \forall k \in K, p \in P \tag{48}$$

Equation (46) defines the demand deviation $\Delta DP_{k,p,t}$ as the difference between the uncertain value of the product demand $\widetilde{DP}_{k,p,t}$ and the nominal value of the demand $\widehat{DP}_{k,p,t}$. Equation (47) models the symmetrical and bounded range for the deviations of the product demand $\Delta DP_{k,p,t}$. Equation (48) ensures that the sum of the normalized absolute values of the demand deviations for each pair of location and product, across all time points t, must not exceed the user-defined budget of uncertainty $\Gamma_{k,p}$. Notice that the normalized absolute value corresponds to the quotient between the absolute and maximum values of the demand deviation, while the uncertain value of the product demand refers to the unknown realization of the product demand. Appendix E includes an example showing how to define the vertices of a polyhedral budget uncertainty set.

4.4. The Adaptive Robust Formulation

In this section, the above robust formulation, that is, the set of Equations (8)–(48), is reformulated to be fully adaptive to the realizations of the uncertain deviations of the product demands $\Delta DP_{k,p,t}$. In order to transform the prior robust formulation into a single-level approach, the worst-case recourse profit $Q_{\mathcal{U}}(\cdot)$ may be denoted by an auxiliary variable α, which must be lower or equal to the difference among the financial items—see Equation (49). Moreover, the set of recourse decisions of the second-stage problem $Q_{\mathcal{U}}(\cdot)$ is reformulated as a function of the uncertain deviations of the product demands $\Delta DP_{k,p,t}$, which are defined by the uncertainty set $\mathcal{U}$ (Equations (46)–(48)). Because this uncertainty set

$\mathcal{U}$ defines a polyhedron with infinite number of points, this reformulation would result in a recourse problem with an infinite number of variables and equations. For this reason, the uncertainty set $\mathcal{U}$ must be partitioned into K parts in order to make the resulting formulation tractable. According to Bertsimas et al. [25], only an extreme point (i.e., a vertex) of the defined polyhedron can be part of the optimal solution of the recourse problem $Q_{\mathcal{U}}(\cdot)$. Hence, only the set of vertices $v \in V$ of the polyhedron defined through the uncertainty set $\mathcal{U}$ is considered in this adjustable robust formulation. Therefore, the adjustable robust formulation is defined in the following set of Equations (49)–(86):

$$\max_{\mathcal{R},\mathcal{H},\alpha} WCP = -\sum_{i\in I, t\in T} XCO_{i,t} \times PO_t + \alpha \tag{49}$$

$$s.t. XCO_{i,t} \le DO_{i,t} \forall i \in I, t \in T \tag{50}$$

$$\alpha \le \min_{\Delta DP_{k,p,t} \in \mathcal{U}} \max_{\mathcal{R} \in \mathcal{F}(\mathcal{H}, \Delta DP_{k,p,t}^{\max})} \left(\sum_{i\in I, t\in T} RR_{i,t,v} + \sum_{j\in J, p\in P, t\in T} MD_{j,p,t,v} \right.$$
$$+ \sum_{k\in K, p\in P, t\in T} MC_{k,p,t,v} - \sum_{i\in I, p\in P, t\in T} CE_{i,p,t,v} - \sum_{h\in H, p\in P, t\in T} CI_{h,p,t,v}$$
$$- \sum_{i\in I, p\in P, t\in T} CPD_{i,p,t,v} - \sum_{j\in J, p\in P, t\in T} CSD_{j,p,t,v} - \sum_{i\in I, t\in T} CIO_{i,t,v}$$
$$- \sum_{i\in I, p\in P, t\in T} CIR_{i,p,t,v} - \sum_{j\in J, p\in P, t\in T} CID_{j,p,t,v} - \sum_{k\in K, p\in P, t\in T} CIM_{k,p,t,v}$$
$$\left. - \sum_{k\in K, p\in P, t\in T} CUD_{k,p,t,v} \right)$$
$$\forall v \in V \tag{51}$$

$$RR_{i,t,v} = \sum_{p\in P} \left[XR_{i,p,t,v} \times \left(PP_{a,p,t} - TN_{r,p} \right) \right] \quad \forall i \in I, t \in T, v \in V, a = r = refinery \tag{52}$$

$$MD_{j,p,t,v} = \sum_{(k,m)\in Route_{i,j,m}} XS_{j,k,m,p,t,v} \times \left(PP_{a,p,t} - TN_{r,p} \right) - \sum_{(i,m)\in Route_{i,j,m}} XP_{i,j,m,p,t,v} \times PP_{a_2,p,t}$$
$$\forall j \in J, p \in P, t \in T, v \in V, a = r = depot, a_2 = refinery \tag{53}$$

$$ML_{k,p,t,v} = \left[\left(XRP_{k,p,t,v} - XU_{k,p,t,v} \right) \times \left(PP_{a_1,p,t} - TN_{r,p} \right) \right]$$
$$- \left[\left(\sum_{(i,m)\in Route_{i,k,m}} XP_{i,k,m,p,t,v} \right) \times PP_{a_2,p,t} + \left(\sum_{(j,m)\in Route_{j,k,m}} XS_{j,k,m,p,t,v} \right) \times PP_{a_3,p,t} \right]$$
$$\forall k \in K, p \in P, t \in T, v \in V, a_1 = r = retail, a_2 = refinery, a_3 = depot \tag{54}$$

$$CE_{i,p,t,v} = XE_{i,p,t,v} \times TN_{r,p} \quad \forall i \in I, p \in P, t \in T, v \in V, r = export \tag{55}$$

$$CI_{h,p,t,v} = XI_{h,p,t,v} \times \left(PP_{a,p} + TN_{r,p} \right) \quad \forall h \in H, p \in P, t \in T, v \in V, a = r = import \tag{56}$$

$$CPD_{i,p,t,v} = \sum_{(l,m)\in Route_{i,l,m}} \left(XP_{i,l,m,p,t,v} \times CT_{m,p} \times Dist_{i,l,m} \right) \quad \forall i \in I, p \in P, t \in T, v \in V \tag{57}$$

$$CSD_{j,p,t,v} = \sum_{(k,m)\in Route_{j,k,m}} \left(XS_{j,k,m,p,t,v} \times CT_{m,p} \times Dist_{j,k,m} \right) \quad \forall j \in J, p \in P, t \in T, v \in V \tag{58}$$

$$CIO_{i,t,v} = CKI \times IO_{i,t,v} \times PO_t \quad \forall i \in I, t \in T, v \in V, p = oil \tag{59}$$

$$CIR_{i,p,t,v} = CKI \times IR_{i,p,t,v} \times PP_{a,p,t} \quad \forall i \in I, p \in P, t \in T, v \in V, a = refinery \tag{60}$$

$$CID_{j,p,t,v} = CKI \times ID_{j,p,t,v} \times PP_{a,p,t} \quad \forall j \in J, p \in P, t \in T, v \in V, a = depot \tag{61}$$

$$CIM_{k,p,t,v} = CKI \times IM_{k,p,t,v} \times PP_{a,p,t} \quad \forall k \in K, p \in P, t \in T, v \in V, a = retail \tag{62}$$

$$CUD_{k,p,t,v} = XU_{k,p,t,v} \times TN_{r,p} \quad \forall k \in K, p \in P, t \in T, v \in V, r = unmet \tag{63}$$

$$XOP_{i,t,v} \leq PC_i \quad \forall i \in I, t \in T, v \in V \tag{64}$$

$$XR_{i,p,t,v} = XOP_{i,t,v} \times YF_{i,p} \quad \forall i \in I, p \in P, t \in T, v \in V \tag{65}$$

$$IO_{i,t,v} = ISO_i + XCO_{i,t} - XOP_{i,t,v} \quad \forall i \in I, v \in V, t \in T, t = t_1 \tag{66}$$

$$IO_{i,t,v} = IO_{i,t-1,v} + XCO_{i,t} - XOP_{i,t,v} \quad \forall i \in I, v \in V, t \in T, t > t_1 \tag{67}$$

$$IR_{i,p,t,v} = ISP_{i,p} + XR_{i,p,t,v} + XI_{i,p,t,v} - \sum_{(l,m)\in Route_{i,l,m}} XP_{i,l,m,p,t,v} - XE_{i,p,t,v}$$
$$\forall i \in I, p \in P, v \in V, t \in T, t = t_1 \tag{68}$$

$$IR_{i,p,t,v} = IR_{i,p,t-1,v} + XR_{i,p,t,v} + XI_{i,p,t,v} - \sum_{(l,m)\in Route_{i,l,m}} XP_{i,l,m,p,t,v} - XE_{i,p,t,v}$$
$$\forall i \in I, p \in P, v \in V, t \in T, t > t_1 \tag{69}$$

$$ID_{j,p,t,v} = ISP_{j,p} + XI_{j,p,t,v} + \sum_{(i,m)\in Route_{i,j,m}} XP_{i,j,m,p,t,v} - \sum_{(k,m)\in Route_{j,k,m}} XS_{j,k,m,p,t,v}$$
$$\forall j \in J, p \in P, v \in V, t \in T, t = t_1 \tag{70}$$

$$ID_{j,p,t,v} = ID_{j,p,t-1,v} + XI_{j,p,t,v} + \sum_{(i,m)\in Route_{i,j,m}} XP_{i,j,m,p,t,v} - \sum_{(k,m)\in Route_{j,k,m}} XS_{j,k,m,p,t,v}$$
$$\forall j \in J, p \in P, v \in V, t \in T, t > t_1 \tag{71}$$

$$IM_{k,p,t,v} = IM_{k,p,t-1,v} + \sum_{(i,m)\in Route_{i,k,m}} XP_{i,k,m,p,t,v} + \sum_{(j,m)\in Route_{j,k,m}} XS_{j,k,m,p,t,v}$$
$$+ XU_{k,p,t,v} - XRP_{k,p,t,v} + \Delta DP_{k,p,t,v} \tag{72}$$
$$\forall k \in K, p \in P, v \in V, t \in T, t > t_1 \tag{73}$$

$$XRP_{k,p,t,v} \leq \left(\widehat{DP}_{k,p,t,v} + \Delta DP_{k,p,t,v} \right) \quad \forall k \in K, p \in P, t \in T, v \in V \tag{74}$$

$$SSO_i \times SCO_i \leq IO_{i,t,v} \leq SCO_i \quad \forall i \in I, t \in T, v \in V \tag{75}$$

$$SSP_i \times SC_{i,p} \leq IR_{i,p,t,v} \leq SC_{i,p} \quad \forall i \in I, p \in P, t \in T, v \in V \tag{76}$$

$$SSP_j \times SC_{j,p} \leq ID_{j,p,t,v} \leq SC_{j,p} \quad \forall j \in J, p \in P, t \in T, v \in V \tag{77}$$

$$SSP_k \times SC_{k,p} \leq IM_{k,p,t,v} \leq SC_{k,p} \quad \forall k \in K, p \in P, t \in T, v \in V \tag{78}$$

$$\sum_{l \in Route_{i,l,m}} XP_{i,l,m,p,t,v} \leq ASPD_{i,m,p} \quad \forall i \in I, m \in M, p \in P, t \in T, v \in V \tag{79}$$

$$\sum_{k \in Route_{j,k,m}} XS_{j,k,m,p,t,v} \leq ASSD_{j,m,p} \quad \forall j \in J, m \in M, p \in P, t \in T, v \in V \tag{80}$$

$$\sum_{p \in P} XP_{i,l,m,p,t,v} \leq ACPD_{i,l,m} \quad \forall (i,l,m) \in Route_{i,l,m}, t \in T, v \in V \tag{81}$$

$$\sum_{p \in P} XS_{j,k,m,p,t,v} \leq ACSD_{j,k,m} \quad \forall (j,k,m) \in Route_{j,k,m}, t \in T, v \in V \tag{82}$$

$$\sum_{(i,m)\in Route_{i,j,m}} XP_{i,j,m,p,t,v} + XI_{j,p,t,v} \leq SC_{j,p} \times TCM_{j,p} \quad \forall j \in J, p \in P, t \in T, v \in V \tag{83}$$

$$\sum_{(i,m)\in Route_{i,k,m}} XP_{i,k,m,p,t,v} + \sum_{(j,m)\in Route_{j,k,m}} XS_{j,k,m,p,t,v} \leq SC_{k,p} \times TCM_{k,p} \quad \forall k \in K, p \in P, t \in T, v \in V \tag{84}$$

$$XP_{i,l,m,p,t,v} = 0 \quad \forall i \in I, l \in L, p \in P, t \in T, v \in V, m = road, \mu_{i,j,m} > MTD \tag{85}$$

$$XS_{j,k,m,p,t,v} = 0 \quad \forall j \in J, k \in K, p \in P, t \in T, v \in V, m = road, \mu_{j,k,m} > MTD \tag{86}$$

5. Case Study

The ARO model is applied to a real case study on a refined products network distribution in the Portuguese oil industry, which was originally characterized by Lima et al. [1]. Figure 1 represents the network under study.

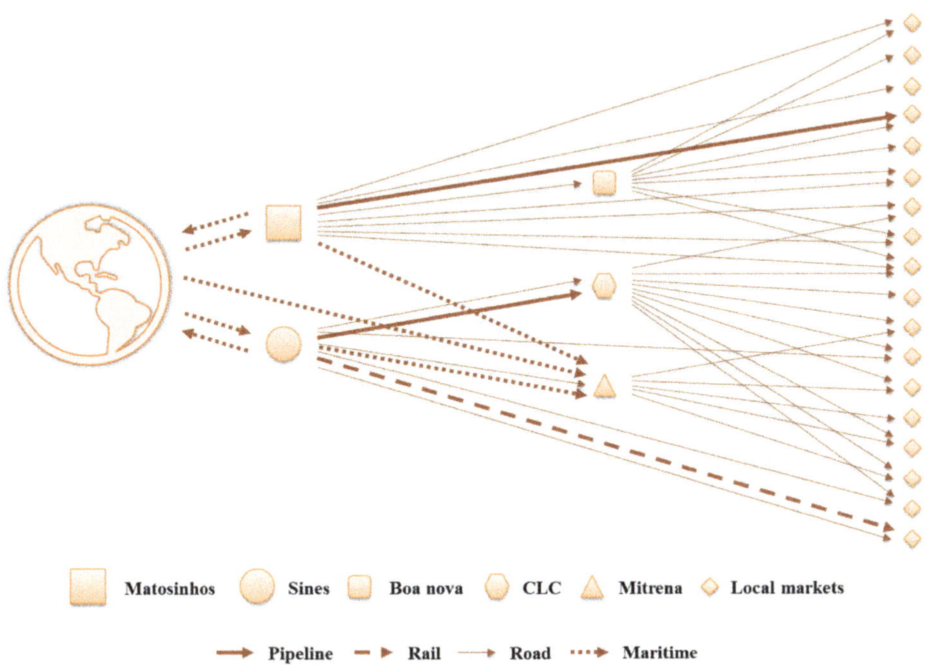

Figure 1. Portuguese downstream oil supply chain (from Lima et al., 2018).

The network comprises two oil refineries (Sines and Matosinhos), three distribution centers (Boa nova, CLC and Mitrena), 278 local markets (Portuguese cities) and four transportation modes (pipeline transport, railway, roadway and waterway). In this network, the oil refineries import the same type of crude oil and produce eight refined products, namely, jet fuel, diesel, propane, butane, fuel oil, gas oil, gasoline 95 and gasoline 98. These refined products are then transported using pipelines, tanker ships, tank wagons and tank trucks from the oil refineries, and using only tank trucks from the distribution centers. Roadway is only considered for distances less than or equal to 250 km. The planning horizon considered is 6 months, which is discretized into monthly cycles. For the sake of comparison, the same database employed by Lima et al. [1] is used in the case study.

6. Results and Discussion

The proposed two-stage ARO model is implemented in GAMS 24.5.6 and solved using CPLEX 12.6 on an Intel(R) Xeon(R) processor CPU E5-2660 v3 @ with 2.60 GHz (two processors) and 64 GB RAM memory.

6.1. Setup of the ARO Model

In the practical application, the ARO model was run under a specific configuration of the budget uncertainty set (Equations (46)–(48)). In other words, the ARO model was solved after specifying the parameters of the polyhedral uncertainty set, namely, the nominal value $\widehat{DP}_{k,p,t}$ and the maximum deviation in absolute value $\Delta DP^{max}_{k,p,t}$ of the uncertain product demand $\widetilde{DP}_{k,p,t}$, besides the budget of uncertainty $\Gamma_{k,p}$. The values of these parameters were estimated from the dataset described by Lima et al. [1]. Therefore, the maximum deviation $\Delta DP^{max}_{k,p,t}$ was estimated as 10% of the nominal value $\widehat{DP}_{k,p,t}$, while the budget of uncertainty $\Gamma_{k,p}$ was set to the average of the total relative deviations over the time horizon for the pairs of local market and product. To be more precise, for each pair of location and product (k,p), the absolute value of the product demand deviation $\Delta DP_{k,p,t}$ is normalized to the associated maximum deviation $\Delta DP^{max}_{k,p,t}$ at each time point. We then sum of all the so normalized deviations over the time horizon. Finally, we average these normalized values over all pairs of products and locations, and the result (let's denote it by $\widehat{\Gamma}_{k,p}$) is considered as the value of the budget of uncertainty, that is, $\widehat{\Gamma}_{k,p}$ for all (k,p). This results in $\widehat{\Gamma}_{k,p} = 0.895$, for all (k,p), which leads to uncertainty sets (i.e., Equations (46)–(48)) consisting in polyhedrons of 12 vertices.

6.2. Comparison with the Developed Approaches

In this section, we compare the results given by the proposed adjustable robust optimization (ARO) model with those obtained using the equivalent non-adjustable robust optimization (NARO), stochastic programming (SP) and the deterministic (DM) models. The mathematical formulations, the model notations and the specific settings for the DM, NARO and SP models can be found in Appendix B, Appendix C, Appendix D, respectively.

6.2.1. Computational Performance

Table 1 presents the problem size, economic performance and solving time for all the optimization models.

Table 1. Statistics.

Cases	Scenarios	Tree Nodes	Binary	Continuous	Equations	Profit (€)	Solution Time (s)
ARO	1	-	-	4,721,774	2,291,857	2,775,311,076	95.300
NARO	1	-	1	395,730	214,388	2,753,378,686	11.880
SP	243	364	-	7,249,759	3,052,522	2,779,767,979	236.800
DM	1	-	-	393,493	190,999	2,791,797,659	1.484

From Table 1, we can conclude that the economic performance (i.e., the profit) declines when uncertainty in demand for refined products is taken into consideration by the optimization model. On the other hand, a model in which all the parameters are considered as deterministic (DM) is not realistic, and it should consider random parameters to provide robustness and consistency to the decision maker.

In this way, the SP model aims at maximizing the expected profit for a set of scenarios, while the robust approaches maximize the total profit of the worst-case scenario. Compared to the ARO approach, the profit is higher about 0.16% for the SP approach. When comparing both robust approaches, the ARO model shows a better economic performance, providing a profit that is almost 0.80% higher than the one returned by the NARO model, which is the most conservative approach.

The problem size is also different for all the considered formulations. That is, the size of the deterministic model is smaller than the other methods, and thus it requires less computational time (1.484 CPU seconds). The NARO model does not increase too much the problem size and the computational time, because its size raises as more dual variables and robust constraints are incorporated into the problem, and such an increase is at the same scale as the number of the added random parameters, which in this case is only the demand for refined products. Nevertheless, much more computational burden is demanded to solve the much bigger ARO and SP models (95.3 and

236.8 CPU seconds, respectively). The size of the ARO model relies on the size of the polyhedral uncertainty set (Equations (46)–(48)), which in turn is determined by the budget of uncertainty $\Gamma_{k,p}$ in Equation (48). The size of the scenario-based SP model depends on the number of the considered scenarios, which in this case amounts to 243, in order to model the uncertainty in the stochastic parameter. All in all, the ARO approach does not increase the problem size as much as the SP approach when compared with the nominal problem (i.e., DM), having better solution performance, while it approximates better the economic result of the SP when comparing to the NARO approach.

6.2.2. Insights about the Network Profitability

The contribution of the different activities to the network profitability is illustrated in Figure 2 for all the optimization models. The oil procurement costs and the refinery revenues are all included into the margins of the refinery sector, so that the comparisons can be possible with the margins of the distribution and market sectors. In general, the profit breakdown is not too affected when the case study is solved by the different models. As the only difference in these approaches is how uncertainty is modeled, the reason comes from that. Firstly, the same uncertainty set $\mathcal{U}$ is considered by both robust optimization approaches. Secondly, the stochastic approach provides an average result of the considered scenarios, which are properly defined in accordance with the level of uncertainty considered by the robust approaches. Lastly, the product demand uncertainty does not actually interfere too much in the profit breakdown.

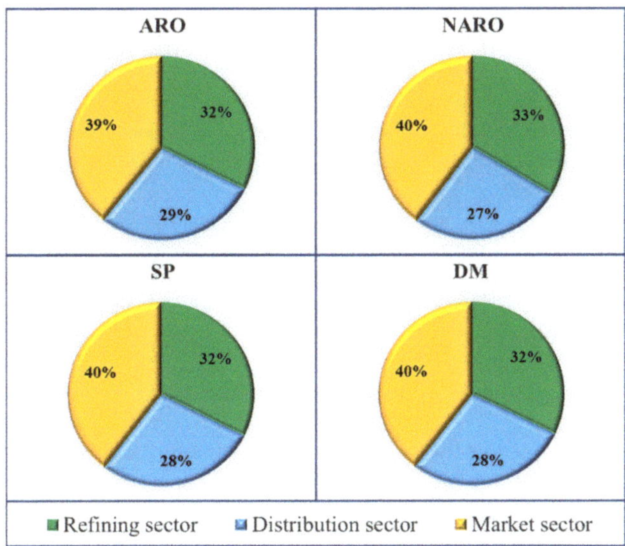

Figure 2. Contribution margins of the network echelons to the profitability.

As it can be seen in Figure 2, the market sector delivers the largest contribution to the network profitability, followed by the refining and distribution sectors, respectively. When comparing both robust approaches, the profit breakdown differs. The NARO model increases the shares of the refining and market sectors and decreases the contribution of the distribution sector when compared with the ARO model. The SP and DM models result in the same profit breakdown, which is slightly different from the one determined by the ARO model, where the market contribution rises 1% and the distribution sector diminishes 1%, whereas the share of the refining sector remains unaltered.

Figure 3 compares the effective profitability contribution of the network sectors among the different optimization models. All the optimization approaches provide quite similar results for the refining sector. The same fact is not observed in the distribution and market sectors, where the NARO model returns the lowest economic performance. These facts corroborate that the NARO model is the most conservative approach to handle uncertainty in this case study.

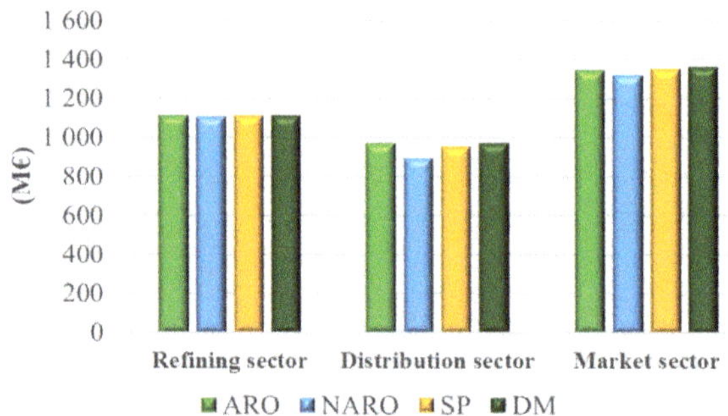

Figure 3. Comparison of profitability contributions among the optimization models.

In order to get more insights about the network profitability, the cost breakdown information is also investigated. As abovementioned, refinery margins comprise the revenues for selling oil products and the cost for acquiring crude oil. The oil procurement cost is the major cost item, and accounts for at least 78.53% of the overall cost in all the developed approaches, while the other costs together account for at most 21.47% of the same total. The oil procurement cost amounts to 78.53%, 80.92%, 78.94% and 78.54% of the total network cost in the ARO, NARO, SP and DM approaches, respectively. The cost breakdown information of the studied models is illustrated in Figure 4. Even though the exportation and lost demand costs are included in the estimation of the cost breakdown, their slices of the overall cost are quite small, i.e., almost 0%, and hence they have been omitted to enhance the readability of the charts without loss of generality.

As shown in Figure 4, the total network cost is divided into the same proportions for both ARO and DM approaches, while the proportions are much different for the other approaches. For instance, the SP and NARO approaches have increased the oil procurement cost by 1% and 3%, respectively, and have decreased the total of the other costs by 1% and 3%, respectively, when compared with the previous approaches. In most cases, the importation cost is the second most relevant cost item, followed by the secondary and primary transportation costs, respectively. The inventory cost aggregates the costs for keeping inventories of oil and refined products at the network facilities and amounts to 1% of the overall cost.

Figure 5 compares the absolute values of the cost items among the optimization methods. Notice that the oil procurement cost is not displayed, because it is much larger than the other cost items, and assumes quite similar values in all the methods, i.e., €2,408,811,163 on average. As it can be seen in Figure 5, the ARO and DM methods have much similar network costs, which are usually higher than the cost items for the NARO and SP models. Notice that the export and unmet demand costs are much smaller than the other cost items, and hence they are displayed in a different scale. Even though the NARO presents the best performance with regards to the network costs, its poorer performance in the network margins makes it as the most conservative approach. In turn, the SP model is the less conservative model, but it is not so efficient as the others in order to fulfill the required demand, and thus the lost demand is larger.

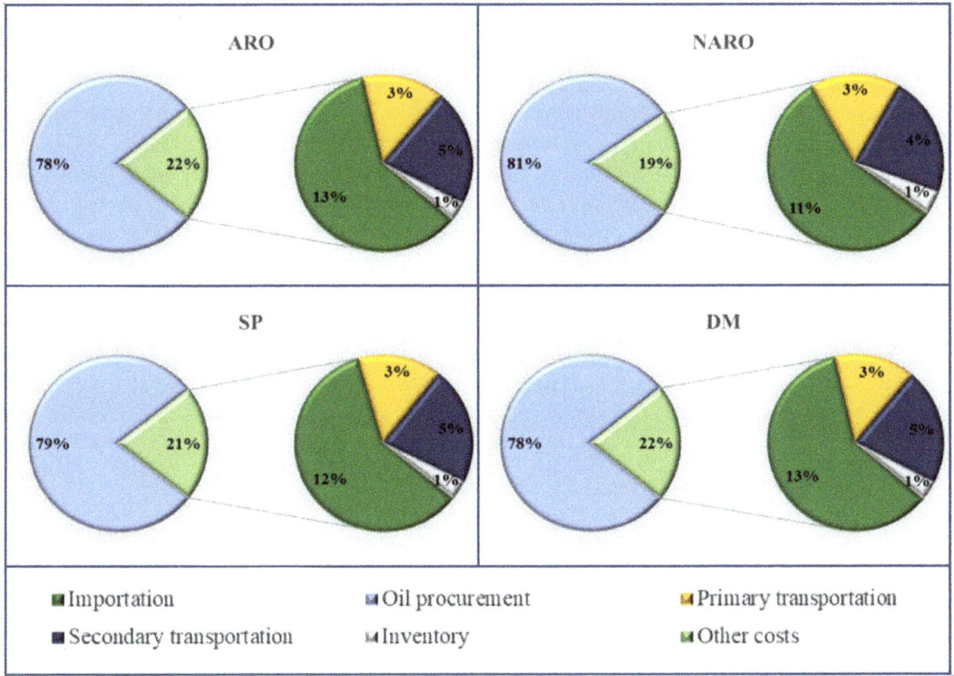

Figure 4. Cost breakdown for the optimization models.

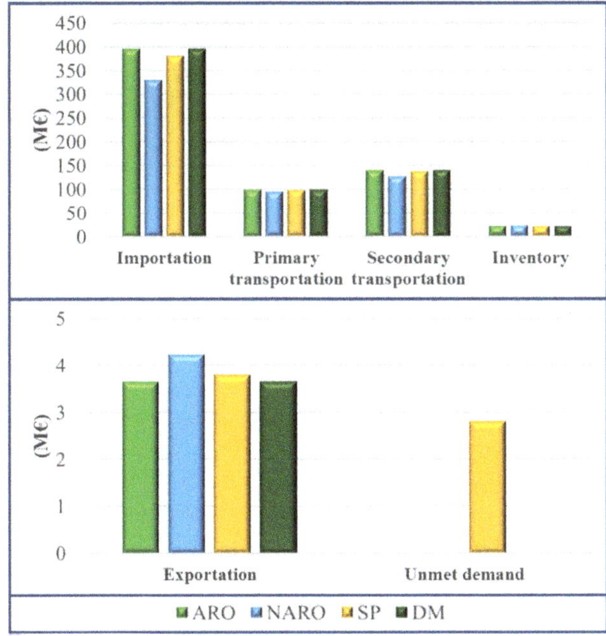

Figure 5. Cost items for the optimization models.

6.2.3. Network Planning for the Refined Products Distribution

Figure 6 shows how the total volume, which is transported over the network, is divided into different proportions by using the optimization models. As it can be observed, the ARO and DM models return the same volume portions, what is fully consistent with the cost breakdown information in Figure 4. However, the SP and NARO models provide different results when compared with the previous models, but some similarities are also observed. For instance, the amount of oil processed also corresponds to the largest slice of the total network volume, and the associated percentages are 26% and 28% in the SP and DM models, respectively. The oil delivery refers to the second biggest piece of the total volume, and accounts for 23% in the ARO, SP and DM models and 24% in the NARO model. Notice that, at all the volume charts, the oil processed is higher than the oil delivery due to the initial oil inventory at oil refineries in the first-time point, which is consumed throughout the planning horizon.

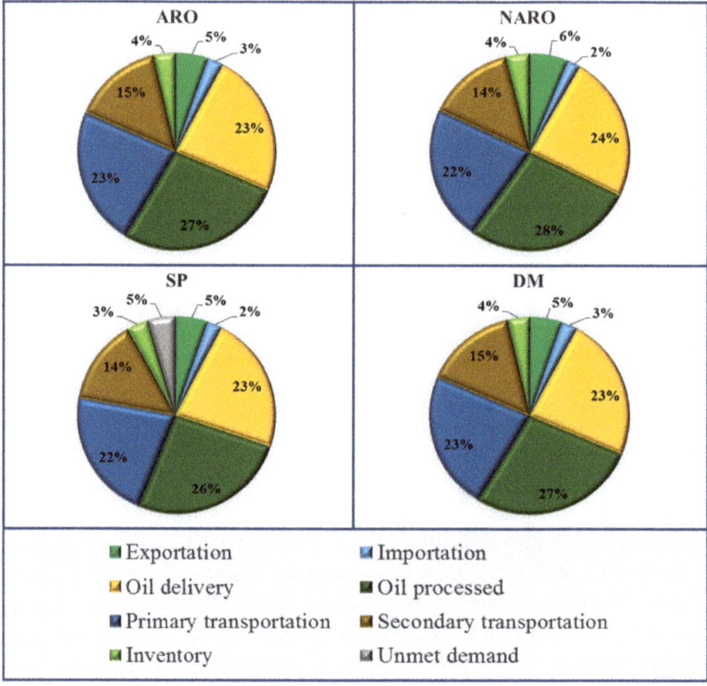

Figure 6. Proportions of the total network volumes among the optimization models.

As it can be seen in Figure 6, the portion of volumes conveyed through primary distribution is considerably bigger than the piece of volumes distributed via secondary distribution. The secondary distribution costs are much bigger than the primary distributions cost as displayed in Figures 4 and 5, because the former is only performed via roadway, i.e., the most expensive way to convey refined products in the network, while the latter can be undertaken by any transportation mode. Similarly, the importation volumes are lower than the exportation volumes, but they generate much bigger costs due to the importation tariffs that are paid to bring the refined product into the Portuguese network. The inventory volumes also present a certain relevance in the network and account for 3% or 4% of the total network volume.

Figure 7 displays the actual volumes determined by solving the case study using the optimization approaches. The NARO model presents the most different network flows among the optimization models, as well as the worst overall performance. Although the NARO model defines to purchase and process the same volumes of crude oil seen in the other models, the network flows through the primary and secondary distributions are significantly shorter, besides the inventories are sensibly higher. Hence, the NARO model does not provide a network that is so profitable as the other models do — see Table 1. Notice that these other models present much similar performances. However, the SP model is the only one that has a so evident lost demand.

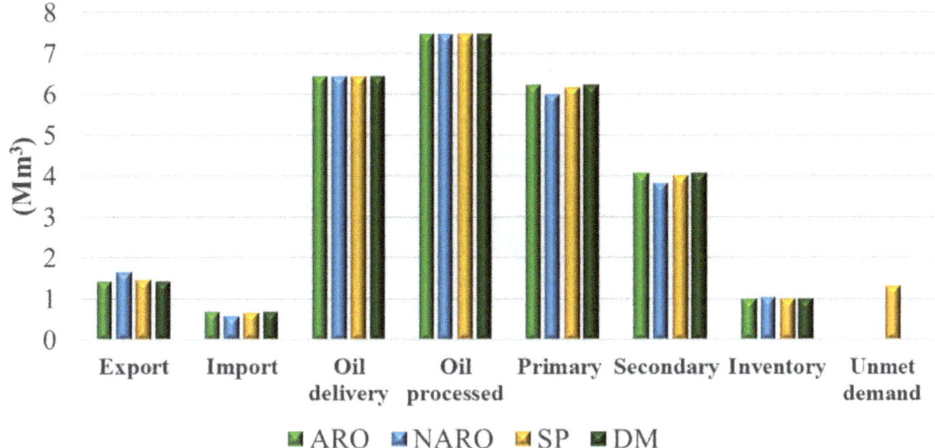

Figure 7. Network flows by solving the case study with different optimization models.

6.2.4. General Aspects about the Developed Modeling Frameworks

All the proposed approaches are useful to handle uncertainty in DOSC problems, and some conclusions can be withdrawn from the previous analyses. Under a second-stage stochastic programming approach, the decision variables are separated into two different groups, i.e., the first- and second-stage variables. This is a troublesome task, which depends on the decision maker's knowledge about the problem under study [8]. Nonetheless, once this separation is successfully performed, it may improve the model robustness against infeasibilities caused by the realization of the random parameters, because the second-stage variables might be properly adjusted to any particular realization of uncertainty [6]. In general, stochastic programming with recourse might be a good option when the probability distribution of the random parameters can be obtained from the historical data, such that a set of scenarios, i.e., a scenario tree, can be generated to represent the underlying uncertainty [32]. The decision maker can precisely model uncertainty by eliminating the undesirable scenarios and specifying the most adequate scenarios. However, the assignment of probabilities to scenarios, as well as the definition of scenario tree frameworks, could not be easy. Additionally, a wide range of scenarios should be considered to model uncertainty adequately, which could result in large-scale or even intractable mathematical programs [31]. Hence, the use of decomposition methods and approximation schemes for their solution are usually employed to solve this class of optimization problems [33].

In contrast to stochastic programming, robust optimization does not assume that uncertainty has a probability distribution [27], but alternatively it assumes that uncertainty is represented through uncertainty sets [29]. In this way, the decision maker can represent uncertainty in the random parameters by defining their nominal values and variation amplitudes from the historical data. Another advantage of employing robust optimization is the computational tractability for solving numerous classes of uncertainty sets and problem types [25]. In this methodology, the decision maker aims at establishing a feasible solution for any realization of the random parameters in a given uncertainty set [34], while the decision maker can control the trade-off between robustness and performance by using a budget of uncertainty that is introduced in the prescribed set [35]. However, a single-stage (non-adjustable) robust optimization approach tends to be very conservative because all decisions are made before uncertainty is revealed [27] while a two-stage adjustable robust optimization has a higher modeling capability, in which the second-stage problem models decision making after the first-stage decisions are made and uncertainty is disclosed [28].

Generally, the proposed modeling frameworks showed to be efficient and effective to cope with DOSC problems under uncertainty. These methodologies present different goals and consider specific assumptions and simplifications to represent uncertainty. Therefore, the most adequate methodology depends on the considered problem, the dataset and the decision maker preferences.

7. Conclusions

In this paper, a two-stage adjustable robust optimization model is developed to deal with demand uncertainty in the tactical planning of refined products distribution in a downstream oil supply chain. The adjustable robust model is then compared with its non-adjustable, stochastic and deterministic counterparts, whose objectives are different, that is, the robust approaches concern to maximize the profit at the worst-case scenario, the stochastic approach aims at maximizing the expected profit for a set of scenarios, and the deterministic approach at maximizing the total profit of the nominal problem. However, all the optimization approaches provide comparable results in terms of economic performance and material flows, and the major discrepancies occur with regards to problem sizes and computational properties.

Specifically, the obtained results show that the non-adjustable model is the most conservative, while the stochastic model is in turn the least conservative. However, the main drawback of the stochastic approach is the limitation of problem size due to the computational burden, while the major advantage of the non-adjustable robust approach is that the problem size is not overly enlarged with regards to the nominal deterministic approach. In comparison, the adjustable robust model establishes a problem that is not as short as its non-adjustable counterpart, as well as is not so big as its stochastic counterpart, so that the model tractability issues are reasonable. All the optimization approaches provide different network flows, but too comparable. The adjustable approach presents the best performance in this respect among the developed approaches to cope with uncertainty, because it provides the highest service level in order to fulfill the required demand for refined products. In contrast, the non-adjustable approach has the most inferior performance over the supply chain, whose network flows are majorly lower in the primary and secondary distributions, for example.

In summary, all the developed optimization approaches are valuable to deal adequately with refined products distribution under uncertainty. Each approach has a specific objective and assumes distinct assumptions and simplifications in order to model and represent adequately uncertainty. Hence, the most appropriate method depends on the problem under study, as well as on the available dataset and on the decision maker preferences.

As future work, and as a direct extension of the present work, the two-stage ARO model could be further explored to include other uncertainty sets, as well as modeling more than one uncertainty type simultaneously, ex. crude oil and product prices. Also, other approaches to deal with uncertainty could be explored such as fuzzy programming and chance-constrained programming. In addition, Markov chain and game theory might be investigated and employed to cope with the stochastic parameters. Finally, the studied approaches to deal with uncertainty could be also applied to the strategic and tactical problem of the downstream oil network allowing for the design and planning of such system.

Author Contributions: Writing—original draft, C.L.; Major revision—review and edition, S.R. and A.B.-P.; Mathematical model analysis—significant comments and suggestions, J.M.M.

Funding: This research received no external funding.

Conflicts of Interest: The authors declare no conflicts of interest.

Appendix A. Nomenclature

Table A1. Model notation.

Sets		
	$a \in A$	Set of developed activities
	$d \in D$	Set of route distances
	$m \in M$	Set of transportation modes
	$n, o \in N$	Set of all network nodes
	$p \in P$	Set of products
	$r \in R$	Set of resources and network stages
	$t, \tau \in T$	Set of time points
	$v \in V$	Set of vertices of the polyhedral uncertainty set
Subsets		
	$i \in I \subseteq N$	Set of refineries
	$j \in J \subseteq N$	Set of depots
	$k \in K \subseteq N$	Set of markets
Subset unions		
	$h \in H = I \cup J$	Set of refineries and depots
	$l \in L = J \cup K$	Set of depots and markets
Parameters		
	$ACPD_{n,o,m}$	Arc capacity between nodes n and o when transportation mode m is considered at primary distribution
	$ACSD_{n,o,m}$	Arc capacity between nodes n and o when transportation mode m is considered at secondary distribution
	$ASPD_{i,m,p}$	Availability of supplying product p from refinery i by transportation mode m through the primary distribution
	$ASSD_{j,m,p}$	Availability of supplying product p from depot j by transportation mode m through the secondary distribution
	CKI	Cost of keeping inventory defined as a percentage of the inventory value
	$CT_{m,p}$	Transportation cost per transportation mode m and product p
	$Dist_{n,o,m}$	Distance between nodes n and o depending on transportation mode m
	$DO_{i,t}$	Demand of oil at refinery i at time point t
	$\widetilde{DP}_{k,p,t,v}$	True value of demand of product p per market k at time point t and vertice v
	$\overline{DP}_{k,p,t,v}$	Nominal value of demand of product p per market k at time point t and vertice v
	ISO_i	Initial stock of oil at refinery i
	$ISP_{n,p}$	Initial stock of product p at node n
	MTD	Maximum travel distance in meters allowed in the road transportation mode
	PC_i	Processing capacity at refinery i
	PO_t	Price of oil at time point t
	$PP_{a,p,t}$	Price of product p at activity a at time point t
	$Route_{n,o,m}$	Route between nodes n and o connected by transportation mode m
	$SC_{n,p}$	Storage capacity of product p at node n
	SCO_i	Storage capacity of oil at refinery i
	SSO_i	Safety stock of oil at refinery i defined as a percentage of the overall oil storage capacity
	SSP_n	Safety stock of products at node n defined as a percentage of the overall storage capacity
	$TN_{r,p}$	Tariff per network stage r and product p
	$TCM_{n,p}$	Throughput capacity multiplier per node n and product p
	$YF_{i,p}$	Yield fractions by refinery i of product p per cubic meters of oil
	$\Gamma_{k,p}$	Budget of uncertainty for deviations of the demand for product p at market k
	$\Delta DP_{k,p,t,v}$	Deviation of the demand of product p at local market k at time point t and vertice v
	$\Delta DP^{max}_{k,p,t,v}$	Maximum deviation of the demand of product p at local market k at time point t and vertice v

Table A1. *Cont.*

Positive continuous variables	
$CE_{i,p,t,v}$	Cost of exporting product p by refinery i at time point t and vertice v
$CI_{h,p,t,v}$	Cost of importing product p by refinery or depot h at time point t and vertice v
$CID_{j,p,t,v}$	Cost of inventory at depot j for product p at time point t and vertice v
$CIM_{k,p,t,v}$	Cost of inventory at market k for product p at time point t and vertice v
$CIO_{i,t,v}$	Cost of inventory for oil at refinery i at time point t and vertice v
$CIR_{i,p,t,v}$	Cost of inventory at refinery i for product p at time point t and vertice v
$CPD_{i,p,t,v}$	Cost of primary distribution from refinery i for product p at time point t and vertice v
$CSD_{j,p,t,v}$	Cost of secondary distribution from depot j for product p at time point t and vertice v
$CUD_{k,p,t,v}$	Cost of unsatisfied demand for product p at market k at time point t and vertice v
$ID_{j,p,t,v}$	Inventory of product p at depot j at time point t and vertice v
$IM_{k,p,t,v}$	Inventory of product p at market k at time point t and vertice v
$IO_{i,t,v}$	Inventory of oil at refinery i at time point t and vertice v
$IR_{i,p,t,v}$	Inventory of product p at refinery i at time point t and vertice v
$XCO_{i,t,v}$	Volume of crude oil received by refinery i at time point t and vertice v
$XE_{i,p,t,v}$	Volume of product p exported by refinery i at time point t and vertice v
$XI_{h,p,t,v}$	Volume of product p imported by refinery or depot h at time point t and vertice v
$XOP_{i,t,v}$	Volume of oil processed by refinery i at time point t and vertice v
$XP_{i,l,m,p,t,v}$	Volume of product p sent by refinery i to location l by transportation mode m at time point t and vertice v
$XR_{i,p,t,v}$	Volume of product p yielded by refinery i at time point t and vertice v
$XRP_{k,p,t,v}$	Volume of product p delivered to market k at time point t and vertice v
$XS_{j,k,m,p,t,v}$	Volume of product p sent by depot j to market k by transportation mode m at time point t and vertice v
$XU_{k,p,t,v}$	Volume of unsatisfied demand per market k and product p at time point t and vertice v
Continuous variables	
$MD_{j,p,t,v}$	Margin per depot j and product p at time point t and vertice v
$MC_{k,p,t,v}$	Margin per local market k and product p at time point t and vertice v
$RR_{i,t,v}$	Revenue per refinery i at time point t and vertice v
$Q_u(\cdot)$	The worst-case recourse profit
WCP	The worst-case profit for the downstream oil network

Appendix B. Deterministic Mathematical Formulation

The deterministic formulation is briefly reported below through its objective function (Equations (A1)–(A13)), network equations (Equations (A14)–(A37)) and model notation (Table A2). For the sake of brevity, the descriptions of the equations are omitted, but the adjustable robust optimization (ARO) counterpart can be consulted within the full paper for further details.

Objective function

$$
\begin{aligned}
\underset{\Theta}{\text{maximize}}\, \pi = {} & \sum_{i \in I, t \in T} MR_{i,t} + \sum_{j \in J, p \in P, t \in T} MD_{j,p,t} + \sum_{k \in K, p \in P, t \in T} MC_{k,p,t} - \sum_{i \in I, p \in P, t \in T} CE_{i,p,t} \\
& - \sum_{h \in H, p \in P, t \in T} CI_{h,p,t} - \sum_{i \in I, p \in P, t \in T} CPD_{i,p,t} - \sum_{j \in J, p \in P, t \in T} CSD_{j,p,t} \\
& - \sum_{i \in I, t \in T} CIO_{i,t} - \sum_{i \in I, p \in P, t \in T} CIR_{i,p,t} - \sum_{j \in J, p \in P, t \in T} CID_{j,p,t} \\
& - \sum_{k \in K, p \in P, t \in T} CIM_{k,p,t} - \sum_{k \in K, p \in P, t \in T} CUD_{k,p,t}
\end{aligned}
\tag{A1}
$$

$$
MR_{i,t} = \sum_{p \in P} \left[XR_{i,p,t} \times \left(PP_{a,p,t} - TN_{r,p} \right) \right] - \left[XCO_{i,t} \times PO_t \right] \quad \forall i \in I, t \in T, a = r = refinery
\tag{A2}
$$

$$
MD_{j,p,t} = \sum_{(k,m) \in Route_{i,j,m}} XS_{j,k,m,p,t} \times \left(PP_{a,p,t} - TN_{r,p} \right) - \sum_{(i,m) \in Route_{i,j,m}} XP_{i,j,m,p,t} \times PP_{a_2,p,t}
\tag{A3}
$$
$$
\forall j \in J, p \in P, t \in T, a = r = depot, a_2 = refinery
$$

$$MC_{k,p,t} = \left[\left(XRP_{k,p,t} - XU_{k,p,t}\right) \times \left(PP_{a_1,p,t} - TN_{r,p}\right)\right] - \left[\left(\sum_{(i,m)\in Route_{i,k,m}} XP_{i,k,m,p,t}\right)\right.$$
$$\left. \times PP_{a_2,p,t} + \left(\sum_{(j,m)\in Route_{j,k,m}} XS_{j,k,m,p,t}\right) \times PP_{a_3,p,t}\right] \tag{A4}$$
$$\forall k \in K, p \in P, t \in T, a_1 = r = retail, a_2 = refinery, a_3 = depot$$

$$CE_{i,p,t} = XE_{i,p,t} \times TN_{r,p} \quad \forall i \in I, p \in P, t \in T, r = export \tag{A5}$$

$$CI_{h,p,t} = XI_{h,p,t} \times \left(PP_{a,p} + TN_{r,p}\right) \quad \forall h \in H, p \in P, t \in T, a = r = import \tag{A6}$$

$$CPD_{i,p,t} = \sum_{(l,m)\in Route_{i,l,m}} \left(XP_{i,l,m,p,t} \times CT_{m,p} \times Dist_{i,l,m}\right) \quad \forall i \in I, p \in P, t \in T \tag{A7}$$

$$CSD_{j,p,t} = \sum_{(k,m)\in Route_{j,k,m}} \left(XS_{j,k,m,p,t} \times CT_{m,p} \times Dist_{j,k,m}\right) \quad \forall j \in J, p \in P, t \in T \tag{A8}$$

$$CIO_{i,t} = CKI \times IO_{i,t} \times PO_t \quad \forall i \in I, t \in T, p = oil \tag{A9}$$

$$CIR_{i,p,t} = CKI \times IR_{i,p,t} \times PP_{a,p,t} \quad \forall i \in I, p \in P, t \in T, a = refinery \tag{A10}$$

$$CID_{j,p,t} = CKI \times ID_{j,p,t} \times PP_{a,p,t} \quad \forall j \in J, p \in P, t \in T, a = depot \tag{A11}$$

$$CIM_{k,p,t} = CKI \times IM_{k,p,t} \times PP_{a,p,t} \quad \forall k \in K, p \in P, t \in T, a = retail \tag{A12}$$

$$CUD_{k,p,t} = XU_{k,p,t} \times TN_{r,p} \quad \forall k \in K, p \in P, t \in T, r = unmet \tag{A13}$$

Network equations

$$XOP_{i,t} \leq PC_i \quad \forall i \in I, t \in T \tag{A14}$$

$$XR_{i,p,t} = XOP_{i,t} \times YF_{i,p} \quad \forall i \in I, p \in P, t \in T \tag{A15}$$

$$XCO_{i,t} \leq DO_{i,t} \quad \forall i \in I, t \in T \tag{A16}$$

$$IO_{i,t} = ISO_i + XCO_{i,t} - XOP_{i,t} \quad \forall i \in I, t \in T, t = t_1 \tag{A17}$$

$$IO_{i,t} = IO_{i,t-1} + XCO_{i,t} - XOP_{i,t} \quad \forall i \in I, t \in T, t > t_1 \tag{A18}$$

$$IR_{i,p,t} = ISP_{i,p} + XR_{i,p,t} + XI_{i,p,t} - \sum_{(l,m)\in Route_{i,l,m}} XP_{i,l,m,p,t} - XE_{i,p,t} \quad \forall i \in I, p \in P, t \in T, t = t_1 \tag{A19}$$

$$IR_{i,p,t} = IR_{i,p,t-1} + XR_{i,p,t} + XI_{i,p,t} - \sum_{(l,m)\in Route_{i,l,m}} XP_{i,l,m,p,t} - XE_{i,p,t} \quad \forall i \in I, p \in P, t \in T, t > t_1 \tag{A20}$$

$$ID_{j,p,t} = ISP_{j,p} + XI_{j,p,t} + \sum_{(i,m)\in Route_{i,j,m}} XP_{i,j,m,p,t} - \sum_{(k,m)\in Route_{j,k,m}} XS_{j,k,m,p,t}$$
$$\forall j \in J, p \in P, t \in T, t = t_1 \tag{A21}$$

$$ID_{j,p,t} = ID_{j,p,t-1} + XI_{j,p,t} + \sum_{(i,m)\in Route_{i,j,m}} XP_{i,j,m,p,t} - \sum_{(k,m)\in Route_{j,k,m}} XS_{j,k,m,p,t}$$
$$\forall j \in J, p \in P, t \in T, t > t_1 \tag{A22}$$

$$IM_{k,p,t} = ISP_{k,p} + \sum_{(i,m)\in Route_{i,k,m}} XP_{i,k,m,p,t} + \sum_{(j,m)\in Route_{j,k,m}} XS_{j,k,m,p,t} + XU_{k,p,t} - XRP_{k,p,t}$$
$$\forall k \in K, p \in P, t \in T, t = t_1 \tag{A23}$$

$$IM_{k,p,t} = IM_{k,p,t-1} + \sum_{(i,m)\in Route_{i,k,m}} XP_{i,k,m,p,t} + \sum_{(j,m)\in Route_{j,k,m}} XS_{j,k,m,p,t} + XU_{k,p,t} - XRP_{k,p,t}$$
$$\forall k \in K, p \in P, t \in T, t > t_1 \tag{A24}$$

$$XRP_{k,p,t} \leq DP_{k,p,t} \quad \forall k \in K, p \in P, t \in T \tag{A25}$$

$$SSO_i \times SCO_i \leq IO_{i,t} \leq SCO_i \quad \forall i \in I, t \in T \tag{A26}$$

$$SSP_i \times SC_{i,p} \leq IR_{i,p,t} \leq SC_{i,p} \quad \forall i \in I, p \in P, t \in T \tag{A27}$$

$$SSP_j \times SC_{j,p} \leq ID_{j,p,t} \leq SC_{j,p} \quad \forall j \in J, p \in P, t \in T \tag{A28}$$

$$SSP_k \times SC_{k,p} \leq IM_{k,p,t} \leq SC_{k,p} \quad \forall k \in K, p \in P, t \in T \tag{A29}$$

$$\sum_{l \in Route_{i,l,m}} XP_{i,l,m,p,t} \leq ASPD_{i,m,p} \quad \forall i \in I, m \in M, p \in P, t \in T \tag{A30}$$

$$\sum_{k \in Route_{j,k,m}} XS_{j,k,m,p,t} \leq ASSD_{j,m,p} \quad \forall j \in J, m \in M, p \in P, t \in T \tag{A31}$$

$$\sum_{p \in P} XP_{i,l,m,p,t} \leq ACPD_{i,l,m} \quad \forall (i,l,m) \in Route_{i,l,m}, t \in T \tag{A32}$$

$$\sum_{p \in P} XS_{j,k,m,p,t} \leq ACSD_{j,k,m} \quad \forall (j,k,m) \in Route_{j,k,m}, t \in T \tag{A33}$$

$$\sum_{(i,m) \in Route_{i,j,m}} XP_{i,j,m,p,t} + XI_{j,p,t} \leq SC_{j,p} \times TCM_{j,p} \quad \forall j \in J, p \in P, t \in T \tag{A34}$$

$$\sum_{(i,m) \in Route_{i,k,m}} XP_{i,k,m,p,t} + \sum_{(j,m) \in Route_{j,k,m}} XS_{j,k,m,p,t} \leq SC_{k,p} \times TCM_{k,p} \quad \forall k \in K, p \in P, t \in T \tag{A35}$$

$$XP_{i,l,m,p,t} = 0 \quad \forall i \in I, l \in L, p \in P, t \in T, m = road, \mu_{i,j,m} > MTD \tag{A36}$$

$$XS_{j,k,m,p,t} = 0 \quad \forall j \in J, k \in K, p \in P, t \in T, m = road, \mu_{j,k,m} > MTD \tag{A37}$$

Table A2. Model notation.

Sets		
	$a \in A$	Set of developed activities
	$d \in D$	Set of route distances
	$m \in M$	Set of transportation modes
	$n,o \in N$	Set of all network nodes
	$p \in P$	Set of products
	$r \in R$	Set of resources and network stages
	$t, \tau \in T$	Set of time points
	$\Theta \in \mathbb{R}_+$	Set of optimization variables: $\Theta = \{XR, XOP, XCO, XP, XS, XRP, XU, XE, XI, IO, IR, ID, IM\}$
Subsets		
	$i \in I \subseteq N$	Set of refineries
	$j \in J \subseteq N$	Set of depots
	$k \in K \subseteq N$	Set of markets
Subset unions		
	$h \in H = I \cup J$	Set of refineries and depots
	$l \in L = J \cup K$	Set of depots and markets
Parameters		
	$ACPD_{n,o,m}$	Arc capacity between nodes n and o when transportation mode m is considered at primary distribution
	$ACSD_{n,o,m}$	Arc capacity between nodes n and o when transportation mode m is considered at secondary distribution
	$ASPD_{i,m,p}$	Availability of supplying product p from refinery i by transportation mode m through the primary distribution
	$ASSD_{j,m,p}$	Availability of supplying product p from depot j by transportation mode m through the secondary distribution
	CKI	Cost of keeping inventory defined as a percentage of the inventory value
	$CT_{m,p}$	Transportation cost per transportation mode m and product p
	$Dist_{n,o,m}$	Distance between nodes n and o depending on transportation mode m
	$DO_{i,t}$	Demand of oil at refinery i at time point t
	$DP_{k,p,t}$	Demand of product p per market k at time point t
	ISO_i	Initial stock of oil at refinery i
	$ISP_{n,p}$	Initial stock of product p at node n
	MTD	Maximum travel distance in meters allowed in the road transportation mode
	PC_i	Processing capacity at refinery i
	PO_t	Price of oil at time point t
	$PP_{a,p,t}$	Price of product p at activity a at time point t
	$Route_{n,o,m}$	Route between nodes n and o connected by transportation mode m
	$SC_{n,p}$	Storage capacity of product p at node n
	SCO_i	Storage capacity of oil at refinery i
	SSO_i	Safety stock of oil at refinery i defined as a percentage of the overall oil storage capacity
	SSP_n	Safety stock of products at node n defined as a percentage of the overall storage capacity
	$TN_{r,p}$	Tariff per network stage r and product p
	$TCM_{n,p}$	Throughput capacity multiplier per node n and product p
	$YF_{i,p}$	Yield fractions by refinery i of product p per cubic meters of oil

Table A2. *Cont.*

Positive continuous variables	
$CE_{i,p,t}$	Cost of exporting product p by refinery i at time point t
$CI_{h,p,t}$	Cost of importing product p by refinery or depot h at time point t
$CID_{j,p,t}$	Cost of inventory at depot j for product p at time point t
$CIM_{k,p,t}$	Cost of inventory at market k for product p at time point t
$CIO_{i,t}$	Cost of inventory for oil at refinery i at time point t
$CIR_{i,p,t}$	Cost of inventory at refinery i for product p at time point t
$CPD_{i,p,t}$	Cost of primary distribution from refinery i for product p at time point t
$CSD_{j,p,t}$	Cost of secondary distribution from depot j for product p at time point t
$CUD_{k,p,t}$	Cost of unsatisfied demand for product p at market k at time point t
$ID_{j,p,t}$	Inventory of product p at depot j at time point t
$IM_{k,p,t}$	Inventory of product p at market k at time point t
$IO_{i,t}$	Inventory of oil at refinery i at time point t
$IR_{i,p,t}$	Inventory of product p at refinery i at time point t
$XCO_{i,t}$	Volume of crude oil received by refinery i at time point t
$XE_{i,p,t}$	Volume of product p exported by refinery i at time point t
$XI_{h,p,t}$	Volume of product p imported by refinery or depot h at time point t
$XOP_{i,t}$	Volume of oil processed by refinery i at time point t
$XP_{i,l,m,p,t}$	Volume of product p sent by refinery i to depot or market l by transportation mode m at time point t
$XR_{i,p,t}$	Volume of product p yielded by refinery i at time point t
$XRP_{k,p,t}$	Volume of product p delivered to market k at time point t
$XS_{j,k,m,p,t}$	Volume of product p sent by depot j to market k by transportation mode m at time point t
$XU_{k,p,t}$	Volume of unsatisfied demand per market k and product p at time point t
Continuous variables	
$MD_{j,p,t}$	Margin per depot j and product p at time point t
$MC_{k,p,t}$	Margin per consumer market k and product p at time point t
$MR_{i,p,t}$	Margin per refinery i at time point t
π	Deterministic objective function

Appendix C. Non-Adjustable Robust Optimization (NARO) Mathematical Formulation

The non-adjustable robust formulation considers the same polyhedral budget uncertainty set of its equivalent ARO model, presented in the full paper. Hereinafter, the NARO model is introduced through its objective function (Equations (A38)–(A50)), network equations (Equations (A51)–(A77)) and model notation (Table A3). The presentation of the ARO model in the full paper must be consulted for details on the descriptions of the equations.

Objective function

$$\underset{\substack{profit,\Theta,\\ \xi^{+demand}_{k,p,t},\,\xi^{-demand}_{k,p,t},\,\eta^{demand}_{k,p}}}{\text{maximize}} \quad profit$$

$$
\begin{aligned}
\text{s.t. } profit &- \sum_{i\in I, t\in T} MR_{i,t} + \sum_{j\in J, p\in P, t\in T} MD_{j,p,t} + \sum_{k\in K, p\in P, t\in T} MC_{k,p,t} - \sum_{i\in I, p\in P, t\in T} CE_{i,p,t}\\
&- \sum_{h\in H, p\in P, t\in T} CI_{h,p,t} - \sum_{i\in I, p\in P, t\in T} CPD_{i,p,t} - \sum_{j\in J, p\in P, t\in T} CSD_{j,p,t}\\
&- \sum_{i\in I, t\in T} CIO_{i,t} - \sum_{i\in I, p\in P, t\in T} CIR_{i,p,t} - \sum_{j\in J, p\in P, t\in T} CID_{j,p,t}\\
&- \sum_{k\in K, p\in P, t\in T} CIM_{k,p,t} - \sum_{k\in K, p\in P, t\in T} CUD_{k,p,t} \leq 0
\end{aligned}
\tag{A38}
$$

$$MR_{i,t} = \sum_{p\in P}\left[XR_{i,p,t} \times \left(PP_{a,p,t} - TN_{r,p}\right)\right] - \left[XCO_{i,t} \times PO_t\right] \forall i \in I, t \in T, a = r = refinery \tag{A39}$$

$$
\begin{aligned}
MD_{j,p,t} &= \sum_{(k,m)\in Route_{i,j,m}} XS_{j,k,m,p,t} \times \left(PP_{a,p,t} - TN_{r,p}\right) - \sum_{(i,m)\in Route_{i,j,m}} XP_{i,j,m,p,t} \times PP_{a_2,p,t}\\
&\forall j \in J, p \in P, t \in T, a = r = depot, a_2 = refinery
\end{aligned}
\tag{A40}
$$

$$
\begin{aligned}
MC_{k,p,t} &= \left[\left(XRP_{k,p,t} - XU_{k,p,t}\right) \times \left(PP_{a_1,p,t} - TN_{r,p}\right)\right] \\
&\quad - \left[\left(\sum_{(i,m)\in Route_{i,k,m}} XP_{i,k,m,p,t}\right) \times PP_{a_2,p,t} + \left(\sum_{(j,m)\in Route_{j,k,m}} XS_{j,k,m,p,t}\right)\right] \\
&\quad\quad \times PP_{a_3,p,t}
\end{aligned}
$$

$$\forall k \in K, p \in P, t \in T, a_1 = r = retail, a_2 = refinery, a_3 = depot \tag{A41}$$

$$CE_{i,p,t} = XE_{i,p,t} \times TN_{r,p} \quad \forall i \in I, p \in P, t \in T, r = export \tag{A42}$$

$$CI_{h,p,t} = XI_{h,p,t} \times \left(PP_{a,p} + TN_{r,p}\right) \quad \forall h \in H, p \in P, t \in T, a = r = import \tag{A43}$$

$$CPD_{i,p,t} = \sum_{(l,m)\in Route_{i,l,m}} \left(XP_{i,l,m,p,t} \times CT_{m,p} \times Dist_{i,l,m}\right) \quad \forall i \in I, p \in P, t \in T \tag{A44}$$

$$CSD_{j,p,t} = \sum_{(k,m)\in Route_{j,k,m}} \left(XS_{j,k,m,p,t} \times CT_{m,p} \times Dist_{j,k,m}\right) \quad \forall j \in J, p \in P, t \in T \tag{A45}$$

$$CIO_{i,t} = CKI \times IO_{i,t} \times PO_t \quad \forall i \in I, t \in T, p = oil \tag{A46}$$

$$CIR_{i,p,t} = CKI \times IR_{i,p,t} \times PP_{a,p,t} \quad \forall i \in I, p \in P, t \in T, a = refinery \tag{A47}$$

$$CID_{j,p,t} = CKI \times ID_{j,p,t} \times PP_{a,p,t} \quad \forall j \in J, p \in P, t \in T, a = depot \tag{A48}$$

$$CIM_{k,p,t} = CKI \times IM_{k,p,t} \times PP_{a,p,t} \quad \forall k \in K, p \in P, t \in T, a = retail \tag{A49}$$

$$CUD_{k,p,t} = XU_{k,p,t} \times TN_{r,p} \quad \forall k \in K, p \in P, t \in T, r = unmet \tag{A50}$$

Network equations

$$XOP_{i,t} \leq PC_i \quad \forall i \in I, t \in T \tag{A51}$$

$$XR_{i,p,t} = XOP_{i,t} \times YF_{i,p} \quad \forall i \in I, p \in P, t \in T \tag{A52}$$

$$XCO_{i,t} \leq DO_{i,t} \quad \forall i \in I, t \in T \tag{A53}$$

$$IO_{i,t} = ISO_i + XCO_{i,t} - XOP_{i,t} \quad \forall i \in I, t \in T, t = t_1 \tag{A54}$$

$$IO_{i,t} = IO_{i,t-1} + XCO_{i,t} - XOP_{i,t} \quad \forall i \in I, t \in T, t > t_1 \tag{A55}$$

$$IR_{i,p,t} = ISP_{i,p} + XR_{i,p,t} + XI_{i,p,t} - \sum_{(l,m)\in Route_{i,l,m}} XP_{i,l,m,p,t} - XE_{i,p,t} \quad \forall i \in I, p \in P, t \in T, t = t_1 \tag{A56}$$

$$IR_{i,p,t} = IR_{i,p,t-1} + XR_{i,p,t} + XI_{i,p,t} - \sum_{(l,m)\in Route_{i,l,m}} XP_{i,l,m,p,t} - XE_{i,p,t} \quad \forall i \in I, p \in P, t \in T, t > t_1 \tag{A57}$$

$$
\begin{aligned}
ID_{j,p,t} &= ISP_{j,p} + XI_{j,p,t} + \sum_{(i,m)\in Route_{i,j,m}} XP_{i,j,m,p,t} - \sum_{(k,m)\in Route_{j,k,m}} XS_{j,k,m,p,t} \\
&\quad \forall j \in J, p \in P, t \in T, t = t_1
\end{aligned}
\tag{A58}
$$

$$
\begin{aligned}
ID_{j,p,t} &= ID_{j,p,t-1} + XI_{j,p,t} + \sum_{(i,m)\in Route_{i,j,m}} XP_{i,j,m,p,t} - \sum_{(k,m)\in Route_{j,k,m}} XS_{j,k,m,p,t} \\
&\quad \forall j \in J, p \in P, t \in T, t > t_1
\end{aligned}
\tag{A59}
$$

$$
\begin{aligned}
IM_{k,p,t} &= ISP_{k,p} + \sum_{(i,m)\in Route_{i,k,m}} XP_{i,k,m,p,t} + \sum_{(j,m)\in Route_{j,k,m}} XS_{j,k,m,p,t} + XU_{k,p,t} - XRP_{k,p,t} \\
&\quad \forall k \in K, p \in P, t \in T, t = t_1
\end{aligned}
\tag{A60}
$$

$$
\begin{aligned}
IM_{k,p,t} &= IM_{k,p,t-1} + \sum_{(i,m)\in Route_{i,k,m}} XP_{i,k,m,p,t} + \sum_{(j,m)\in Route_{j,k,m}} XS_{j,k,m,p,t} + XU_{k,p,t} - XRP_{k,p,t} \\
&\quad \forall k \in K, p \in P, t \in T, t > t_1
\end{aligned}
\tag{A61}
$$

$$XRP_{k,p,t} - DP_{k,p,t}Y_g + DP_{k,p,t}^{max}\left(\xi_{k,p,t}^{+demand} + \xi_{k,p,t}^{-demand}\right) + \Gamma_{k,p}^{demand}\eta_{k,p}^{demand} \leq 0 \tag{A62}$$
$$\forall k \in K, p \in P, t \in T, g = |I| + 1$$

$$\xi_{k,pt}^{+demand} + \frac{\eta_{k,p}^{demand}}{DP_{k,p,t}^{max}} \geq +Y_g \quad \forall i \in I, g = |I| + 1, t \in T \tag{A63}$$

$$\xi_{k,p,t}^{-demand} + \frac{\eta_{k,p}^{demand}}{DP_{k,p,t}^{max}} \geq -Y_g \quad \forall i \in I, g = |I| + 1, t \in T \tag{A64}$$

$$Y_g = 1g = |I| + 1 \tag{A65}$$

$$SSO_i \times SCO_i \leq IO_{i,t} \leq SCO_i \quad \forall i \in I, t \in T \tag{A66}$$

$$SSP_i \times SC_{i,p} \leq IR_{i,p,t} \leq SC_{i,p} \quad \forall i \in I, p \in P, t \in T \tag{A67}$$

$$SSP_j \times SC_{j,p} \leq ID_{j,p,t} \leq SC_{j,p} \quad \forall j \in J, p \in P, t \in T \tag{A68}$$

$$SSP_k \times SC_{k,p} \leq IM_{k,p,t} \leq SC_{k,p} \quad \forall k \in K, p \in P, t \in T \tag{A69}$$

$$\sum_{l \in Route_{i,l,m}} XP_{i,l,m,p,t} \leq ASPD_{i,m,p} \quad \forall i \in I, m \in M, p \in P, t \in T \tag{A70}$$

$$\sum_{k \in Route_{j,k,m}} XS_{j,k,m,p,t} \leq ASSD_{j,m,p} \quad \forall j \in J, m \in M, p \in P, t \in T \tag{A71}$$

$$\sum_{p \in P} XP_{i,l,m,p,t} \leq ACPD_{i,l,m} \quad \forall (i,l,m) \in Route_{i,l,m}, t \in T \tag{A72}$$

$$\sum_{p \in P} XS_{j,k,m,p,t} \leq ACSD_{j,k,m} \quad \forall (j,k,m) \in Route_{j,k,m}, t \in T \tag{A73}$$

$$\sum_{(i,m) \in Route_{i,j,m}} XP_{i,j,m,p,t} + XI_{j,p,t} \leq SC_{j,p} \times TCM_{j,p} \quad \forall j \in J, p \in P, t \in T \tag{A74}$$

$$\sum_{(i,m) \in Route_{i,k,m}} XP_{i,k,m,p,t} + \sum_{(j,m) \in Route_{j,k,m}} XS_{j,k,m,p,t} \leq SC_{k,p} \times TCM_{k,p} \quad \forall k \in K, p \in P, t \in T \tag{A75}$$

$$XP_{i,l,m,p,t} = 0 \quad \forall i \in I, l \in L, p \in P, t \in T, m = road, \mu_{i,j,m} > MTD \tag{A76}$$

$$XS_{j,k,m,p,t} = 0 \quad \forall j \in J, k \in K, p \in P, t \in T, m = road, \mu_{j,k,m} > MTD \tag{A77}$$

Table A3. Model notation.

Sets		
	$a \in A$	Set of developed activities
	$d \in D$	Set of route distances
	$m \in M$	Set of transportation modes
	$n, o \in N$	Set of all network nodes
	$p \in P$	Set of products
	$r \in R$	Set of resources and network stages
	$t, \tau \in T$	Set of time points
	$\Theta \in \mathbb{R}_+$	Set of optimization variables: $\Theta = \{XR, XOP, XCO, XP, XS, XRP, XU, XE, XI, IO, IR, ID, IM\}$
Subsets		
	$i \in I \subseteq N$	Set of refineries
	$j \in J \subseteq N$	Set of depots
	$k \in K \subseteq N$	Set of markets
	$g = |I| + 1$	Auxiliary set used in the robust formulation
Subset unions		
	$h \in H = I \cup J$	Set of refineries and depots
	$l \in L = J \cup K$	Set of depots and markets

Table A3. *Cont.*

Parameters	
$ACPD_{n,o,m}$	Arc capacity between nodes n and o when transportation mode m is considered at primary distribution
$ACSD_{n,o,m}$	Arc capacity between nodes n and o when transportation mode m is considered at secondary distribution
$ASPD_{i,m,p}$	Availability of supplying product p from refinery i by transportation mode m through the primary distribution
$ASSD_{j,m,p}$	Availability of supplying product p from depot j by transportation mode m through the secondary distribution
CKI	Cost of keeping inventory defined as a percentage of the inventory value
$CT_{m,p}$	Transportation cost per transportation mode m and product p
$Dist_{n,o,m}$	Distance between nodes n and o depending on transportation mode m
$DO_{i,t}$	Demand of oil at refinery i at time point t
$DP_{k,p,t}$	Demand of product p per market k at time point t
ISO_i	Initial stock of oil at refinery i
$ISP_{n,p}$	Initial stock of product p at node n
MTD	Maximum travel distance in meters allowed in the road transportation mode
PC_i	Processing capacity at refinery i
PO_t	Price of oil at time point t
$PP_{a,p,t}$	Price of product p at activity a at time point t
$Route_{n,o,m}$	Route between nodes n and o connected by transportation mode m
$SC_{n,p}$	Storage capacity of product p at node n
SCO_i	Storage capacity of oil at refinery i
SSO_i	Safety stock of oil at refinery i defined as a percentage of the overall oil storage capacity
SSP_n	Safety stock of products at node n defined as a percentage of the overall storage capacity
$TN_{r,p}$	Tariff per network stage r and product p
$TCM_{n,p}$	Throughput capacity multiplier per node n and product p
$YF_{i,p}$	Yield fractions by refinery i of product p per cubic meters of oil
Positive continuous variables	
$CE_{i,p,t}$	Cost of exporting product p by refinery i at time point t
$CI_{h,p,t}$	Cost of importing product p by refinery or depot h at time point t
$CID_{j,p,t}$	Cost of inventory at depot j for product p at time point t
$CIM_{k,p,t}$	Cost of inventory at market k for product p at time point t
$CIO_{i,t}$	Cost of inventory for oil at refinery i at time point t
$CIR_{i,p,t}$	Cost of inventory at refinery i for product p at time point t
$CPD_{i,p,t}$	Cost of primary distribution from refinery i for product p at time point t
$CSD_{j,p,t}$	Cost of secondary distribution from depot j for product p at time point t
$CUD_{k,p,t}$	Cost of unsatisfied demand for product p at market k at time point t
$ID_{j,p,t}$	Inventory of product p at depot j at time point t
$IM_{k,p,t}$	Inventory of product p at market k at time point t
$IO_{i,t}$	Inventory of oil at refinery i at time point t
$IR_{i,p,t}$	Inventory of product p at refinery i at time point t
$XCO_{i,t}$	Volume of crude oil received by refinery i at time point t
$XE_{i,p,t}$	Volume of product p exported by refinery i at time point t
$XI_{h,p,t}$	Volume of product p imported by refinery or depot h at time point t
$XOP_{i,t}$	Volume of oil processed by refinery i at time point t
$XP_{i,l,m,p,t}$	Volume of product p sent by refinery i to depot or market l by transportation mode m at time point t
$XR_{i,p,t}$	Volume of product p yielded by refinery i at time point t
$XRP_{k,p,t}$	Volume of product p delivered to market k at time point t
$XS_{j,k,m,p,t}$	Volume of product p sent by depot j to market k by transportation mode m at time point t
$XU_{k,p,t}$	Volume of unsatisfied demand per market k and product p at time point t
Continuous variables	
$Profit$	Profit for the downstream oil supply chain over the planning horizon
$MD_{j,p,t}$	Margin per depot j and product p at time point t
$MC_{k,p,t}$	Margin per consumer market k and product p at time point t
$MR_{i,p,t}$	Margin per refinery i at time point t
Binary variable	
Y_g	Auxiliary variable to aid the robust formulation to handle product demand uncertainty, where $g = |I| + 1$

Table A3. *Cont.*

Robust parameters	
$\Gamma_{k,p}^{demand}$	Budget parameter to adjust the robustness of product demand
$DP_{k,p,t}^{max}$	Maximum variation in product demand for market k and product p at time point t
Robust dual variables	
$\eta_{k,p}^{demand}$	Dual variable associated with the establishment of the budget parameter of product demand
$\xi_{k,p,t}^{+demand}$	Quantify the sensitivity to positive deviation in product demand for market k and product p at time point t
$\xi_{k,p,t}^{-demand}$	Quantify the sensitivity to negative deviation in product demand for market k and product p at time point t

Appendix D. Stochastic Mathematical Programming Formulation

The two-stage stochastic programming (SP) model is formulated using node-variable formulation, where the decision variables of the optimization problem are associated with the nodes of the scenario tree. The objective function (Equations (A78)–(A92)), network equations (Equations (A93)–(A117)) and model notation (Table A4) of the SP model are presented below. The scenario tree to represent the evolution of the product demands is established considering an optimistic growth of 5% with 0.35 probability, a realistic and unchangeable occurrence of 0% with 0.50 probability and a pessimistic decrease of 10% with 0.15 probability, in accordance with the reasoning developed by Fernandes et al. [36]. For more details about the SP formulation, the interested reader is referred to Lima et al. [1].

Objective function

$$\underset{XCO}{\text{maximize}} - \sum_{i \in I, t \in T} XCO_{i,t} PO_t + Q(XCO) \tag{A78}$$

$$\text{s.t. } XCO_{i,t} \leq DO_{i,t} \forall i \in I, t \in T \tag{A79}$$

$$\begin{aligned}
Q(XCO) = \underset{XCO,\Theta}{\text{maximize}} \sum_{s \in S} P_s \Bigg(& \sum_{i \in I, t \in T} RR_{i,t} + \sum_{j \in J, p \in P, t \in T} MD_{j,p,t} + \sum_{k \in K, p \in P, t \in T} MC_{k,p,t} \\
& - \sum_{i \in I, p \in P, t \in T} CE_{i,p,t} - \sum_{h \in H, p \in P, t \in T} CI_{h,p,t} - \sum_{i \in I, p \in P, t \in T} CPD_{i,p,t} \\
& - \sum_{j \in J, p \in P, t \in T} CSD_{j,p,t} - \sum_{i \in I, t \in T} CIO_{i,t} - \sum_{i \in I, p \in P, t \in T} CIR_{i,p,t} \\
& - \sum_{j \in J, p \in P, t \in T} CID_{j,p,t} - \sum_{k \in K, p \in P, t \in T} CIM_{k,p,t} - \sum_{k \in K, p \in P, t \in T} CUD_{k,p,t} \Bigg)
\end{aligned} \tag{A80}$$

$$RR_{i,t} = \sum_{p \in P} \left[XR_{i,p,t,s} \times \left(PP_{a,p,t} - TN_{r,p} \right) \right] \quad \forall i \in I, (t,s) \in TS, a = r = refinery \tag{A81}$$

$$\begin{aligned}
MD_{j,p,t,s} = & \left[\sum_{(k,m) \in Route_{j,k,m}} XS_{j,k,m,p,t,s} \times \left(PP_{a,p,t} - TN_{r,p} \right) \right] - \left[\sum_{(i,m) \in Route_{i,j,m}} XP_{i,j,m,p,t,s} \times PP_{a_2,p,t} \right] \\
& \forall j \in J, p \in P, (t,s) \in TS, a = r = depot, a_2 = refinery
\end{aligned} \tag{A82}$$

$$\begin{aligned}
MC_{k,p,t,s} = & \left[\left(XRP_{k,p,t,s} - XU_{k,p,t,s} \right) \times \left(PP_{a_1,p,t} - TN_{r,p} \right) \right] - \left[\left(\sum_{(i,m) \in Route_{i,k,m}} XP_{i,k,m,p,t,s} \right) \right. \\
& \left. \times PP_{a_2,p,t} + \left(\sum_{(j,m) \in Route_{j,k,m}} XS_{j,k,m,p,t,s} \right) \times PP_{a_3,p,t} \right] \\
& \forall k \in K, p \in P, (t,s) \in TS, a_1 = r = retail, a_2 = refinery, a_3 = depot
\end{aligned} \tag{A83}$$

$$CE_{i,p,t,s} = XE_{i,p,t,s} \times TN_{r,p} \quad \forall i \in I, p \in P, (t,s) \in TS, r = export \tag{A84}$$

$$CI_{h,p,t,s} = XI_{h,p,t,s} \times \left(PP_{a,p,s} + TN_{r,p} \right) \quad \forall h \in H, p \in P, (t,s) \in TS, a = r = import \tag{A85}$$

$$CPD_{i,p,t,s} = \sum_{(l,m) \in Route_{i,l,m}} \left(XP_{i,l,m,p,t,s} \times CT_{m,p} \times Dist_{i,l,m} \right) \quad \forall i \in I, p \in P, (t,s) \in TS \tag{A86}$$

$$CSD_{j,p,t,s} = \sum_{(k,m)\in Route_{j,k,m}} \left(XS_{j,k,m,p,t,s} \times CT_{m,p} \times Dist_{j,k,m} \right) \quad \forall j \in J, p \in P, (t,s) \in TS \tag{A87}$$

$$CIO_{i,t,s} = CKI \times IO_{i,t,s} \times PO_{a,s} \quad \forall i \in I, (t,s) \in TS, a = procurement \tag{A88}$$

$$CIR_{i,p,t,s} = CKI \times IR_{i,p,t,s} \times PP_{a,p,t} \quad \forall i \in I, p \in P, (t,s) \in TS, a = refinery \tag{A89}$$

$$CID_{j,p,t,s} = CKI \times ID_{j,p,t,s} \times PP_{a,p,t} \quad \forall j \in J, p \in P, (t,s) \in TS, a = depot \tag{A90}$$

$$CIM_{k,p,t,s} = CKI \times IM_{k,p,t,s} \times PP_{a,p,t} \quad \forall k \in K, p \in P, (t,s) \in TS, a = retail \tag{A91}$$

$$CUD_{k,p,t,s} = XU_{k,p,t,s} \times TN_{r,p} \quad \forall k \in K, p \in P, (t,s) \in TS, r = unmet \tag{A92}$$

Network equations

$$XOP_{i,t,s} \leq PC_i \quad \forall i \in I, (t,s) \in TS \tag{A93}$$

$$XR_{i,p,t,s} = XOP_{i,t,s} \times YF_{i,p} \quad \forall i \in I, p \in P, (t,s) \in TS \tag{A94}$$

$$IO_{i,t,s} = ISO_i + DO_{i,t} - XOP_{i,t,s} \quad \forall i \in I, (t,s) \in TS, t = t_1 \tag{A95}$$

$$IO_{i,t,s} = IO_{i,t-1,\bar{s}} + DO_{i,t} - XOP_{i,t,s} \quad \forall i \in I, (t,s) \in TS, t > t_1, s \in S\bar{S} \tag{A96}$$

$$IR_{i,p,t,s} = ISP_{i,p} + XR_{i,p,t,s} + XI_{i,p,t,s} - \sum_{(l,m)\in Route_{i,l,m}} XP_{i,l,m,p,t,s} - XE_{i,p,t,s}$$
$$\forall i \in I, p \in P, (t,s) \in TS, t = t_1 \tag{A97}$$

$$IR_{i,p,t,s} = IR_{i,p,t-1,\bar{s}} + XR_{i,p,t,s} + XI_{i,p,t,s} - \sum_{(l,m)\in Route_{i,l,m}} XP_{i,l,m,p,t,s} - XE_{i,p,t,s}$$
$$\forall i \in I, p \in P, (t,s) \in TS, t > t_1, s \in S\bar{S} \tag{A98}$$

$$ID_{j,p,t,s} = ISP_{j,p} + XI_{j,p,t,s} + \sum_{(i,m)\in Route_{i,j,m}} XP_{i,j,m,p,t,s} - \sum_{(k,m)\in Route_{j,k,m}} XS_{j,k,m,p,t,s}$$
$$\forall j \in J, p \in P, (t,s) \in TS, t = t_1 \tag{A99}$$

$$ID_{j,p,t,s} = ID_{j,p,t-1,\bar{s}} + XI_{j,p,t,s} + \sum_{(i,m)\in Route_{i,j,m}} XP_{i,j,m,p,t,s} - \sum_{(k,m)\in Route_{j,k,m}} XS_{j,k,m,p,t,s}$$
$$\forall j \in J, p \in P, (t,s) \in TS, t > t_1, s \in S\bar{S} \tag{A100}$$

$$IM_{k,p,t,s} = ISP_{k,p} + \sum_{(i,m)\in Route_{i,k,m}} XP_{i,k,m,p,t,s} + \sum_{(j,m)\in Route_{j,k,m}} XS_{j,k,m,p,t,s} + XU_{k,p,t,s} - XRP_{k,p,t,s}$$
$$\forall k \in K, p \in P, (t,s) \in TS, t = t_1 \tag{A101}$$

$$IM_{k,p,t,s} = IC_{k,p,t-1,\bar{s}} + \sum_{(i,m)\in Route_{i,k,m}} XP_{i,k,m,p,t,s} + \sum_{(j,m)\in Route_{j,k,m}} XS_{j,k,m,p,t,s} + XU_{k,p,t,s} - XRP_{k,p,t,s}$$
$$\forall k \in K, p \in P, (t,s) \in TS, t > t_1, s \in S\bar{S} \tag{A102}$$

$$DPR_{k,p,t,s} = DP_{k,p} \quad \forall k \in K, p \in P, (t,s) \in TS, t = t_1 \tag{A103}$$

$$DPR_{k,p,t,s} = DPR_{k,p,t-1,\bar{s}} \chi^p \psi^s \quad \forall k \in K, p \in P, (t,s) \in TS, t > t_1, s \in S\bar{S} \tag{A104}$$

$$XRP_{k,p,t,s} \leq DPR_{k,p,t,s} \quad \forall k \in K, p \in P, (t,s) \in TS \tag{A105}$$

$$SSO_i \times SCO_i \leq IO_{i,t,s} \leq SCO_i \quad \forall i \in I, (t,s) \in TS \tag{A106}$$

$$SSP_i \times SC_{i,p} \leq IR_{i,p,t,s} \leq SC_{i,p} \quad \forall i \in I, p \in P, (t,s) \in TS \tag{A107}$$

$$SSP_j \times SC_{j,p} \leq ID_{j,p,t,s} \leq SC_{j,p} \quad \forall j \in J, p \in P, (t,s) \in TS \tag{A108}$$

$$SSP_k \times SC_{k,p} \leq IM_{k,p,t,s} \leq SC_{k,p} \quad \forall k \in K, p \in P, (t,s) \in TS \tag{A109}$$

$$\sum_{l\in Route_{i,l,m}} XP_{i,l,m,p,t,s} \leq ASPD_{i,m,p} \quad \forall i \in I, m \in M, p \in P, (t,s) \in TS \tag{A110}$$

$$\sum_{k\in Route_{j,k,m}} XS_{j,k,m,p,t,s} \leq ASSD_{j,m,p} \quad \forall j \in J, m \in M, p \in P, (t,s) \in TS \tag{A111}$$

$$\sum_{p \in P} XP_{i,l,m,p,t,s} \leq ACPD_{i,l,m} \quad \forall (i,l,m) \in Route_{i,l,m}, (t,s) \in TS \tag{A112}$$

$$\sum_{p \in P} XS_{j,k,m,p,t,s} \leq ACSD_{j,k,m} \quad \forall (j,k,m) \in Route_{j,k,m}, (t,s) \in TS \tag{A113}$$

$$\sum_{(i,m) \in Route_{i,j,m}} XP_{i,j,m,p,t,s} + XI_{j,p,t,s} \leq SC_{j,p} \times TCM_{j,p} \quad \forall j \in J, p \in P, (t,s) \in TS \tag{A114}$$

$$\sum_{(i,m) \in Route_{i,k,m}} XP_{i,k,m,p,t,s} + \sum_{(j,m) \in Route_{j,k,m}} XS_{j,k,m,p,t,s} \leq SC_{k,p} \times TCM_{k,p} \quad \forall k \in K, p \in P, (t,s) \in TS \tag{A115}$$

$$XP_{i,l,m,p,t,s} = 0 \quad \forall i \in I, l \in L, p \in P, (t,s) \in TS, m = road, Dist_{i,j,m} > MTD \tag{A116}$$

$$XS_{j,k,m,p,t,s} = 0 \quad \forall j \in J, k \in K, p \in P, (t,s) \in TS, m = road, Dist_{j,k,m} > MTD \tag{A117}$$

Table A4. Model notation.

Sets		
	$a \in A$	Set of activities developed
	$d \in D$	Set of route distances
	$m \in M$	Set of transportation modes
	$n, o \in N$	Set of all network nodes
	$p \in P$	Set of products
	$r \in R$	Set of resources and network stages
	$s, \bar{s} \in S$	Set of nodes/states in the scenario tree
	$t \in T$	Set of time points
	$\Theta \in \mathbb{R}_+$	Set of optimization variables: $\Theta = \{XR, XOP, XCO, XP, XS, XRP, XU, XE, XI, IO, IR, ID, IM\}$
Subsets		
	$i \in I \subseteq N$	Set of refineries
	$j \in J \subseteq N$	Set of depots
	$k \in K \subseteq N$	Set of markets
Subset unions		
	$h \in H = I \cup J$	Set of refineries and depots
	$l \in L = J \cup K$	Set of depots and markets
	$Route_{n,o,m}$	Possible route combination between network nodes n and o connected by transportation mode m
	$S\bar{S}$	Set of predecessors $\bar{s}$ of nodes/states s in the scenario tree: $S\bar{S} = \{(s,\bar{s}) : s \in S(t), \bar{s} \in S(t-1)\}$
	TS	Set of nodes/states s that belong to each time point t: $TS = \{(t,s) : t \in T, s \in S(t)\}$
Parameters		
	$ACPD_{n,o,m}$	Arc capacity between network nodes n and o when transportation mode m is considered at primary distribution
	$ACSD_{n,o,m}$	Arc capacity between network nodes n and o when transportation mode m is considered at secondary distribution
	$ASPD_{i,m,p}$	Availability of supplying product p from refinery i by transportation mode m through the primary distribution
	$ASSD_{j,m,p}$	Availability of supplying product p from depot j by transportation mode m through the secondary distribution
	CKI	Cost of keeping inventory defined as a percentage of the inventory value
	$CT_{m,p}$	Transportation cost per transportation mode m and product p
	$Dist_{n,o,m}$	Distance between network nodes n and o depending on transportation mode m
	$DO_{i,t}$	Demand of oil at refinery i at time point t
	$DP_{k,p}$	Demand of product p per market k
	$DPR_{k,p,t,s}$	Demand realization of product p for market k in time point t and state s
	ISO_i	Initial stock of oil at refinery i
	$ISP_{n,p}$	Initial stock of product p at network node n
	MTD	Maximum travel distance in meters allowed in the road transportation mode
	NTP	Number of time points
	P_s	Probability of each state s in the scenario tree approach
	PC_i	Processing capacity at refinery i
	PO_t	Price of oil at activity a at time point t
	$PP_{a,p,t}$	Price of product p at activity a and time point t
	$SC_{n,p}$	Storage capacity of product p at network node n
	SCO_i	Storage capacity of oil at refinery i
	SSO_i	Safety stock of oil at refinery i defined as a percentage of the overall oil storage capacity
	SSP_n	Safety stock of products at network node n defined as a percentage of the overall storage capacity
	$TN_{r,p}$	Tariff per network stage r and product p
	$TCM_{n,p}$	Throughput capacity multiplier per network node n and product p
	$YF_{i,p}$	Yield fractions by refinery i of product p per cubic meters of oil
	χ^p	Market tendency per product p
	ψ^s	Market tendency per state s

Table A4. *Cont.*

Positive continuous variables	
$CE_{i,p,t,s}$	Cost of exporting product p by refinery i at time point t and state s
$CI_{h,p,t,s}$	Cost of importing product p by refinery or depot h at time point t and state s
$CID_{j,p,t,s}$	Cost of inventory at depot j for product p in time point t and state s
$CIM_{k,p,t,s}$	Cost of inventory at market k for product p in time point t and state s
$CIO_{i,t,s}$	Cost of inventory for oil at refinery i in time point t and state s
$CIR_{i,p,t,s}$	Cost of inventory at refinery i for product p in time point t and state s
$CPD_{i,p,t,s}$	Cost of primary transportation from refinery i for product p in time point t and state s
$CSD_{j,p,t,s}$	Cost of secondary transportation from depot j for product p in time point t and state s
$CUD_{k,p,t,s}$	Cost of unsatisfied demand for product p at market k in time point t and state s
$ID_{j,p,t,s}$	Inventory of product p at depot j in time point t and state s
$IM_{k,p,t,s}$	Inventory of product p at market k in time point t and state s
$IO_{i,t,s}$	Inventory of oil at refinery i in time point t and state s
$IR_{i,p,t,s}$	Inventory of product p at refinery i in time point t and state s
$XCO_{i,t}$	Volume of crude oil received by refinery i at time point t
$XE_{i,p,t,s}$	Volume of product p exported by refinery i at time point t and state s
$XI_{h,p,t,s}$	Volume of product p imported by refinery or depot h at time point t and state s
$XOP_{i,t,s}$	Volume of oil processed by refinery i at time point t and state s
$XP_{i,l,m,p,t,s}$	Volume of product p sent by refinery i to depot or market l by transportation mode m at time point t and state s
$XR_{i,p,t,s}$	Volume of product p yielded by refinery i at time point t and state s
$XRP_{k,p,t,s}$	Volume of product p delivered to market k at time point t and state s
$XS_{j,k,m,p,t,s}$	Volume of product p sent by depot j to market k by transportation mode m at time point t and state s
$XU_{k,p,t}$	Volume of unsatisfied demand per market k and product p at time point t and state s
Continuous variables	
$MD_{j,p,t,s}$	Margin per storage depot j and product p at time point t and state s
$MC_{k,p,t,s}$	Margin per consumer market k and product p at time point t and state s
$RR_{i,p,t,s}$	Revenue per refinery i at time point t and state s

Appendix E. Considerations about a Typical Polyhedral Budget Uncertainty Set

In this part, we demonstrate how to enumerate all the possible vertices $v \in V$ in the budget uncertainty set $\mathcal{U}$ for a generic pair of location k and product p, and a specific budget of uncertainty $\Gamma_{k,p}$ over a time horizon T. Consider the Equations (A118) and (A119) below:

$$\left| \Delta DP_{k,p,t} \right| \leq \Delta DP_{k,p,t}^{max}, \ \forall k \in K, p \in P, t \in T \tag{A118}$$

$$\sum_t \frac{\left| \Delta DP_{k,p,t} \right|}{\Delta DP_{k,p,t}^{max}} \leq \Gamma_{k,p}, \ \forall k \in K, p \in P \tag{A119}$$

Equation (A118) determines symmetrical intervals for the deviation of the product demand from the nominal value, while the total deviation across all time points are limited by the budget of uncertainty in Equation (A119). We can omit the indices k and p in Equation (A120) once they could refer to any pair of location and product. Consider a budget of uncertainty $\Gamma_{k,p} = 1$ and a time horizon covering two time points as follows:

$$\frac{\left| \Delta DP_1 \right|}{\Delta DP_1^{max}} + \frac{\left| \Delta DP_2 \right|}{\Delta DP_2^{max}} \leq 1 \tag{A120}$$

Equation (A120) ensures that if the deviation of product demand at time point $t = 1$ is at the lower or upper bound of the range defined by Equation (A118), i.e., $|\Delta DP_1| = \Delta DP_1^{max}$, the deviation of the product demand at time point $t = 2$ will be necessarily zero, $|\Delta DP_2| = 0$. Inversely, when the deviation of product demand at time point $t = 2$ is at the lower or upper bound of the range defined by Equation (A118), i.e., $|\Delta DP_2| = \Delta DP_2^{max}$, the deviation of the product demand at time point $t = 1$ will be certainly zero, $|\Delta DP_1| = 0$. In such a way, we have enumerated all the possible scenarios $v \in V$ in $\mathcal{U}$, i.e., four vertices, where $V = \{(\Delta DP_1, 0); (-\Delta DP_1, 0); (0, -\Delta DP_2); (0, \Delta DP_2)\}$. Note that the size of the set of vertices V depends on the value of the budget of uncertainty $\Gamma_{k,p}$, which takes values in the range $[0; 2]$. When $\Gamma_{k,p} = 1.5$, there are eight vertices within the set V, as shown below:

$$V = \left\{ \begin{array}{l} \left(\Delta DP_1, -\frac{\Delta DP_2}{2}\right); \left(\Delta DP_1, \frac{\Delta DP_2}{2}\right); \left(\frac{\Delta DP_1}{2}, -\Delta DP_2\right); \left(\frac{\Delta DP_1}{2}, \Delta DP_2\right); \\ \left(-\Delta DP_1, -\frac{\Delta DP_2}{2}\right); \left(-\Delta DP_1, \frac{\Delta DP_2}{2}\right); \left(-\frac{\Delta DP_1}{2}, -\Delta DP_2\right); \left(-\frac{\Delta DP_1}{2}, \Delta DP_2\right) \end{array} \right\}$$

We can generalize that when $\Gamma_{k,p} = 0$, the uncertainty set $\mathcal{U}$ has a just one vertex, corresponding to the nominal deterministic case. As $\Gamma_{k,p}$ increases, the size of the uncertainty set $\mathcal{U}$ enlarges. As shown before, when $\Gamma_{k,p}$ takes any value in the interval $[0.01; 1]$, the polyhedron will have four vertices, while if $\Gamma_{k,p}$ assumes any value within the interval $[1.01; 1, 99]$, the polyhedron will have eight vertices. On the other hand, when $\Gamma_{k,p} = 2$, the polyhedron will have four vertices again.

It is important to highlight that this is just a generic illustration to show how to enumerate the vertices of a budget uncertainty set, and it was not used in the case study shown in the full paper. On the other hand, such instance can easily be extended to include a longer time horizon, such that the vertices of more complex polyhedral uncertainty sets can be determined.

References

1. Lima, C.; Relvas, S.; Barbosa-Póvoa, A.P. Stochastic programming approach for the optimal tactical planning of the downstream oil supply chain. *Comput. Chem. Eng.* **2018**, *108*, 314–336. [CrossRef]
2. Oliveira, F.; Gupta, V.; Hamacher, S.; Grossmann, I.E. A Lagrangean decomposition approach for oil supply chain investment planning under uncertainty with risk considerations. *Comput. Chem. Eng.* **2013**, *50*, 184–195. [CrossRef]
3. Lima, C.; Relvas, S.; Barbosa-Póvoa, A.P. Downstream oil supply chain management: A critical review and future directions. *Comput. Chem. Eng.* **2016**, *92*, 78–92. [CrossRef]
4. Escudero, L.F.; Quintana, F.J.; Salmerón, J. CORO, a modeling and an algorithmic framework for oil supply, transformation and distribution optimization under uncertainty. *Eur. J. Oper. Res.* **1999**, *114*, 638–656. [CrossRef]
5. Birge, J.R.; Louveaux, F. *Introduction to Stochastic Programming*, 2nd ed.; Series in Operations Research and Financial Engineering; Springer: New York, NY, USA, 2011.
6. Sahinidis, N.V. Optimization under uncertainty: State-of-the-art and opportunities. *Comput. Chem. Eng.* **2004**, *28*, 971–983. [CrossRef]
7. Lorca, A.; Sun, X.A. Adaptive robust optimization with dynamic uncertainty sets for multi-period economic dispatch under significant wind. *IEEE Trans. Power Syst.* **2015**, *30*, 1702–1713. [CrossRef]
8. Tong, K.; Feng, Y.; Rong, G. Planning under demand and yield uncertainties in an oil supply chain. *Ind. Eng. Chem. Res.* **2011**, *51*, 814–834. [CrossRef]
9. Ning, C.; You, F. Data-driven decision making under uncertainty integrating robust optimization with principal component analysis and kernel smoothing methods. *Comput. Chem. Eng.* **2018**, *112*, 190–210. [CrossRef]
10. Ning, C.; You, F. Data-driven adaptive nested robust optimization: General modeling framework and efficient computational algorithm for decision making under uncertainty. *AIChE J.* **2017**, *63*, 3790–3817. [CrossRef]
11. Sahebi, H.; Nickel, S.; Ashayeri, J. Strategic and tactical mathematical programming models within the crude oil supply chain context—A review. *Comput. Chem. Eng.* **2014**, *68*, 56–77. [CrossRef]
12. Dempster, M.A.H.; Pedrón, N.H.; Medova, E.A.; Scott, J.E.; Sembos, A. Planning logistics operations in the oil industry. *J. Oper. Res. Soc.* **2000**, *51*, 1271–1288. [CrossRef]
13. Al-Othman, W.B.E.; Lababidi, H.M.S.; Alatiqi, I.M.; Al-Shayji, K. Supply chain optimization of petroleum organization under uncertainty in market demands and prices. *Eur. J. Oper. Res.* **2008**, *189*, 822–840. [CrossRef]
14. MirHassani, S.A. An operational planning model for petroleum products logistics under uncertainty. *Appl. Math. Comput.* **2008**, *196*, 744–751. [CrossRef]

15. Carneiro, M.C.; Ribas, G.P.; Hamacher, S. Risk management in the oil supply chain: A CVaR approach. *Ind. Eng. Chem. Res.* **2010**, *49*, 3286–3294. [CrossRef]
16. Ribas, G.P.; Hamacher, S.; Street, A. Optimization under uncertainty of the integrated oil supply chain using stochastic and robust programming. *Int. Trans. Oper. Res.* **2010**, *17*, 777–796. [CrossRef]
17. MirHassani, S.A.; Noori, R. Implications of capacity expansion under uncertainty in oil industry. *J. Petrol. Sci. Eng.* **2011**, *77*, 194–199. [CrossRef]
18. Oliveira, F.; Hamacher, S. Optimization of the petroleum product supply chain under uncertainty: A case study in Northern Brazil. *Ind. Eng. Chem. Res.* **2012**, *51*, 4279–4287. [CrossRef]
19. Leiras, A.; Ribas, G.; Hamacher, S. Tactical and operational planning of multirefinery networks under uncertainty: An iterative integration approach. *Ind. Eng. Chem. Res.* **2013**, *52*, 8507–8517. [CrossRef]
20. Oliveira, F.; Grossmann, I.E.; Hamacher, S. Accelerating Benders stochastic decomposition for the optimization under uncertainty of the petroleum product supply chain. *Comput. Oper. Res.* **2014**, *49*, 47–58. [CrossRef]
21. Tong, K.; Gong, J.; Yue, D.; You, F. Stochastic programming approach to optimal design and operations of integrated hydrocarbon biofuel and petroleum supply chains. *ACS Sustain. Chem. Eng.* **2014**, *2*, 49–61. [CrossRef]
22. Ghatee, M.; Hashemi, S.M. Optimal network design and storage management in petroleum distribution network under uncertainty. *Eng. Appl. Artif. Intell.* **2009**, *22*, 796–807. [CrossRef]
23. Tong, K.; Gleeson, M.J.; Rong, G.; You, F. Optimal design of advanced drop-in hydrocarbon biofuel supply chain integrating with existing petroleum refineries under uncertainty. *Biomass Bioenergy* **2014**, *60*, 108–120. [CrossRef]
24. Tong, K.; You, F.; Rong, G. Robust design and operations of hydrocarbon biofuel supply chain integrating with existing petroleum refineries considering unit cost objective. *Comput. Chem. Eng.* **2014**, *68*, 128–139. [CrossRef]
25. Bertsimas, D.; Litvinov, E.; Sun, X.A.; Zhao, J.; Zheng, T. Adaptive robust optimization for the security constrained unit commitment problem. *IEEE Trans. Power Syst.* **2013**, *28*, 52–63. [CrossRef]
26. Ben-Tal, A.; Nemirovski, A. Robust optimization–methodology and applications. *Math. Program.* **2002**, *92*, 453–480. [CrossRef]
27. Bertsimas, D.; Brown, D.B.; Caramanis, C. Theory and applications of robust optimization. *SIAM Rev.* **2011**, *53*, 464–501. [CrossRef]
28. Zeng, B.; Zhao, L. Solving two-stage robust optimization problems using a column-and-constraint generation method. *Oper. Res. Lett.* **2013**, *41*, 457–461. [CrossRef]
29. Gorissen, B.L.; Yanıkoğlu, İ.; Hertog, D. A practical guide to robust optimization. *Omega* **2015**, *53*, 124–137. [CrossRef]
30. Bertsimas, D.; Sim, M.; Zhang, M. Adaptive distributionally robust optimization. *Manag. Sci.* **2018**, *65*, 604–618. [CrossRef]
31. Zugno, M.; Conejo, A.J. A robust optimization approach to energy and reserve dispatch in electricity markets. *Eur. J. Oper. Res.* **2015**, *247*, 659–671. [CrossRef]
32. Escudero, L.F.; Garín, A.; Merino, M.; Pérez, G. The value of the stochastic solution in multistage problems. *Top* **2007**, *15*, 48–64. [CrossRef]
33. Liu, M.L.; Sahinidis, N.V. Optimization in process planning under uncertainty. *Ind. Eng. Chem. Res.* **1996**, *35*, 4154–4165. [CrossRef]
34. Bertsimas, D.; Brown, D.B. Constructing uncertainty sets for robust linear optimization. *Oper. Res.* **2009**, *57*, 1483–1495. [CrossRef]
35. Bertsimas, D.; Sim, M. The price of robustness. *Oper. Res.* **2004**, *52*, 35–53. [CrossRef]
36. Fernandes, L.J.; Relvas, S.; Barbosa-Póvoa, A.P. Downstream petroleum supply chain planning under uncertainty. In *Proceedings of PSE 2015 ESCAPE 25; May 31-June 4 2015; Copenhagen, Denmark. Computer Aided Chemical Engineering*; Gernaey, K.V., Huusom, J.K., Gani, R., Eds.; Elsevier: Amsterdam, The Netherlands, 2015; pp. 1889–1894.

 *processes*

Article

Modern Modeling Paradigms Using Generalized Disjunctive Programming [†]

Qi Chen and Ignacio Grossmann *

Center for Advanced Process Decision-Making, Carnegie Mellon University, Pittsburgh, PA 15213, USA; qichen@andrew.cmu.edu

* Correspondence: grossmann@cmu.edu; Tel.: +1-412-268-3642

† This paper is dedicated to the memory of Professor Roger Sargent, an intellectual leader in process systems engineering.

Received: 14 October 2019; Accepted: 6 November 2019; Published: 10 November 2019

Abstract: Models involving decision variables in both discrete and continuous domain spaces are prevalent in process design. Generalized Disjunctive Programming (GDP) has emerged as a modeling framework to explicitly represent the relationship between algebraic descriptions and the logical structure of a design problem. However, fewer formulation examples exist for GDP compared to the traditional Mixed-Integer Nonlinear Programming (MINLP) modeling approach. In this paper, we propose the use of GDP as a modeling tool to organize model variants that arise due to characterization of different sections of an end-to-end process at different detail levels. We present an illustrative case study to demonstrate GDP usage for the generation of model variants catered to process synthesis integrated with purchasing and sales decisions in a techno-economic analysis. We also show how this GDP model can be used as part of a hierarchical decomposition scheme. These examples demonstrate how GDP can serve as a useful model abstraction layer for simplifying model development and upkeep, in addition to its traditional usage as a platform for advanced solution strategies.

Keywords: process design; process modeling; mathematical programming; MINLP; generalized disjunctive programming

---

1. Introduction

Mathematical programming is a powerful tool for process design and optimization, allowing the modeler to consider both continuous and discrete decisions. In process design, discrete degrees of freedom often determine topological structure (selection/activation/ordering of nodes and edges) while continuous variables determine system states such as flow rates or qualities. In the general case, these process design problems can involve nonlinear variable relationships and are addressed as Mixed-Integer Nonlinear Programming (MINLP) problems.

$$\min z = f(x, y)$$
$$\text{s.t.} \quad g(x, y) \leq 0$$
$$h(x, y) = 0 \qquad \text{(MINLP)}$$
$$x \in X \subseteq \mathbb{R}^n$$
$$y \in Y \subseteq \mathbb{Z}^m$$

The general form for these optimization models is given in (MINLP). An objective function $f(x, y)$ is minimized by selecting values for continuous variables x and integer variables y, subject to satisfying inequality constraints $g(x, y) \leq 0$ and equality constraints $h(x, y) = 0$. In processes, the continuous variables usually represent flows, pressures, and temperatures. The integer variables are commonly 0–1 variables for the selection of units, but can also represent the number of units. The inequalities usually describe process and equipment limitations while equality constraints describe physical property relationships. An abundant literature exists for the formulation and solution of MINLP models [1–4]. Particularly in chemical engineering, many models are now formulated using algebraic relationships. Such equation-oriented modeling is becoming progressively more common [5], with differential equations used to describe temporal and spatial dynamics. For process design problems, postulation of alternatives is also an important consideration, with several approaches described in literature [6–15]. However, even with the growth of more complex models, there has been limited emphasis on the link between algebraic relationships and model logic.

Generalized Disjunctive Programming (GDP) represents one effort to systematize the relationship between algebraic relations and logical clauses [2,16,17], in pursuit of a framework to simplify both model formulation and solution of the eventual mathematical programming problem. GDP can be seen as the extension of theoretical work in disjunctive programming from the operations research community [18,19] to formulations involving nonlinear algebraic relationships. GDP gives the modeler a mathematical framework to express high-level logical statements without needing to immediately translate them into algebraic form. The general form for GDP optimization models is given in (GDP).

$$
\begin{aligned}
\min\ obj &= f(x,z) \\
s.t.\quad & g(x,z) \leq 0 \\
& \bigvee_{i \in D_k} \begin{bmatrix} Y_{ik} \\ r_{ik}(x,z) \leq 0 \end{bmatrix} \quad \forall k \in K \\
& \underline{\bigvee}_{i \in D_k} Y_{ik} \quad \forall k \in K \\
& \Omega(Y) = True \\
& x \in X \subseteq \mathbb{R}^n \\
& Y_{ik} \in \{True, False\} \quad \forall i \in D_k, \forall k \in K \\
& z \in Z \subseteq \mathbb{Z}^m
\end{aligned}
\tag{GDP}
$$

As with the MINLP formulation, an objective function $f(x,z)$ is minimized. Continuous decisions variables are still represented by x, but Boolean variables Y now describe selection among discrete alternatives. Remaining integer variables are denoted by z. This is preferable, as the conditional constraints $r_{ik}(x,z) \leq 0$ corresponding to selection of alternative Y_{ik} can be grouped together and separated from the globally valid constraints $g(x,z) \leq 0$ that must hold true for any selection of alternatives. Note that equality constraints are implicitly captured in (GDP) through the use of two inequality constraints. We term these groupings of a Boolean indicator variable with relevant conditional constraints a "disjunct", as they each constitute one term of a disjunction $\vee$ (logical "OR" relationship). Next, we state that for each disjunction $k \in K$, exactly one of the disjuncts $i \in D_k$ will be selected, a generalization of the logical XOR $\underline{\vee}$. Finally, GDP also allows the explicit specification of logical propositions $\Omega(Y) = True$ to describe logical relationships between selection of the discrete alternatives. These logical propositions are key to the modeling strategies addressed later in this work.

GDP offers two major advantages over the traditional MINLP modeling approach. First, it facilitates more intuitive modeling of process decision-making by allowing explicit specification of logical relationships [20]. The grouping of related constraints in disjuncts also helps to keep GDP models more organized. Second, by exploiting explicit logical structure provided by GDP models [21], advanced solution algorithms can reap benefits in convergence speed and robustness [22–24]. In this work, we focus on the modeling implications of GDP use.

While long-time practitioners of MINLP modeling approaches may sometimes find logical propositions too verbose, we contend that explicit logic is more readable and better preserves a modeler's original intent. Take for example the logical statement in Equation (1), which may correspond to the following process specification: if we purchase the cheap feed ($Y_1 = True$), then we will need to install a pretreater ($Y_2 = True$) or install an additional separation unit ($Y_3 = True$). That is, Y_1 implies Y_2 or Y_3.

$$Y_1 \implies (Y_2 \vee Y_3) \tag{1}$$

This resolves to the equivalent algebraic constraint in Equation (2), using binary variables y [20,25].

$$-y_1 + y_2 + y_3 \geq 0 \tag{2}$$

Both descriptions are valid, but the logical statement is self-documenting and is much clearer to a new modeler. Expert MINLP modelers are accustomed to automatically preprocessing their formulation to encode relevant problem logic in the algebraic constraints. With GDP, this mental overhead and the associated potential for human error is eliminated. As a result, GDP models promise to be easier to develop and maintain.

Compounding this effect is the fact that mathematical programming is frequently an analysis tool in a larger process design procedure [26]. Business needs and customer expectations are not always initially stated in a form amenable to formulation as an algebraic objective or constraint [27]. A process design problem therefore involves several iterations of reassessing assumptions and adjusting constraints to match new business needs or revised customer expectations. This means that neither the data nor the structure for a model formulation can be regarded as static throughout the design workflow. As a result, the optimization model may be rewritten or revised several times in the course of tackling a single process design problem [28]. Considerations such as environmental/social impacts, safety, and operability may also be added to the model as constraints or secondary objectives. Moreover, multiple versions of the same model are often necessary to trade-off process detail versus model tractability for different process sections, increasing the number of model formulations that must be developed and maintained. By separating model logical structure from the underlying algebraic descriptions, GDP reduces the work necessary to revise a model. In doing so, it aims to advance the state-of-the-art in process design [26,29–32]. Later, we also show how GDP can help manage model variants.

GDP can be viewed in a broader context as a logic-based model abstraction layer, facilitating intuitive expression of the discrete decision spaces. These abstractions are a necessary response to complexity [33], to make mathematical programming capabilities accessible to a broader range of process modelers. The ubiquity of commercial chemical process simulators is attributable to their ability to abstract large-scale mathematical computations from chemical engineering decision-making. They provide a drag-and-drop interface to assemble a process structure and a drop-down menu to pick among standard physical property packages. Similar efforts to provide high-level modeling capabilities underpinned by mathematical programming—such as Egret [34] for power systems design, ICAS [35] for process and product design, MIPSYN [36] for process synthesis, and IDAES [37] for advanced energy systems design—can benefit from the logical abstraction provided by GDP. Note that modeling in GDP does not preclude the use of MINLP solution methods. Instead, it can offer a systematic yet flexible approach to generate the appropriate MINLP formulation via reformulation. For instance, Castro and Grossmann [38] derive several traditional scheduling formulations using standardized reformulations of GDP models.

The two most popular ways to reformulate a GDP model as an MINLP model are the Big-M (BM) [16] and Hull Reformulation (HR) [39] methods. BM and HR trade-off problem size (number of variables and constraints) and the tightness (quality) of the continuous relaxation. Other reformulations are also possible [40–43] with different tradeoffs in problem size, relaxation tightness, and computational cost to generate the reformulation. In general, multiple valid MINLP formulations exist to describe the same problem logic, and there exists no general way to determine a priori the most tractable formulation [17]. Direct formulation as an MINLP requires the modeler to commit to a single formulation approach, while GDP allows multiple algebraic formulations to be systematically generated from a single logical description, so that the most advantageous variant may be utilized.

Despite theoretical progress, lack of computational tools to support GDP modeling has hindered its adoption in both academic and industrial settings. The GAMS algebraic modeling language [44] provides support for GDP models through its Extended Mathematical Programming syntax, with the ability to generate the BM and HR reformulations. Prior to version 23.7, GAMS also supported solution of GDP models using the LOGMIP 1.0 solver [45]. However, as a closed-source commercial platform, academic interest has been limited. More recently, Pyomo.GDP [46] has emerged as an open-source ecosystem for GDP modeling and development, built on top of the Pyomo algebraic modeling language [47] in Python. As an open-source platform, it has been able to incorporate recent innovations in reformulation strategies [48] and logic-based solution algorithms [22]. Powerful options now exist for formulating and solving GDP models. However, compared to the MINLP literature, relatively few formulation examples exist for GDP. This paper aims to address that gap.

In this paper, we focus on GDP as a modeling tool to manage model variants. We demonstrate its use for two modeling use cases: (1) end-to-end analysis with focus on various portions of the overall process, and (2) a single solution scheme involving use of models at different detail levels. In Section 2, we describe these use cases and their modeling challenges in detail. In Section 3, we discuss application of these techniques on an illustrative example and the resulting implications. We present concluding remarks in Section 4.

2. Problem Statement

We examine two scenarios in which GDP is useful as a model management platform.

- Case 1: Generate model variants that focus on various portions of an end-to-end process;
- Case 2: Use higher-level (approximate) models to do preliminary analysis, and drill down into increasing model detail for promising options as part of the solution scheme.

In the first case, complex value chains often result in a process being subdivided among major process sections, assigned to different modeling teams. Each of these modeling teams may develop specialized, detailed models that describe decision points and specifications relevant to their portion of the overall process. However, as sequential optimizations of process sections may yield a suboptimal overall result, coordination is necessary. Each section therefore needs to model the secondary impact of their decisions on the rest of the process. Ideally, the detailed models for each section could simply be linked together to produce a single optimization formulation. However, this formulation is often intractably large. Therefore, a less-detailed surrogate is often employed to model nonfocal portions of the process.

Consider an illustrative chemical production process consisting of three sections: procurement, production, and sales, displayed in Figure 1. We examine this process in more detail in Section 3. Multiple discrete options are available as decision variables for each section, denoted by the Boolean variables Y_i, Y_j, and Y_k for each section, respectively. For procurement, these discrete decisions may describe selection among several available supply contracts from various vendors. For sales, there might exist several sales opportunities corresponding to different customers. In production, selection among various production modes and capital purchases are often key decisions. For our illustrative process, $Y_{12} = True$ describes the selection of the second procurement contract, for example.

In typical industrial practice, models at varying detail levels are often developed separately from each other, with interoperability suffering as a result. By making use of GDP, the choice of modeling detail can be integrated within a single framework, allowing related models to be developed in proximity to each other. Note that care should still be taken to define appropriate interconnections between the process sections so that relevant phenomena can be described (e.g., time dependence).

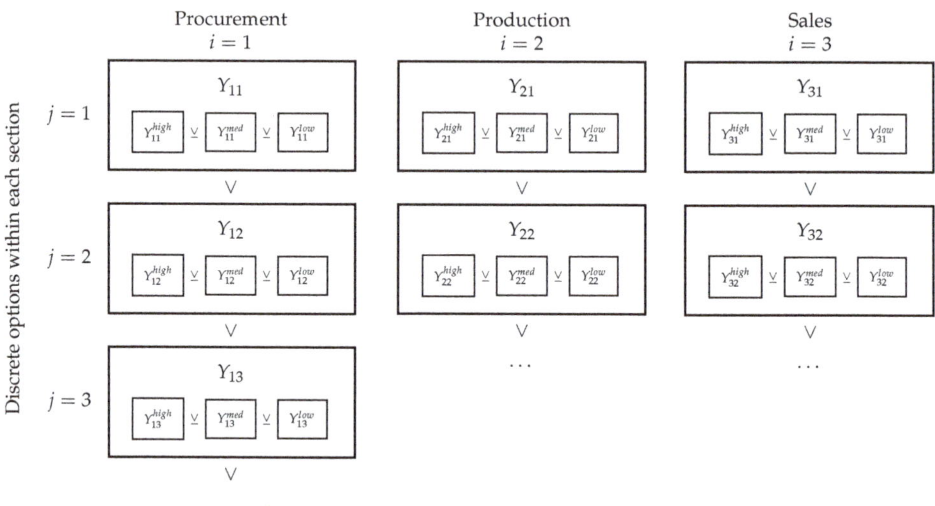

Figure 1. Example process illustrating the embedding of multiple detail levels within discrete options for each process section.

In the second case, a modeler may wish to use approximate models in a preliminary analysis to identify promising candidates among numerous alternatives, then drill down into progressively more detailed representations for remaining alternatives. For chemical processes, making assumptions that restrict temperatures, transport phenomena, or thermodynamic complexity can greatly simplify a model representation. However, variations in degrees of freedom and relevant physical phenomena should be subsequently revisited. In specialized simulation software, provisions for changing the thermodynamic assumptions of a chemical process are commonplace. However, the ability to do so is less common in equation-oriented optimization frameworks. Instead, a new model must frequently be developed at the desired complexity level. With GDP, the imposition or relaxation of these assumptions may be made by setting the value of Boolean variables. For a broader perspective, by imposing the implication that selection of unit u requires its modeling at a low detail level, $Y_u \implies Y_u^{low}$, we can restrict consideration to the approximate models. Conversely, for a deeper perspective, we can consider only more rigorously modeled alternatives by imposing the use of high detail models $Y_u \implies Y_u^{high}$. Note that at high fidelity, some alternatives may themselves involve selection among a discrete decision space. For example, selection of a distillation column in a chemical process may involve deciding on the number of trays. This discrete decision could itself be treated at various levels of modeling fidelity. Fortunately, with GDP, this decision can simply be nested within the higher-level selection as a nested disjunction [49].

3. Case Study

To demonstrate the principles of GDP as a tool for model management, we propose an illustrative end-to-end methanol synthesis process example adapted from literature [22]. As previously stated, the process consists of three sections: procurement, production, and sales (see Figure 2). Procurement must source the syngas from one of two different vendors, each of which offers a different purity-dependent cost schedule per unit feed. Production must select the optimal equipment configuration and operating conditions (temperatures, pressures, flows, and compositions) to convert syngas to methanol. This includes the discrete decision between single and two-stage compression for both the feed and recycle streams, as well as the choice between a higher-conversion, higher-cost reactor and a cheaper variant. Sales then contracts with one of two different customers, who are willing to pay a unit price that depends on the product purity. Each of these sections may be modeled at a high, medium, and low level of detail. The production section superstructure appears in Figure 3, with unit 9 as the cheap reactor alternative with low conversion and unit 10 denoting the expensive reactor with high conversion. Amendments to the literature methanol process synthesis model are presented in Appendix C.

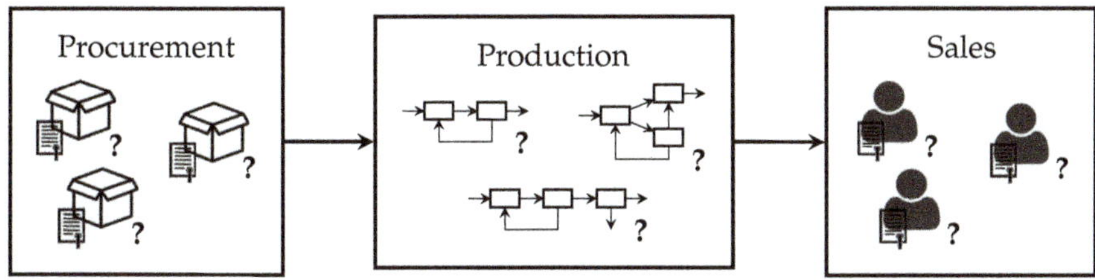

Figure 2. Simplified process diagram for the illustrative example. Three sections exist: procurement, production, and sales. In each section, decisions must be made in both discrete and continuous variable domains.

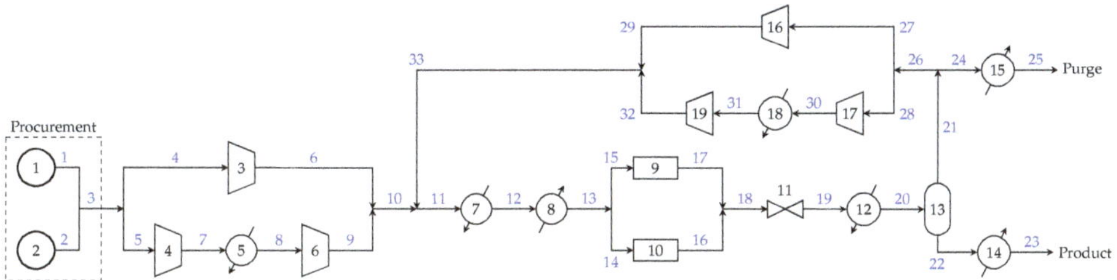

Figure 3. Methanol process flowsheet superstructure, adapted from [22], showing stream numbers in blue.

Below, we present a generic formulation for an end-to-end techno-economic analysis model and demonstrate how it may be adapted to our methanol synthesis example. In the following subsections, we lay out the relevant sets, parameters, decision variables, and functions before presenting the formulation. We then discuss manipulations of the formulation for both use cases from Section 2.

3.1. Sets

The chemical process involves a set of components C, which include the relevant raw materials, reaction intermediates, inerts, and products. For the methanol process, we have feed components H_2 and CO, inert CH_4, and product CH_3OH. The process is subdivided into three major process sections I: procurement, production, and sales. Of all possible process section alternatives J, a subset $J_i \subseteq J$ is available for selection for each section $i \in I$. In the production section of the methanol process, two of these alternatives are single-stage feed compression, and two-stage feed compression. The process also involves a set of streams K that describe flows of material between process sections. Finally, for each process alternative, a set of modeling detail levels $l \in L$ are available. We consider in our example three detail levels: "low", "medium", and "high."

$c \in C$ Set of components (feeds, intermediates, inerts, and products)
$i \in I$ Set of end-to-end process sections
$j \in J$ Set of process section alternatives
J_i Set of alternatives available for process section i
$k \in K$ Set of streams
$l \in L$ Set of detail levels

3.2. Variables

The decision variables include characterization of each process stream $k \in K$: total molar flowrate F_k and component molar flowrate f_{kc} for each component $c \in C$, as well as the stream temperature T_k and pressure P_k. A profit or cost (negative) contribution from each section $i \in I$ is given by z_i. This, in turn, may be influenced by the contribution ζ_{ij} from selection of alternative $j \in J_i$ for section $i \in I$. Other continuous state variables x may also be relevant for internal calculations within the alternatives. In the methanol case study, these variables include conversion rates in the reactors and shaft work required in the compressors. Finally, the Boolean variables Y govern selection among the process alternatives and modeling detail levels. Y_{ij} determines whether alternative $j \in J_i$ is active for section $i \in I$. Y_i^l determines whether process section $i \in I$ is modeled at detail level $l \in L_i$. Finally, for each process alternative $j \in J_i$ in section $i \in I$, Boolean Y_{ij}^l determines the modeling detail level $l \in L_{ij}$.

f_{ck}	molar flow of component c on stream k
F_k	total molar flow of stream k
T_k	temperature of stream k
P_k	pressure of stream k
z_i	profit or cost contribution from section i
ζ_{ij}	profit or cost contribution from alternative j in section i
x	other continuous state variables
Y_{ij}	Boolean selection of process alternative j for section i
Y_i^l	Boolean selection of detail level l for modeling process section i
Y_{ij}^l	Boolean selection of detail level l for modeling process alternative j in section i

3.3. Functions

The problem-specific variable relationships for the end-to-end process are represented by several functions. The globally relevant constraints $g(f, F, T, P, x)$ describe variable relationships that must be satisfied regardless of discrete selections of the process alternatives or modeling detail levels. These include the linking constraints that equate stream flow properties between different process sections. That is, the exit stream from the procurement section should be equivalent to the inlet stream to the production section. The constraints $r_{ij}(f, F, T, P, x)$ describe the relationships that are enforced regardless of the selected detail level when alternative $j \in J_i$ is selected for section $i \in I$. For each section $i \in I$, the constraints $h_i^l(f, F, T, P, x)$ describe variable interactions at each detail level $l \in L_i$ that are relevant regardless of the selected process alternatives. These constraints include potential equality relationships that link different process alternatives in a section with each other. For each of these alternatives $j \in J_i$, the constraints $s_{ij}^l(f, F, T, P, x)$ describe the variable relationships at detail level $l \in L_{ij}$. Here are included the kinetic calculations for the reactor conversion, or the shaft work calculation for the compressors. The cost functions are computed using $\phi_i^l(f, F, T, P, x, \zeta)$ at the section level, and $\psi_{ij}^l(f, F, T, P, x)$ for each process alternative. One common interpretation of $\phi_i^l(f, F, T, P, x, \zeta)$ is given in Equation (3), where the section cost is simply equal to the sum of the contributions ζ_{ij} from each alternative $j \in J_i$, but more complex relationships are possible.

$g(f, F, T, P, x)$	globally relevant constraints
$r_{ij}(f, F, T, P, x)$	constraints relevant to selection of alternative $j \in J_i$ for process section $i \in I$ for any detail level
$h_i^l(f, F, T, P, x)$	constraints describing process section $i \in I$ at detail level $l \in L_i$
$s_{ij}^l(f, F, T, P, x)$	constraints describing alternative $j \in J_i$ for process section $i \in I$ at detail level $l \in L_{ij}$
$\phi_i^l(f, F, T, P, x, \zeta)$	calculation of profit or cost contribution for section $i \in I$
$\psi_{ij}^l(f, F, T, P, x)$	calculation of profit or cost contribution for selecting alternative $j \in J_i$ in section $i \in I$
$\Omega(Y)$	Logical propositions between Boolean selections

$$z_i = \phi_i^l(f, F, T, P, x, \zeta) = \sum_{j \in J_i} \zeta_{ij} \tag{3}$$

3.4. Formulation

The overall generic problem formulation is given in Problem (P1). The objective is to maximize the profit, denoted by Z, equal to the summation of profit (or negative cost) contributions from each section $i \in I$. In the methanol process, we consider revenue from the methanol sales, purchase costs from the syngas feed, utility costs for the heaters and coolers, electricity costs for the compressors, fuel credit for the purge stream, and annualized capital costs for equipment purchases.

$$\max Z = \sum_{i \in I} z_i \tag{P1.1}$$

$$\text{s.t.} \quad g(f, F, T, P, x) \leq 0 \tag{P1.2}$$

$$\bigvee_{j \in J_i} \left[\begin{array}{c} Y_{ij} \\ r_{ij}(f, F, T, P, x) \leq 0 \end{array} \right] \qquad \forall i \in I \tag{P1.3}$$

$$\bigvee_{l \in L_i} \left[\begin{array}{c} Y_i^l \\ h_i^l(f, F, T, P, x) \leq 0 \\ z_i = \phi_i^l(f, F, T, P, x, \zeta) \end{array} \right] \qquad \forall i \in I \tag{P1.4}$$

$$Y_{ij} \implies \bigvee_{l \in L_{ij}} \left[\begin{array}{c} Y_{ij}^l \\ s_{ij}^l(f, F, T, P, x) \leq 0 \\ \zeta_{ij} = \psi_{ij}^l(f, F, T, P, x) \end{array} \right] \qquad \forall i \in I, \forall j \in J_i \tag{P1.5}$$

$$\underline{\bigvee_{l \in L_i}} Y_i^l \qquad \forall i \in I \tag{P1.6}$$

$$Y_{ij} \implies \underline{\bigvee_{l \in L_{ij}}} Y_{ij}^l \qquad \forall i \in I, \forall j \in J_i \tag{P1.7}$$

$$Y_{ij}^l \implies Y_{ij} \qquad \forall i \in I, \forall j \in J_i, \forall l \in L_{ij} \tag{P1.8}$$

$$\Omega(Y) = True \tag{P1.9}$$

$$z_i \in \mathbb{R} \qquad \forall i \in I \tag{P1.10}$$

$$z_{ij} \in \mathbb{R} \qquad \forall i \in I, \forall j \in J_i \tag{P1.11}$$

$$f \in \mathbb{R}^{|C||K|} \tag{P1.12}$$

$$F \in \mathbb{R}^{|K|} \tag{P1.13}$$

$$T \in \mathbb{R}^{|K|} \tag{P1.14}$$

$$P \in \mathbb{R}^{|K|} \tag{P1.15}$$

$$x \in X \subseteq \mathbb{R}^{n_X} \tag{P1.16}$$

$$Y \in \{True, False\}^p \tag{P1.17}$$

(P1.2) describes global constraints that are enforced independent of any alternative or detail selection. Disjunction (P1.3) governs the selection among the process alternatives $j \in J_i$ for each section $i \in I$. Disjunction (P1.4) gives the detail level $l \in L_i$ at which the major process sections $i \in I$ are modeled. Implication (P1.5) states that the selection of an alternative $j \in J_i$ implies the choice of a detail level $l \in L_{ij}$. For each section, the exclusive-OR relationship (P1.6) states that exactly one detail level $l \in L_i$ is used to model alternative-independent interactions. Similarly, for a selection of process alternative $j \in J_i$, Implication (P1.7) governs selection of exactly one detail level $l \in L_{ij}$ for modeling each alternative. Implication (P1.8) enforces that selection of a modeling detail level for an alternative implies that the alternative is selected. Other logical propositions are expressed using $\Omega(Y)$. Finally, the continuous variable definitions are given in lines (P1.10)–(P1.17).

Note that this illustrative example is meant to give a sense of the complexity that is possible to represent in a process design problem using GDP modeling techniques. GDP modeling easily supports augmentation of the model to consider, for example, methane reforming as another process section. Other process relationships and logical expressions are also possible to include, as the problem demands.

3.5. Discussion

The GDP model in Section 3.4 captures both the choice among discrete process alternatives as well as the level of modeling detail used to describe each alternative. The observant reader may note that the MINLP resulting from the reformulation of this GDP is more complex than simply modeling the entire process in high detail. While true, the power of GDP lies in the ability to systematically activate or deactivate entire blocks of related constraints. The intention of formulation (P1) is not to solve the monolithic GDP model, but rather to systematically generate models that trade off fidelity and tractability for different analyses from a single source of truth by imposing the relevant logical implications. As a result, Problem (P1) could be regarded as a parametric optimization in which the modeler prespecifies the values of Y_i^l to satisfy Equation (P1.6) for each section $i \in I$ and provides logical implications to tie selection of an alternative Y_{ij} to selection of the desired detail level Y_{ij}^l for each alternative $j \in J_i$.

Once these decisions are made, model simplifications driven by logical inference are applied to generate a process model at the desired level of detail for each of its constituent sections. For example, the production team may want to adopt a simplified view of the procurement and sales sections while preserving a high-fidelity view of production section alternatives. To accomplish this, the logical statements in Equation (4) may be appended to the model.

$$
\begin{aligned}
Y_i &\implies Y_i^{low}, & i &\in \{procurement, sales\} \\
Y_i &\implies Y_i^{high}, & i &\in \{production\} \\
Y_{ij} &\implies Y_{ij}^{low}, & i &\in \{procurement, sales\}, \forall j \in J_i \\
Y_{ij} &\implies Y_{ij}^{high}, & i &\in \{production\}, \forall j \in J_i
\end{aligned}
\tag{4}
$$

Due to the exclusive-OR-type relationship between different levels of modeling detail established by logical statements (P1.6) and (P1.7), this forces the implied level of detail to be selected for its corresponding alternative or section. Applying standard principles of logical inference [21], we arrive at the model shown in Problem (P2), with the sets $I^{low} = \{procurement, sales\}$ and $I^{high} = \{production\}$.

$$
\max Z = \sum_{i \in I} z_i \tag{P2.1}
$$

$$
s.t. \quad g(f, F, T, P, x) \leq 0 \tag{P2.2}
$$

$$
\bigvee_{j \in J_i} \begin{bmatrix} Y_{ij} \\ r_{ij}(f, F, T, P, x) \leq 0 \end{bmatrix} \qquad \forall i \in I \tag{P2.3}
$$

$$
\left.\begin{aligned}
Y_i^{low} &= True \\
h_i^{low}(f, F, T, P, x) &\leq 0 \\
z_i &= \phi_i^{low}(f, F, T, P, x, \zeta)
\end{aligned}\right\} \qquad \forall i \in I^{low} \tag{P2.4}
$$

$$Y_i^l = False \qquad \forall i \in I^{low}, \forall l \in L_i \setminus \{low\} \tag{P2.5}$$

$$\left. \begin{array}{l} Y_i^{high} = True \\ h_i^{high}(f, F, T, P, x) \leq 0 \\ z_i = \phi_i^{high}(f, F, T, P, x, \zeta) \end{array} \right\} \qquad \forall i \in I^{high} \tag{P2.6}$$

$$Y_i^l = False \qquad \forall i \in I^{high}, \forall l \in L_i \setminus \{high\} \tag{P2.7}$$

$$Y_{ij} \implies \left\{ \begin{array}{l} Y_{ij}^{low} \\ s_{ij}^{low}(f, F, T, P, x) \leq 0 \\ \zeta_{ij} = \psi_{ij}^{low}(f, F, T, P, x) \end{array} \right\} \qquad \forall i \in I^{low}, \forall j \in J_i \tag{P2.8}$$

$$Y_{ij}^l = False \qquad \forall i \in I^{low}, \forall j \in J_i, \forall l \in L_i \setminus \{low\} \tag{P2.9}$$

$$Y_{ij} \implies \left\{ \begin{array}{l} Y_{ij}^{high} \\ s_{ij}^{high}(f, F, T, P, x) \leq 0 \\ \zeta_{ij} = \psi_{ij}^{high}(f, F, T, P, x) \end{array} \right\} \qquad \forall i \in I^{high}, \forall j \in J_i \tag{P2.10}$$

$$Y_{ij}^l = False \qquad \forall i \in I^{high}, \forall j \in J_i, \forall l \in L_i \setminus \{high\} \tag{P2.11}$$

$$Y_{ij}^l \implies Y_{ij} \qquad \forall i \in I, \forall j \in J_i, \forall l \in L_{ij} \tag{P2.12}$$

$$\Omega(Y) = True \tag{P2.13}$$

$$z_i \in \mathbb{R} \qquad \forall i \in I \tag{P2.14}$$

$$z_{ij} \in \mathbb{R} \qquad \forall i \in I, \forall j \in J_i \tag{P2.15}$$

$$f \in \mathbb{R}^{|C||K|} \tag{P2.16}$$

$$F \in \mathbb{R}^{|K|} \tag{P2.17}$$

$$T \in \mathbb{R}^{|K|} \tag{P2.18}$$

$$P \in \mathbb{R}^{|K|} \tag{P2.19}$$

$$x \in X \subseteq \mathbb{R}^{n_X} \tag{P2.20}$$

$$Y \in \{True, False\}^p \tag{P2.21}$$

Problem (P2) now describes the decision space of the overall process with a focus on the production section. Changing the logical implications in Equation (4), we can easily shift focus in model fidelity to other sections of the process. By imposing different logical relationships on the general GDP model and applying easily automated principles of logical inference, we are able to derive multiple model variants from a single source of truth. Different levels of detail can also be evaluated as a post-solve solution quality check. The modeler can hold constant the production section decisions and increase modeling detail in the other sections to examine impacts of their decision-making on other sections. While this type of analysis may be possible under other engineered frameworks, GDP offers the formalism of an end-to-end perspective that is tied together by mathematical theory.

3.6. Solution Strategies

After generating a variant such as Problem (P2), a solution approach may be selected to obtain the optimal decision values. As previously introduced, the BM and HR reformulations to MINLP are the most popular approaches [2], trading off formulation size versus tightness of the continuous relaxation. The BM formulation for Problem (P2) may be found in Appendix A. For BM, equality constraints in the disjuncts are replaced by their corresponding two inequalities to facilitate relaxation of the constraint. BM results in a smaller problem size, as it does not require the introduction of new variables. However, the use of the Big-M parameter M results in a looser continuous relaxation.

Note that the relaxation may be improved by selecting unique values of M for each constraint [48]. For a tighter continuous relaxation, the HR formulation found in Appendix B may be used. HR requires additional disaggregated variables to be defined for each disjunct, so the problem size may be significantly larger than BM. However, the tighter HR formulation may require fewer iterations to converge. Once the reformulation to MINLP is made, the model can be sent to the user's solver of choice. Direct logic-based decomposition approaches [22,39] are also possible for solving GDP models, with implementations available in Pyomo.GDP via the GDPopt solver [46].

We solve the methanol synthesis model described in Problem (P2) and obtain a solution with annual profit of $1.8 million using two-stage feed compression, the cheap reactor, and single-stage recycle compression, see Figure 4. We can now fix the production section discrete decisions and evaluate the solution at varying detail levels for the procurement and sales sections. The results are given in Table 1.

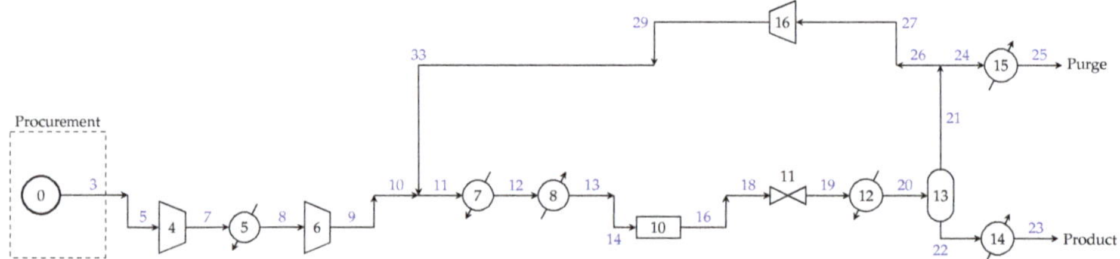

Figure 4. Solution flowsheet for Problem (P2), using two-stage feed compression, the cheap reactor, and single-stage recycle compression. At low procurement modeling detail, no feed selection decisions are made.

Table 1. Profit (1000 USD) at the fixed production section design of two-stage feed compression, the cheap reactor, and single-stage recycle compression compared to the profit achievable when the design is allowed to vary.

Procurement Detail	Sales Detail	Fixed Design	Best Design	Difference
low	low	1793	1793	
low	med	1564	1614	single-stage feed compression
low	high	2617	2667	single-stage feed compression
med	low	1793	1793	
med	med	1564	1614	single-stage feed compression
med	high	2617	2667	single-stage feed compression
high	low	1709	1832	single-stage feed compression
high	med	1746	1850	single-stage feed compression
high	high	3133	3183	single-stage feed compression

Notice that the solution profit tends to increase at higher modeling details for the feed and sales. This is due to the additional flexibility in adjusting coordinating purity levels across procurement, production, and sales. In the high-detail solutions, a higher purity syngas is purchased to enable production of a higher purity methanol product. The supplier and customer contracts are also selected to facilitate the purity decision. The procurement contract is selected with a higher base cost, but lower incremental cost for improved purity. Conversely, the customer contract is chosen where a higher purity is more valued. Thus, despite an increase in feed cost of $4.3 million vs. $3.4 million in the low-detail solution, the product revenue rises to $10.6 million rather than $7.7 million.

The Problem (P2) solution is also compared in Table 1 to the profit possible when the production section configuration is allowed to change at higher procurement and sales modeling detail levels. Here, we see that at higher levels of modeling detail, the single stage feed compressor becomes more advantageous, but only by the $50 thousand margin that accounts for the difference in annualized capital cost. In general, this analysis may not be possible, as the formulation for solving all decision degrees of freedom at a high level of modeling detail may be intractable. However, even at a medium level of sales modeling detail, it is possible to notice that the choice of single-stage feed compression may be relevant in the optimal production configuration.

As an illustration of the flexibility of GDP modeling, another solution approach that can be utilized is to emulate traditional process design strategies. GDP model (P1) is compatible with a design analysis akin to the hierarchical decomposition approach described by Douglas [50], by solving sequentially at different detail levels. First, the overall process (P1) is solved with a low detail level in iteration $iter = 0$, enforcing the implications in Equation (5).

$$\begin{aligned}
Y_i^0 &\implies Y_i^{0,low}, & \forall i \in I \\
Y_{ij}^0 &\implies Y_{ij}^{0,low}, & \forall i \in I, \forall j \in J_i
\end{aligned} \tag{5}$$

The solution to this overview problem gives high-level decisions among the process alternatives and defines the following sets for the iteration $iter = 1$. Let $J_i^1 = \{j \in J_i : Y_{ij}^0 = True\}$ denote the selected alternatives from the overview problem from iteration $iter = 0$. In the next iteration, we enforce for (P1) the implications in Equation (6) such that these alternatives are evaluated at a progressively higher detail level.

$$\begin{aligned}
Y_i^1 &\implies Y_i^{1,med}, & \forall i \in I \\
Y_{ij}^1 &\implies Y_{ij}^{1,med}, & \forall i \in I, \forall j \in J_i^1 \\
Y_{ij}^1 &\implies Y_{ij}^{0,low}, & \forall i \in I, \forall j \in J_i \setminus J_i^1
\end{aligned} \tag{6}$$

The algorithm would terminate at iteration $iter = N$ when all selected alternatives are evaluated at a high detail level: $Y_{ij}^{high} = True, \forall i \in I, \forall j \in J_i$. Note that solutions at lower detail levels can be used to initialize the higher fidelity models for each alternative, aiding in algorithm robustness. The amount of backtracking done by the algorithm to evaluate other alternatives can be tuned by applying a penalty factor to alternatives modeled at lower detail levels.

We apply this algorithm to the methanol synthesis. In the base iteration ($iter = 0$), all alternatives are modeled at a low detail level. The solution gives a configuration of the cheap reactor with two-stage feed compression and single-stage recycle compression at a profit of \$1.1 million. The structure for this solution is identical to that shown in Figure 4. We increase the modeling detail level for these selections and solve iteration 1.

Solving the low-detail overview problem, we obtain a configuration using the cheap reactor, two-stage feed compression, and single-stage recycle compression, yielding a profit of \$1.1 million. We then increase the level of modeling detail for the selected alternatives. Solving again, we obtain a profit of \$0.4 million with the expensive feed alternative, the cheap reactor, two-stage feed compression, single-stage recycle compression, and the high-purity sales option. At this intermediate level of modeling detail, the solution profit decreases because physical constraints are more tightly enforced, but not as many optimization degrees of freedom are made available to the solver yet. For some analyses, it may be advantageous for intermediate detail levels to produce a monotonically tightening approximation of the high-detail representation, but we do not address that consideration in this work. At the next iteration ($iter = 2$), we increase the detail level again, obtaining now a profit of \$3.1 million, using the contract with supplier 1, the cheap reactor, two-stage feed compression, single-stage recycle compression, and the contract with customer 1. At this point, the algorithm terminates, as all selected alternatives are modeled at the highest possible level of detail.

4. Conclusions

In this paper, we present GDP not only as a mathematical modeling framework supporting advanced solution strategies, but also as a modeling tool to organize model variants. We demonstrate through an illustrative case study that GDP is a useful model abstraction that separates algebraic and logical relationships within a process design problem. From a single GDP model capturing discrete design alternatives as well as alternatives in modeling fidelity, variants can be generated to suit the modeling scope and focus for different process sections by specifying appropriate logical implications. As a result, model interoperability is improved, facilitating simplified post-solution validation analysis. Preservation of model logical structure in GDP also offers the modeler flexibility to reformulate the problem as an MINLP or to apply a variety of automated decomposition methods.

Author Contributions: Q.C. wrote the manuscript. I.G. supervised the project.

Funding: We graciously acknowledge funding from the U.S. Department of Energy, Office of Fossil Energy's Crosscutting Research Program through the Institute for the Design of Advanced Energy Systems (IDAES).

Conflicts of Interest: The authors declare no conflict of interest.

Appendix A. Big-M Reformulation (BM)

The Big-M reformulation for GDP model in Problem (P2) is presented in Equations (A1)–(A31). All variables' domains for BM are assumed to be a subset of the non-negative real numbers.

$$\max Z = \sum_{i \in I} z_i \tag{A1}$$

$$s.t. \quad g(f, F, T, P, x) \le 0 \tag{A2}$$

$$r_{ij}(f, F, T, P, x) \le M(1 - y_{ij}) \qquad \forall i \in I, \forall j \in J_i \tag{A3}$$

$$\sum_{j \in J_i} y_{ij} \ge 1 \qquad \forall i \in I \tag{A4}$$

$$y_i^{low} = 1 \qquad \forall i \in I^{low} \tag{A5}$$

$$h_i^{low}(f, F, T, P, x) \le 0 \qquad \forall i \in I^{low} \tag{A6}$$

$$z_i = \phi_i^{low}(f, F, T, P, x, \zeta) \qquad \forall i \in I^{low} \tag{A7}$$

$$y_i^l = 0 \qquad \forall i \in I^{low}, \forall l \in L_i \setminus \{low\} \tag{A8}$$

$$y_i^{high} = 1 \qquad \forall i \in I^{high} \tag{A9}$$

$$h_i^{high}(f, F, T, P, x) \le 0 \qquad \forall i \in I^{high} \tag{A10}$$

$$z_i = \phi_i^{high}(f, F, T, P, x, \zeta) \qquad \forall i \in I^{high} \tag{A11}$$

$$y_i^l = 0 \qquad \forall i \in I^{high}, \forall l \in L_i \setminus \{high\} \tag{A12}$$

$$y_{ij}^{low} = y_{ij} \qquad \forall i \in I^{low}, \forall j \in J_i \tag{A13}$$

$$s_{ij}^{low}(f, F, T, P, x) \le M(1 - y_{ij}^{low}) \qquad \forall i \in I^{low}, \forall j \in J_i \tag{A14}$$

$$\zeta_{ij} \ge \psi_{ij}^{low}(f, F, T, P, x) + M(1 - y_{ij}^{low}) \qquad \forall i \in I^{low}, \forall j \in J_i \tag{A15}$$

$$\zeta_{ij} \le \psi_{ij}^{low}(f, F, T, P, x) - M(1 - y_{ij}^{low}) \qquad \forall i \in I^{low}, \forall j \in J_i \tag{A16}$$

$$y_{ij}^l = 0 \qquad \forall i \in I^{low}, \forall j \in J_i, \forall l \in L_i \setminus \{low\} \tag{A17}$$

$$y_{ij}^{high} = y_{ij} \qquad \forall i \in I^{high}, \forall j \in J_i \tag{A18}$$

$$s_{ij}^{high}(f, F, T, P, x) \le M(1 - y_{ij}^{high}) \qquad \forall i \in I^{high}, \forall j \in J_i \tag{A19}$$

$$\zeta_{ij} \le \psi_{ij}^{high}(f, F, T, P, x) + M(1 - y_{ij}^{high}) \qquad \forall i \in I^{high}, \forall j \in J_i \tag{A20}$$

$$\zeta_{ij} \ge \psi_{ij}^{high}(f, F, T, P, x) - M(1 - y_{ij}^{high}) \qquad \forall i \in I^{high}, \forall j \in J_i \tag{A21}$$

$$y_{ij}^l = 0 \qquad \forall i \in I^{high}, \forall j \in J_i, \forall l \in L_i \setminus \{high\} \tag{A22}$$

$$\hat{\Omega} y = b \tag{A23}$$

$$z_i \in \mathbb{R} \qquad \forall i \in I \tag{A24}$$

$$z_{ij} \in \mathbb{R} \qquad \forall i \in I, \forall j \in J_i \tag{A25}$$

$$f \in \mathbb{R}^{|C||K|} \tag{A26}$$

$$F \in \mathbb{R}^{|K|} \tag{A27}$$

$$T \in \mathbb{R}^{|K|} \tag{A28}$$

$$P \in \mathbb{R}^{|K|} \tag{A29}$$

$$x \in X \subseteq \mathbb{R}^{n_X} \tag{A30}$$

$$y \in \{0,1\}^p \tag{A31}$$

Appendix B. Hull Reformulation (HR)

The Hull Reformulation model for Problem (P2) is presented in Equations (A32)–(A70). Note that in HR, the conditional nonlinear functions, Equations (A34), (A55), (A56), (A59), and (A60) use a perspective function formulation. To avoid the singularity at a zero value of the indicator binary variable, various reformulations have been proposed in literature [39,43,51] and are available in Pyomo.GDP.

$$\max Z = \sum_{i \in I} z_i \tag{A32}$$

$$s.t. \quad g(f, F, T, P, x) \leq 0 \tag{A33}$$

$$y_{ij} r_{ij} \left(\frac{\hat{f}_{ij}}{y_{ij}}, \frac{\hat{F}_{ij}}{y_{ij}}, \frac{\hat{T}_{ij}}{y_{ij}}, \frac{\hat{P}_{ij}}{y_{ij}}, \frac{\hat{x}_{ij}}{y_{ij}} \right) \leq 0 \qquad \forall i \in I, j \in J_i \tag{A34}$$

$$\sum_{j \in J_i} y_{ij} \geq 1 \qquad \forall i \in I \tag{A35}$$

$$f^{lb} y_{ij} \leq \hat{f}_{ij} \leq f^{ub} y_{ij} \qquad \forall i \in I, \forall j \in J_i \tag{A36}$$

$$F^{lb} y_{ij} \leq \hat{F}_{ij} \leq F^{ub} y_{ij} \qquad \forall i \in I, \forall j \in J_i \tag{A37}$$

$$T^{lb} y_{ij} \leq \hat{T}_{ij} \leq T^{ub} y_{ij} \qquad \forall i \in I, \forall j \in J_i \tag{A38}$$

$$P^{lb} y_{ij} \leq \hat{P}_{ij} \leq P^{ub} y_{ij} \qquad \forall i \in I, \forall j \in J_i \tag{A39}$$

$$x^{lb} y_{ij} \leq \hat{x}_{ij} \leq x^{ub} y_{ij} \qquad \forall i \in I, \forall j \in J_i \tag{A40}$$

$$f_{ij} = \sum_{j \in J_i} \hat{f}_{ij} \qquad \forall i \in I \tag{A41}$$

$$F_{ij} = \sum_{j \in J_i} \hat{F}_{ij} \qquad \forall i \in I \tag{A42}$$

$$T_{ij} = \sum_{j \in J_i} \hat{T}_{ij} \qquad \forall i \in I \tag{A43}$$

$$P_{ij} = \sum_{j \in J_i} \hat{P}_{ij} \qquad \forall i \in I \tag{A44}$$

$$x_{ij} = \sum_{j \in J_i} \hat{x}_{ij} \qquad \forall i \in I \tag{A45}$$

$$y_i^{low} = 1 \qquad \forall i \in I^{low} \tag{A46}$$

$$h_i^{low}(f, F, T, P, x) \leq 0 \qquad \forall i \in I^{low} \tag{A47}$$

$$z_i = \phi_i^{low}(f, F, T, P, x, \zeta) \qquad \forall i \in I^{low} \tag{A48}$$

$$y_i^l = 0 \qquad \forall i \in I^{low}, \forall l \in L_i \setminus \{low\} \tag{A49}$$

$$Y_i^{high} = 1 \qquad \forall i \in I^{high} \tag{A50}$$

$$h_i^{high}(f, F, T, P, x) \leq 0 \qquad \forall i \in I^{high} \tag{A51}$$

$$z_i = \phi_i^{high}(f, F, T, P, x, \zeta) \qquad \forall i \in I^{high} \tag{A52}$$

$$y_i^l = 0 \qquad \forall i \in I^{high}, \forall l \in L_i \setminus \{high\} \tag{A53}$$

$$y_{ij}^{low} = y_{ij} \qquad \forall i \in I^{low}, \forall j \in J_i \tag{A54}$$

$$y_{ij} s_{ij}^{low} \left(\frac{\hat{f}_{ij}}{y_{ij}}, \frac{\hat{F}_{ij}}{y_{ij}}, \frac{\hat{T}_{ij}}{y_{ij}}, \frac{\hat{P}_{ij}}{y_{ij}}, \frac{\hat{x}_{ij}}{y_{ij}} \right) \leq 0 \qquad \forall i \in I^{low}, \forall j \in J_i \tag{A55}$$

$$\zeta_{ij} = y_{ij} \psi_{ij}^{low} \left(\frac{\hat{f}_{ij}}{y_{ij}}, \frac{\hat{F}_{ij}}{y_{ij}}, \frac{\hat{T}_{ij}}{y_{ij}}, \frac{\hat{P}_{ij}}{y_{ij}}, \frac{\hat{x}_{ij}}{y_{ij}} \right) \qquad \forall i \in I^{low}, \forall j \in J_i \tag{A56}$$

$$y_{ij}^l = 0 \qquad \forall i \in I^{low}, \forall j \in J_i, \forall l \in L_i \setminus \{low\} \tag{A57}$$

$$y_{ij}^{high} = y_{ij} \qquad \forall i \in I^{high}, \forall j \in J_i \tag{A58}$$

$$y_{ij} s_{ij}^{high} \left(\frac{\hat{f}_{ij}}{y_{ij}}, \frac{\hat{F}_{ij}}{y_{ij}}, \frac{\hat{T}_{ij}}{y_{ij}}, \frac{\hat{P}_{ij}}{y_{ij}}, \frac{\hat{x}_{ij}}{y_{ij}} \right) \leq 0 \qquad \forall i \in I^{high}, \forall j \in J_i \tag{A59}$$

$$\zeta_{ij} = y_{ij} \psi_{ij}^{high} \left(\frac{\hat{f}_{ij}}{y_{ij}}, \frac{\hat{F}_{ij}}{y_{ij}}, \frac{\hat{T}_{ij}}{y_{ij}}, \frac{\hat{P}_{ij}}{y_{ij}}, \frac{\hat{x}_{ij}}{y_{ij}} \right) \qquad \forall i \in I^{high}, \forall j \in J_i \tag{A60}$$

$$y_{ij}^l = 0 \qquad \forall i \in I^{high}, \forall j \in J_i, \forall l \in L_i \setminus \{high\} \tag{A61}$$

$$\hat{\Omega} y = b \tag{A62}$$

$$z_i \in \mathbb{R} \qquad \forall i \in I \tag{A63}$$

$$z_{ij} \in \mathbb{R} \qquad \forall i \in I, \forall j \in J_i \tag{A64}$$

$$f \in \mathbb{R}^{|C||K|} \tag{A65}$$

$$F \in \mathbb{R}^{|K|} \tag{A66}$$

$$T \in \mathbb{R}^{|K|} \tag{A67}$$

$$P \in \mathbb{R}^{|K|} \tag{A68}$$

$$x \in X \subseteq \mathbb{R}^{n_x} \tag{A69}$$

$$y \in \{0,1\}^p \tag{A70}$$

Appendix C. Methanol Model

The methanol synthesis model examined in this paper is adapted from the description in Example 3 of [22]. In this appendix, we describe the main differences between the literature model and the presented variant. We consider the original unit models to be the high-detail version, except for the feed and product description. The original feed model is considered to be the medium-detail description, and the original product sales model is considered the low-detail description. Linking constraints for stream flows and material balance equations are preserved from the original formulation.

Appendix C.1. Feed Procurement

Appendix C.1.1. High Detail

The high-detail feed model involves the choice between two different suppliers, with purity-dependent costs for the syngas feed.

Supplier 1

$$f_{1,H2} = 0.7(F_1 - f_{1,CH_4}) \tag{A71}$$
$$f_{1,CO} = 0.3(F_1 - f_{1,CH_4}) \tag{A72}$$
$$0.03F_1 \leq f_{1,CH_4} \leq 0.20F_1 \tag{A73}$$
$$z_1 = -425.712 + 194.973(\ln f_{1,CH_4} - \ln F_1)F_1 \tag{A74}$$

Supplier 2

$$f_{2,H_2} = 0.7(F_2 - f_{2,CH_4}) \tag{A75}$$
$$f_{2,CO} = 0.3(F_2 - f_{2,CH_4}) \tag{A76}$$
$$0.03F_2 \leq f_{2,CH_4} \leq 0.20F_2 \tag{A77}$$
$$z_1 = -400 + 210(\ln f_{2,CH_4} - \ln F_2)F_2 \tag{A78}$$

Appendix C.1.2. Medium Detail

The medium-detail feed model is the same as in [22], with the choice between a cheaper and a more expensive feed stream.

Cheap Feed

$$f_{1,H_2} = 0.6F_1 \tag{A79}$$
$$f_{1,CO} = 0.25F_1 \tag{A80}$$
$$f_{1,CH_4} = 0.15F_1 \tag{A81}$$
$$z_1 = -795.6F_1 \tag{A82}$$

$$f_{1,H_2} = 0.65F_1 \tag{A83}$$
$$f_{1,CO} = 0.30F_1 \tag{A84}$$
$$f_{1,CH_4} = 0.05F_1 \tag{A85}$$
$$z_1 = -1009.8F_1 \tag{A86}$$

Appendix C.1.3. Low Detail

The low-detail feed model simply defaults to the expensive feed option from the medium-detail case, with no other option available.

$$f_{1,H_2} = 0.65F_1 \tag{A87}$$
$$f_{1,CO} = 0.30F_1 \tag{A88}$$
$$f_{1,CH_4} = 0.05F_1 \tag{A89}$$
$$z_1 = -1009.8F_1 \tag{A90}$$

Appendix C.2. Product Sales

Appendix C.2.1. High Detail

Customer 1

$$0.85F_{23} \leq f_{23,CH_3OH} \leq 0.98F_{23} \tag{A91}$$

$$z_3 = 1682.92 - 2275.6(\ln(f_{23,H_2} + f_{23,CO} + f_{23,CH_4}) - \ln F_{23})F_{23} \tag{A92}$$

Customer 2

$$0.85F_{23} \leq f_{23,CH_3OH} \leq 0.98F_{23} \tag{A93}$$

$$z_3 = 1700 - 2265(\ln(f_{23,H_2} + f_{23,CO} + f_{23,CH_4}) - \ln F_{23})F_{23} \tag{A94}$$

Appendix C.2.2. Medium Detail

Low purity product

$$0.85F_{23} \leq f_{23,CH_3OH} \tag{A95}$$

$$z_3 = 6000F_{23} \tag{A96}$$

High purity product

$$0.95F_{23} \leq f_{23,CH_3OH} \tag{A97}$$

$$z_3 = 8500F_{23} \tag{A98}$$

Appendix C.2.3. Low Detail

$$0.9F_{23} \leq f_{23,CH_3OH} \tag{A99}$$

$$z_3 = 7650F_{23} \tag{A100}$$

Appendix C.3. Compressors

$$0 \leq P_{ratio} \leq 1.74 \tag{A101}$$

$$0 \leq W \leq 50 \tag{A102}$$

Appendix C.3.1. High Detail

$$T_{out} = P_{ratio}T_{in} \tag{A103}$$

$$W = \frac{0.72(P_{ratio} - 1)T_{in}F_{in}}{(10)(0.75)(0.23077)} \tag{A104}$$

$$P_{out}^{0.23077} = P_{ratio}P_{in}^{0.23077} \tag{A105}$$

Appendix C.3.2. Medium Detail

$$P_{ratio} = \begin{cases} 1.3 & \text{if feed compressor} \\ 1.1 & \text{if recycle compressor} \end{cases} \tag{A106}$$

$$T_{out} = \begin{cases} 1.3T_{in} & \text{if feed compressor} \\ 1.1T_{in} & \text{if recycle compressor} \end{cases} \tag{A107}$$

$$W \geq \begin{cases} \dfrac{0.72(0.3)T_{in}F_{in}}{(10)(0.75)(0.23077)} & \text{if feed compressor} \\[2ex] \dfrac{0.72(0.1)T_{in}F_{in}}{(10)(0.75)(0.23077)} & \text{if recycle compressor} \end{cases} \tag{A108}$$

$$P_{out}^{0.23077} = \begin{cases} 1.3P_{in}^{0.23077} & \text{if feed compressor} \\ 1.1P_{in}^{0.23077} & \text{if recycle compressor} \end{cases} \tag{A109}$$

Appendix C.3.3. Low Detail

$$P_{ratio} = \begin{cases} 1.3 & \text{if feed compressor} \\ 1.1 & \text{if recycle compressor} \end{cases} \tag{A110}$$

$$T_{out} = \begin{cases} 1.3T_{in} & \text{if feed compressor} \\ 1.1T_{in} & \text{if recycle compressor} \end{cases} \tag{A111}$$

$$W \geq 0.54F_{in} \tag{A112}$$

$$P_{out} = 3.8P_{in} \tag{A113}$$

Appendix C.4. Expansion Valve

Appendix C.4.1. High Detail

$$T_{out}P_{in}^{0.23077} = T_{in}P_{out}^{0.23077} \tag{A114}$$

$$P_{out} \leq P_{in} \tag{A115}$$

Appendix C.4.2. Medium Detail

$$\begin{bmatrix} T_{out} = 0.94T_{in} \\ 0.75P_{in} \leq P_{out} \leq 0.9P_{in} \end{bmatrix} \vee \begin{bmatrix} T_{out} = 0.9T_{in} \\ 0.6P_{in} \leq P_{out} \leq 0.75P_{in} \end{bmatrix} \tag{A116}$$

Appendix C.4.3. Low Detail

$$T_{out} = 0.94T_{in} \tag{A117}$$

$$P_{out} = 0.77P_{in} \tag{A118}$$

Appendix C.5. Cooler

Appendix C.5.1. High Detail

$$Q = 0.00306(35.0)(F_{in}T_{in} - F_{out}T_{out}) \tag{A119}$$

Appendix C.5.2. Medium Detail

$$\begin{bmatrix} Q = 1.61(T_{in} - T_{out}) \\ 10 \leq F_{in} \leq 20 \end{bmatrix} \vee \begin{bmatrix} Q = 0.80(T_{in} - T_{out}) \\ 5 \leq F_{in} \leq 10 \end{bmatrix} \vee \begin{bmatrix} Q = 0.54(T_{in} - T_{out}) \\ 0 \leq F_{in} \leq 5 \end{bmatrix} \tag{A120}$$

Appendix C.5.3. Low Detail

$$Q = 1.0(T_{in} - T_{out}) \tag{A121}$$

Appendix C.6. Heater

Appendix C.6.1. High Detail

$$Q = 0.00306(35.0)(F_{out}T_{out} - F_{in}T_{in}) \tag{A122}$$

Appendix C.6.2. Medium Detail

$$\begin{bmatrix} Q = 1.8(T_{out} - T_{in}) \\ 10 \leq F_{in} \leq 20 \end{bmatrix} \vee \begin{bmatrix} Q = 0.80(T_{out} - T_{in}) \\ 5 \leq F_{in} \leq 10 \end{bmatrix} \vee \begin{bmatrix} Q = 0.11(T_{out} - T_{in}) \\ 0 \leq F_{in} \leq 5 \end{bmatrix} \tag{A123}$$

Appendix C.6.3. Low Detail

$$Q = 1.0(T_{out} - T_{in}) \tag{A124}$$

Appendix C.7. Reactors

Appendix C.7.1. High Detail

$$P^2 P_{sq,inv} = 1 \tag{A125}$$

$$T T_{inv} = 1 \tag{A126}$$

$$r = \mathcal{X} f_{in,H_2} \tag{A127}$$

$$(F_{in}T_{in} - F_{out}T_{out})(35.0) = 0.01\Delta H_{rxn}r \tag{A128}$$

$$\mathcal{X}_{eq} = 0.415(1 - (26.25e^{-18T_{inv})P_{sq,inv}}) \tag{A129}$$

$$\mathcal{X} F_{in} = \begin{cases} \mathcal{X}_{eq}(1 - e^{-10})(f_{in,H_2} + f_{in,CO} + f_{in,CH_3OH}) & \text{expensive reactor} \\ \mathcal{X}_{eq}(1 - e^{-5})(f_{in,H_2} + f_{in,CO} + f_{in,CH_3OH}) & \text{cheap reactor} \end{cases} \tag{A130}$$

Appendix C.7.2. Medium Detail

$$T = 5.23 \tag{A131}$$

$$P^2 P_{sq,inv} = 1 \tag{A132}$$

$$r = \mathcal{X} f_{in,H_2} \tag{A133}$$

$$\mathcal{X}_{eq} = 0.415(1 - (26.25e^{-18/5.23}P_{sq,inv})) \tag{A134}$$

$$\mathcal{X}F_{in} = \begin{cases} \mathcal{X}_{eq}(1 - e^{-10})(f_{in,H_2} + f_{in,CO} + f_{in,CH_3OH}) & \text{expensive reactor} \\ \mathcal{X}_{eq}(1 - e^{-5})(f_{in,H_2} + f_{in,CO} + f_{in,CH_3OH}) & \text{cheap reactor} \end{cases} \tag{A135}$$

Appendix C.7.3. Low Detail

$$T \geq 5.23 \tag{A136}$$

$$\mathcal{X} = \begin{cases} 0.35 & \text{expensive reactor} \\ 0.30 & \text{cheap reactor} \end{cases} \tag{A137}$$

$$P \geq 14.3 \tag{A138}$$

$$r = \mathcal{X} f_{in,H_2} \tag{A139}$$

Appendix C.8. Flash

Appendix C.8.1. High Detail

$$(13.6333 - \ln(7500.6168 P_{H_2}^{vap}))(100T - 3.19) = 164.9 \tag{A140}$$

$$(14.3686 - \ln(7500.6168 P_{CO}^{vap}))(100T + 13.15) = 530.22 \tag{A141}$$

$$(18.5875 - \ln(7500.6168 P_{CH_3OH}^{vap}))(100T + 34.29) = 3626.55 \tag{A142}$$

$$(15.2243 - \ln(7500.6168 P_{CH_4}^{vap}))(100T + 7.16) = 897.84 \tag{A143}$$

$$\xi_{H_2}(\xi_{CO} P_{H_2}^{vap} + (1 - \xi_{CO})P_{CO}^{vap}) = P_{H_2}^{vap}\xi_{CO} \tag{A144}$$

$$\xi_{H_2}(\xi_{CH_3OH} P_{H_2}^{vap} + (1 - \xi_{CH_3OH})P_{CH_3OH}^{vap}) = P_{H_2}^{vap}\xi_{CH_3OH} \tag{A145}$$

$$\xi_{H_2}(\xi_{CH_4} P_{H_2}^{vap} + (1 - \xi_{CH_4})P_{CH_4}^{vap}) = P_{H_2}^{vap}\xi_{CH_4} \tag{A146}$$

$$f_{21,H_2} = \xi_{H_2} f_{20,H_2} \tag{A147}$$

$$f_{21,CO} = \xi_{CO} f_{20,CO} \tag{A148}$$

$$f_{21,CH_3OH} = \xi_{CH_3OH} f_{20,CH_3OH} \tag{A149}$$

$$f_{21,CH_4} = \xi_{CH_4} f_{20,CH_4} \tag{A150}$$

$$PF_{22} = P_{H_2}^{vap} f_{22,H_2} + P_{CO}^{vap} f_{22,CO} + P_{CH_3OH}^{vap} f_{22,CH_3OH} + P_{CH_4}^{vap} f_{22,CH_4} \tag{A151}$$

Appendix C.8.2. Medium Detail

$$P_{H_2}^{vap} = 73.332 \tag{A152}$$

$$P_{CO}^{vap} = 64.232 \tag{A153}$$

$$P_{CH_3OH}^{vap} = 3.722 \tag{A154}$$

$$P_{CH_4}^{vap} = 60.125 \tag{A155}$$

$$\xi_{H_2}(73.332\xi_{CO} + 64.232(1 - \xi_{CO})) = 73.332\xi_{CO} \tag{A156}$$

$$\xi_{H_2}(73.332\xi_{CH_3OH} + 3.722(1 - \xi_{CH_3OH})) = 73.332\xi_{CH_3OH} \tag{A157}$$

$$\xi_{H_2}(73.332\xi_{CH_4} + 60.125(1 - \xi_{CH_4})) = 73.332\xi_{CH_4} \tag{A158}$$

$$f_{21,H_2} = \xi_{H_2} f_{20,H_2} \tag{A159}$$

$$f_{21,CO} = \xi_{CO} f_{20,CO} \tag{A160}$$

$$f_{21,CH_3OH} = \xi_{CH_3OH} f_{20,CH_3OH} \tag{A161}$$

$$f_{21,CH_4} = \xi_{CH_4} f_{20,CH_4} \tag{A162}$$

$$PF_{22} = 73.332 f_{22,H_2} + 64.232 f_{22,CO} + 3.722 f_{22,CH_3OH} + 60.125 f_{22,CH_4} \tag{A163}$$

Appendix C.8.3. Low Detail

$$\xi_{H_2} = 0.99 \tag{A164}$$

$$\xi_{CO} = 0.99 \tag{A165}$$

$$\xi_{CH_3OH} = 0.85 \tag{A166}$$

$$\xi_{CH_4} = 0.99 \tag{A167}$$

$$f_{21,H_2} = 0.99 f_{20,H_2} \tag{A168}$$

$$f_{21,CO} = 0.99 f_{20,CO} \tag{A169}$$

$$f_{21,CH_3OH} = 0.85 f_{20,CH_3OH} \tag{A170}$$

$$f_{21,CH_4} = 0.99 f_{20,CH_4} \tag{A171}$$

$$T = 4 \tag{A172}$$

References

1. Floudas, C.A.; Gounaris, C.E. A review of recent advances in global optimization. *J. Global Optim.* **2009**, *45*, 3–38. [CrossRef]
2. Trespalacios, F.; Grossmann, I.E. Review of mixed-integer nonlinear and generalized disjunctive programming methods. *Chem. Ing. Tech.* **2014**, *86*, 991–1012. [CrossRef]
3. Lee, J.; Leyffer, S. (Eds.) *Mixed Integer Nonlinear Programming*; The IMA Volumes in Mathematics and its Applications; Springer: New York, NY, USA, 2012; Volume 154. [CrossRef]
4. Kronqvist, J.; Bernal, D.E.; Lundell, A.; Grossmann, I.E. *A Review and Comparison of sSolvers for Convex MINLP*; Springer: New York, NY, USA, 2019; Volume 20, pp. 397–455. [CrossRef]
5. Dowling, A.W.; Biegler, L.T. A framework for efficient large scale equation-oriented flowsheet optimization. *Comput. Chem. Eng.* **2015**, *72*, 3–20. [CrossRef]
6. Kondili, E.; Pantelides, C.; Sargent, R. A general algorithm for short-term scheduling of batch operations—I. MILP formulation. *Comput. Chem. Eng.* **1993**, *17*, 211–227. [CrossRef]
7. Smith, E.; Pantelides, C. Design of reaction/separation networks using detailed models. *Comput. Chem. Eng.* **1995**, *19*, 83–88. [CrossRef]

8. Bagajewicz, M.J.; Manousiouthakis, V. Mass/heat-exchange network representation of distillation networks. *AIChE J.* **1992**, *38*, 1769–1800. [CrossRef]

9. Friedler, F.; Tarjan, K.; Huang, Y.; Fan, L. Combinatorial algorithms for process synthesis. *Comput. Chem. Eng.* **1992**, *16*, S313–S320. [CrossRef]

10. Farkas, T.; Rev, E.; Lelkes, Z. Process flowsheet superstructures: Structural multiplicity and redundancy Part I: Basic GDP and MINLP representations. *Comput. Chem. Eng.* **2005**, *29*, 2180–2197. [CrossRef]

11. D'Anterroches, L.; Gani, R. Group contribution based process flowsheet synthesis, design and modelling. *Fluid Phase Equilibria* **2005**, *228–229*, 141–146. [CrossRef]

12. Lutze, P.; Babi, D.K.; Woodley, J.M.; Gani, R. Phenomena based methodology for process synthesis incorporating process intensification. *Ind. Eng. Chem. Res.* **2013**, *52*, 7127–7144. [CrossRef]

13. Bertran, M.O.; Frauzem, R.; Sanchez-Arcilla, A.S.; Zhang, L.; Woodley, J.M.; Gani, R. A generic methodology for processing route synthesis and design based on superstructure optimization. *Comput. Chem. Eng.* **2017**, *106*, 892–910. [CrossRef]

14. Wu, W.; Henao, C.A.; Maravelias, C.T. A superstructure representation, generation, and modeling framework for chemical process synthesis. *AIChE J.* **2016**, *62*, 3199–3214. [CrossRef]

15. Li, J.; Demirel, S.E.; Hasan, M.M.F. Process Integration Using Block Superstructure. *Ind. Eng. Chem. Res.* **2018**, *57*, 4377–4398. [CrossRef]

16. Raman, R.; Grossmann, I. Modelling and computational techniques for logic based integer programming. *Comput. Chem. Eng.* **1994**, *18*, 563–578. [CrossRef]

17. Grossmann, I.E.; Trespalacios, F. Systematic modeling of discrete-continuous optimization models through generalized disjunctive programming. *AIChE J.* **2013**, *59*, 3276–3295. [CrossRef]

18. Balas, E. Disjunctive programming and a hierarchy of relaxations for discrete optimization problems. *SIAM J. Algebraic Discrete Methods* **1985**, *6*, 466–486. [CrossRef]

19. Balas, E. *Disjunctive Programming*; Springer International Publishing: Cham, Switzerland, 2018. [CrossRef]

20. Raman, R.; Grossmann, I. Relation between MILP modelling and logical inference for chemical process synthesis. *Comput. Chem. Eng.* **1991**, *15*, 73–84. [CrossRef]

21. Hooker, J. *Logic-Based Methods for Optimization*; John Wiley & Sons, Inc.: Hoboken, NJ, USA, 2000. [CrossRef]

22. Türkay, M.; Grossmann, I.E. Logic-based MINLP algorithms for the optimal synthesis of process networks. *Comput. Chem. Eng.* **1996**, *20*, 959–978. [CrossRef]

23. Lee, S.; Grossmann, I.E. Global optimization of nonlinear generalized disjunctive programming with bilinear equality constraints: applications to process networks. *Comput. Chem. Eng.* **2003**, *27*, 1557–1575. [CrossRef]

24. Ruiz, J.P.; Grossmann, I.E. Global optimization of non-convex generalized disjunctive programs: A review on reformulations and relaxation techniques. *J. Global Optim.* **2017**, *67*, 43–58. [CrossRef]

25. Williams, H.P. *Model Building in Mathematical Programming*, 5th ed.; John Wiley & Sons, Ltd.: Hoboken, NJ, USA, 2013.

26. Chen, Q.; Grossmann, I. Recent developments and challenges in optimization-based process synthesis. *Annu. Rev. Chem. Biomol. Eng.* **2017**, *8*, 249–283 [CrossRef] [PubMed]

27. Rolandi, P.A. The Unreasonable Effectiveness of Equations: Advanced Modeling For Biopharmaceutical Process Development. In *Computer Aided Chemical Engineering*; Elsevier B.V.: Amsterdam, The Netherlands, 2019; pp. 137–150. [CrossRef]

28. Siirola, J.D.; Hauan, S. Polymorphic optimization. *Comput. Chem. Eng.* **2007**, *31*, 1312–1325. [CrossRef]

29. Tian, Y.; Demirel, S.E.; Hasan, M.M.; Pistikopoulos, E.N. An overview of process systems engineering approaches for process intensification: State of the art. *Chem. Eng. Process. Process Intensif.* **2018**, *133*, 160–210. [CrossRef]

30. Tsay, C.; Pattison, R.C.; Piana, M.R.; Baldea, M. A survey of optimal process design capabilities and practices in the chemical and petrochemical industries. *Comput. Chem. Eng.* **2018**, *112*, 180–189. [CrossRef]

31. Tula, A.K.; Eden, M.R.; Gani, R. Computer-aided process intensification: Challenges, trends and opportunities. *AIChE J.* **2019**. [CrossRef]

32. Sitter, S.; Chen, Q.; Grossmann, I.E. An overview of process intensification methods. *Curr. Opin. Chem. Eng.* **2019**. [CrossRef]

33. Simon, H.A. The architecture of complexity. In *Facets of Systems Science*; Springer US: Boston, MA, USA, 1991; pp. 457–476. [CrossRef]

34. Knueven, B.; Laird, C.; Watson, J.P.; Bynum, M.; Castillo, A.; US DOE. Egret v. 0.1 (beta), Version v. 0.1 (beta). 2019. [CrossRef]
35. Gani, R.; Hytoft, G.; Jaksland, C.; Jensen, A.K. An integrated computer aided system for integrated design of chemical processes. *Comput. Chem. Eng.* **1997**, *21*, 1135–1146. [CrossRef]
36. Kravanja, Z.; Grossmann, I.E. Prosyn—An automated topology and parameter process synthesizer. *Comput. Chem. Eng.* **1993**, *17*, S87–S94. [CrossRef]
37. Miller, D.C.; Siirola, J.D.; Agarwal, D.; Burgard, A.P.; Lee, A.; Eslick, J.C.; Nicholson, B.; Laird, C.; Biegler, L.T.; Bhattacharyya, D.; et al. Next generation multi-scale process systems wngineering framework. *Comput. Aided Chem. Eng.* **2018**, *44*, 2209–2214. [CrossRef]
38. Castro, P.M.; Grossmann, I.E. Generalized disjunctive programming as a systematic modeling framework to derive scheduling formulations. *Ind. Eng. Chem. Res.* **2012**, *51*, 5781–5792. [CrossRef]
39. Lee, S.; Grossmann, I.E. New algorithms for nonlinear Generalized Disjunctive Programming. *Comput. Chem. Eng.* **2000**, *24*, 2125–2141. [CrossRef]
40. Ruiz, J.P.; Grossmann, I.E. A hierarchy of relaxations for nonlinear convex generalized disjunctive programming. *Eur. J. Operat. Res.* **2012**, *218*, 38–47. [CrossRef]
41. Trespalacios, F.; Grossmann, I.E. Cutting plane algorithm for convex generalized disjunctive programs. *INFORMS J. Comput.* **2016**, *28*, 209–222. [CrossRef]
42. Bogataj, M.; Kravanja, Z. Alternative mixed-integer reformulation of Generalized Disjunctive Programs. *Comput. Aided Chem. Eng.* **2018**, *43*, 549–554. [CrossRef]
43. Furman, K.C.; Sawaya, N.; Grossmann, I. A computationally useful algebraic representation of nonlinear disjunctive convex sets using the perspective function. *Optim. Online* **2017**.
44. Brook, A.; Kendrick, D.; Meeraus, A. GAMS, a user's guide. *ACM SIGNUM Newslett.* **1988**, *23*, 10–11. [CrossRef]
45. Vecchietti, A.; Grossmann, I.E. LOGMIP: A disjunctive 0-1 non-linear optimizer for process system models. *Comput. Chem. Eng.* **1999**, *23*, 555–565. [CrossRef]
46. Chen, Q.; Johnson, E.S.; Siirola, J.D.; Grossmann, I.E. Pyomo.GDP: Disjunctive Models in Python. In *Proceedings of the 13th International Symposium on Process Systems Engineering*; Eden, M.R., Ierapetritou, M.G., Towler, G.P., Eds.; Elsevier B.V.: San Diego, CA, USA, 2018; pp. 889–894. [CrossRef]
47. Hart, W.E.; Laird, C.D.; Watson, J.P.; Woodruff, D.L.; Hackebeil, G.A.; Nicholson, B.L.; Siirola, J.D. *Pyomo—Optimization Modeling in Python*, 2nd ed.; Springer Optimization and Its Applications; Springer International Publishing: Cham, Switzerland, 2017; Volume 67. [CrossRef]
48. Trespalacios, F.; Grossmann, I.E. Improved big-M reformulation for generalized disjunctive programs. *Comput. Chem. Eng.* **2015**, *76*, 98–103. [CrossRef]
49. Vecchietti, A.; Grossmann, I.E. Modeling issues and implementation of language for Disjunctive Programming. *Comput. Chem. Eng.* **2000**, *24*, 2143–2155. [CrossRef]
50. Douglas, J.M. A hierarchical decision procedure for process synthesis. *AIChE J.* **1985**, *31*, 353–362. [CrossRef]
51. Grossmann, I.E.; Lee, S. Generalized convex disjunctive programming: Nonlinear convex hull relaxation. *Comput. Optim. Appl.* **2003**, *26*, 83–100. [CrossRef]

 *processes*

Review

Advances in Energy Systems Engineering and Process Systems Engineering in China—A Review Starting from Sargent's Pioneering Work

Wenhan Qian, Pei Liu * and Zheng Li

State Key Lab of Power Systems, Department of Energy and Power Engineering, Tsinghua University, Beijing 100084, China; qwh5721@126.com (W.Q.); lz-dte@tsinghua.edu.cn (Z.L.)
* Correspondence: liu_pei@tsinghua.edu.cn; Tel.: +86-10-6279-5738

Received: 25 April 2019; Accepted: 3 June 2019; Published: 7 June 2019

Abstract: Process systems engineering (PSE), after being proposed by Sargent and contemporary researchers, has been fast developing in various domains and research communities around the world in the last couple of decades, with energy systems engineering featuring a typical yet still fast propagating domain, and the Chinese PSE community featuring a typical community with its own unique challenges for applying PSE theory and methods. In this paper, development of energy systems engineering and process systems engineering in China is discussed, and Sargent's impacts on these two fields are the main focus. Pioneering work conducted by Sargent is firstly discussed. Then, a venation on how his work and thoughts have motivated later researchers and led to progressive advances is reviewed and analyzed. It shows that Sargent's idea of optimum design and his work on nonlinear programming and superstructure modelling have resulted in well-known methods that are widely adopted in energy systems engineering and PSE applications in tackling problems in China. Following Sargent's pioneering ideas and conceptual design of the PSE mansion, future development directions of energy systems engineering are also discussed.

Keywords: process systems engineering; energy systems engineering; process design; optimization; nonlinear programming

---

1. Introduction

Process systems engineering (PSE) is an interdiscipline of chemical engineering and systems engineering. It aims at providing a systematic methodology for chemical engineering decisions [1,2]. Energy systems engineering (ESE) is a branch of PSE, applying theories from PSE to analyses of energy systems. Since the beginning of PSE and ESE, their application to the industry has produced significant benefits. Researchers are putting more and more emphasis on these two disciplines.

Professor Sargent is entitled the father of PSE, since he had made great contributions to design, boost, and lead the PSE field. His work and impact can almost be found in every corner of PSE, including process design, process control, and others. In addition, his thoughts have a leading role for the development of PSE. A tribute of Sargent, which summarized his important work and his impact had been written by Doherty et al. [3]. However, there seems to be a lack of review work on Sargent's impact on ESE. Moreover, Sargent's work has inspired many research communities around the world with their unique yet challenging PSE problems. Furthermore, a study on Sargent's impact on the Chinese PSE development is chosen in this review.

The remaining part of the paper is organized as follows. Firstly, Sargent's major contributions, especially those with direct impact with energy systems engineering and PSE applications in China, are discussed. Development of energy systems engineering, guided by the framework set by Sargent, and applications in solving local challenging problems of China, inspired by Sargent's work, are then summarized. Last, future directions of energy systems engineering are proposed, following the PSE framework set up by Sargent, and based on characteristics of energy systems which are unique from a generic process system.

2. Contributions of Professor Sargent

Over his research career, Professor Sargent made tremendous contributions to the field of PSE. His work can be found in almost every discipline, including system decomposition, optimal design, process control, solution of nonlinear algebraic equations (NAE), differential algebraic equations (DAE), and others. In addition, it is more creditable that Professor Sargent does not only make great achievements in his own research, but also acts as a leader of PSE researchers, providing guiding light for the development of the whole field. In 1967, he published an inspiring and insightful article, pointing out possible directions for future research [4]. The most important one in this article might be the idea of integrated design, i.e., combining design and optimization instead of regarding them as two separate tasks. Sargent also pointed out the need for future work in better understanding of physical mechanisms of chemical process, in development of model simplification techniques, and in exploration of dynamic behaviour. In 1971, he presented another review article as a supplement of the former one [4], in which he pointed out that much work remained to be done in the solution of large-scale NAEs, stochastic optimal control, nonlinear filtering, and the automatic determination of model structure [5]. Those research directions he had mentioned have almost all become research frontends in the next few decades. With such significant contributions, Professor Sargent is regarded as the father of PSE. In this part, his contributions in several specific disciplines are discussed, whilst some of the significant work are further elaborated in the following sections.

In early years, Sargent had made much effort in system decomposition techniques. In 1964, Sargent and Westerberg developed two algorithms, one for decomposing a system into sequential groups, and the other for determining the computational order within a group, based on dynamic programming [6]. Later Sargent extended the research to systems of equations. In his article in 1977 [7], Sargent studied the connection between decomposition of systems of procedures and systems of equations, and extended Leigh's algorithm [8] to recognizing linear subsets.

In terms of optimization, Sargent mainly focused on solution methods for nonlinear programming (NLP) problems. In 1969, based on the Davidon–Fletcher–Powell (DFP) method for unconstrained nonlinear optimization [9,10] and Rosen's gradient projection method for dealing with constraints [11], Sargent and Murtagh developed the variable-metric projection (VMP) method for NLP [12]. In 1973, they further improved the VMP method in situations of nonlinear constraints [13]. Sargent also did much work to discuss the convergence property of the method, such as Sargent and Murtagh in 1970 [14], Sargent and Sebastian in 1973 [15], and Sargent in 1973 [16]. In 1978, Sargent proposed an efficient implementation of the Lemke algorithm, which could be used to solve quadratic programming (QP) problems [17]. Later this algorithm was used to derive an SQP algorithm in 2001 [18]. Sargent was also one of the first to study process design under uncertainty. In 1978, Grossmann and Sargent proposed an efficient method for NLP problems with uncertain parameters [19]. Their method aimed to design a plant that could meet the specifications over a bounded range of values of parameters, which was essentially the idea of robust optimization.

In process design, a proper representation of the model could make the analysis more convenient. Sargent had done much profound work in this area. In 1976, Sargent and Gaminibandara proposed the superstructure representation for synthesis of distillation columns [20]. In 1993, Sargent and his co-workers introduced the state-task-network (STN) representation for batch scheduling [21]. And in 1998, Sargent further applied the STN representation method to distillation systems [22].

The steady-state simulation of chemical processes usually involved the solution of large systems of NAEs. In 1980, Sargent presented a review of various methods for this problem, such as Newton's method and least-change secant methods [23]. In addition, since solving the whole set of NAEs is rather difficult, development of equation-oriented decomposition techniques was motivated, as discussed above.

When it comes to dynamic simulation, the problem becomes the solution of large systems of DAEs. Sargent and co-workers had studied various solution methods, such as Cameron's work on diagonally implicit Runge–Kutta (DIRK) methods and backward differentiation formulae (BDF) algorithms in 1981 [24]. In 1988, Sargent and his co-workers put this problem in a general mathematical formulation, and they used a BDF-based software to solve the problem of a plate distillation column described by index two DAEs [25].

3. Energy Systems Engineering

Energy systems engineering (ESE) is usually regarded as a branch of PSE, which aims to apply PSE methodologies to analyse energy systems. The concept of ESE was first proposed by Pistikopoulos and co-workers at the sixth European Congress of Chemical Engineering in 2007 [26]. According to their definition, energy systems engineering provides a methodological framework to address the complex energy and environmental problems by an integrated systematic approach, which accounts complexities of very different scales, ranging from technology, plant, to energy supply chain, and megasystem [27]. Typical methodologies include mixed integer programming (MIP), superstructure-based modelling, and others. In this chapter, the development of such techniques as related to Sargent's ideas is discussed, and a venation on how researchers came to realize the usability and helpfulness of those PSE methods in energy systems research is given.

3.1. Development of Optimum Design and Synthesis

The venation on how Sargent's original work in optimum design was gradually developed to commonly-used methods in ESE nowadays is illustrated in Figure 1. Detailed description of the relevant work is given in the following.

Professor Sargent was one of the first researchers to recognize the importance of optimization in PSE. Although his article in 1967 [4] was the most well-known for this idea, Sargent had actually come to it much earlier. In 1964, Sargent and Westerberg published an article describing their SPEED-UP program for analysis and design of chemical processes [6]. The main topic of the article is about decomposition techniques. However, it was mentioned that, at that time, design and optimization were usually treated as two separate tasks, which meant that a feasible design of the system would first be worked out, and then attempts would be made to change the parameters of the model for better performance, and that in optimization, no published work had discussed modifying plant structure to achieve optimum performance. Sargent and co-workers were attempting to develop a program called SPEED-UP, which could automatically make the optimum design of chemical systems, in which both the parameters and the structure would be modifiable. In the 1967 article [4], Sargent expressed this idea in more details. He defined process design as the determination of the fixed parameters for a process so that it meets the specifications as well as being optimum in some sense, and the criterion for being optimum is to maximize (or minimize) an objective function. In addition, he pointed out that it had become common practice to assume a fixed process flow diagram at the outset of design, but with the development of the computer it was hoped that the process structure could be decided automatically.

This idea was soon embodied in Sargent's work. In 1969, Sargent and Murtagh described a program for performance calculation of distillation columns [28], and they pointed out that the program could be used in an integrated optimum design, forming an optimization problem where several degrees of freedom, such as the number of plates and the positions of the feeds, are to be fixed to minimize a specific objective function. This is a pioneering work in optimum design, and since it became necessary to solve optimization problems, usually NLP problems, Sargent began to conduct much research in this field. In 1969, Sargent and Murtagh developed the VMP method for solving NLP problems [12] and further developed it in 1973 [13]. The VMP method was one of the first algorithms for NLP that could handle both linear and nonlinear constraints. However, when applied to large-scale systems, it would show some shortcomings. Motivated by these deficiencies, in 1978, Murtagh and Saunders developed an algorithm for solving large-scale NLP problems with linear constraints, and a code was written based on it, which was now one of the widely used NLP solvers, MINOS [29].

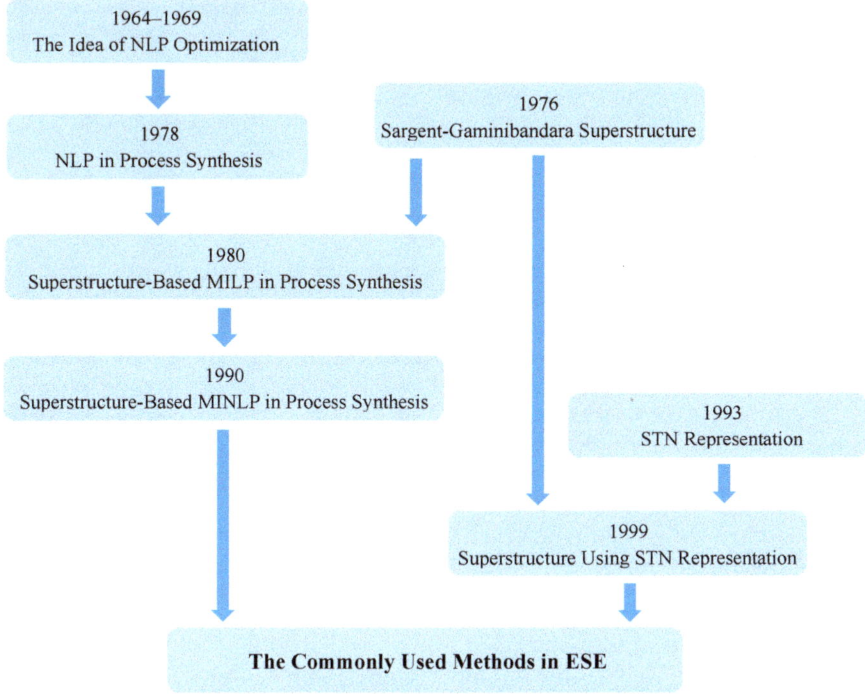

Figure 1. The development of Sargent's work in optimum design to methods used in energy systems engineering (ESE).

For conventional design problems, the process flow diagram is fixed, and equations can be easily written according to the diagram. However, in optimum design, the structure and thus the flow diagram are modifiable, making it difficult for mathematical analysis. Drawing all functionally possible configurations and writing corresponding sets of equations is certainly an effective way, but such a combinatory problem requires too much effort and is of poor efficiency. Faced with this question, Sargent and Gaminibandara proposed the superstructure representation of the system in 1976 [20]. For the design of a separation system with multiple distillation columns, they concluded all possible configurations to one general system—superstructure, from which any particular configuration could be obtained by deleting some of the components in the system. Such deletions could be done through direct optimization of the superstructure, solving continuous NLP problems [30]. With Sargent's idea of superstructure, tedious combinatory problems were avoided, and the optimum design problem could be solved using mathematical programming (MP).

In the meantime, research on the synthesis of heat exchanger network (HEN) emerged. HEN can be defined as a heat recovery system consisting of a set of heat exchangers [31]. By suitable arrangement of heat exchangers, the resulting HEN can have higher heat efficiency and make better use of thermal energy, leading to reduced energy consumption and costs. Around 1970, energy price became higher and higher, and thus researchers came to pay more attention to such a synthesis problem. The first one to tackle this problem was Rudd, who in 1967 presented an article discussing process synthesis with an example about the synthesis of a system of heat exchangers [32]. In addition since the HEN synthesis is for energy-saving purposes, Rudd's work could also be regarded as the origin of ESE. In the current view, there are mainly two kinds of methods for the HEN synthesis problem, one based on thermodynamics and the other based on pure MP [33]. Rudd's method [32] belongs to the latter, in which he formulated the problem as an unconstrained optimization problem. At first, Sargent did not pay much attention to this problem. However, his student Grossmann was obviously very interested in it. As was shown in Sargent's retrospect [34], Grossmann obtained his Ph.D. at Imperial College in 1977, and during his doctoral research, he had involved Sargent in newer areas of the HEN synthesis, batch scheduling, and integer programming. Their cooperative work on optimum design of HEN was published in 1978 [35]. In addition, since Sargent had made great efforts in NLP-based process design, it is not surprising that in this paper NLP optimization was performed to solve the problem, which is a pioneering work in applying MP to the HEN synthesis. This work was also the first step into the MINLP-based HEN synthesis, as discrete variables were introduced to determine the configuration [36].

Since then Grossmann had been studying the solution of MINLP and its application. In 1979, Sargent and Grossmann considered the design of sequential multiproduct batch plants and formulated the problem as a mixed integer nonlinear programming (MINLP) one [37]. However, at that time no efficient methods for MINLP existed, so they used the VMP method to solve a relaxed NLP version of the problem, obtaining a practically satisfactory solution though it might be suboptimal rather than exactly optimal. Since MINLP was difficult to solve, in 1980, Grossmann and Santibanez suggested to reformulate those process synthesis problems as mixed integer linear programming (MILP) problems [38]. In 1983, Papoulias and Grossmann did so to the HEN problem and used the standard Branch & Bound (BB) technique to solve it [39].

The difficulty in MINLP has aroused much research interest. In 1986, Duran and Grossmann proposed the outer-approximation (OA) algorithm for solving MINLP [40]. Their algorithm was designed to solve MINLP problems in which integer variables are in linear form and the nonlinear functions are convex. In addition, on problems of this kind, OA would perform better than generalized benders decomposition (GBD) and BB, since it fully takes advantage of the problem structure. The requirements of OA are often satisfied in process synthesis problems. However, OA cannot explicitly handle nonlinear equality constraints. In the same year, Kocis and Grossmann developed an equality-relaxation (OA/ER) algorithm that could handle such constraints as well as preserve the advantages of OA [41]. In 1989, the two authors further wrote a computer code to implement OA/ER, and showed that it could solve MINLP problems efficiently [42]. Their code, DICOPT, has become one of the most commonly used solvers for MINLP nowadays. As efficient tools for MINLP had been well developed, the process synthesis problems were no longer restricted to MILP forms. In 1990, Grossmann reviewed the current MINLP strategies and algorithms, and illustrated their use in process synthesis [43].

The superstructure representation proposed by Sargent is naturally a method to formulate design problems as MINLP ones, since it covered all possible choices for equipment, pathways and others. As a result, with the rapid development of MINLP methodologies, superstructure-based modelling became more and more widely-known. Grossmann had made great contributions in spreading this method. In 1983, Papoulias and Grossmann proposed a superstructure for process synthesis problems of utility systems, which provide electricity, power and steam for the facilities in the plant [39]. What is worth mentioning is that their superstructure is a more detailed one than the Sargent–Gaminibandara superstructure [20], as the latter only included the possible configurations of the system, while theirs included the configurations as well as the available types of the equipment. However, at that time MINLP techniques were still immature, so the problem was in the form of MILP. In his 1990 paper [43], Grossmann gave a clear scheme for solving process synthesis problems, providing a systematic way to use superstructure representation and MINLP techniques in design problems. The scheme includes three steps, i.e., one should first postulate a superstructure involving all the feasible alternatives, then model the superstructure as an MINLP problem, and finally solve the problem to obtain the optimal design. Later this year, Yee et al. implemented this scheme in simultaneous synthesis of the process and its HEN [21]. Their work was ground-breaking, as it was the first time process synthesis problems were solved using superstructure-based MINLP modelling, showing the feasibility and usefulness of this method. In addition more importantly, as the HEN synthesis is to a large extent an energy problem, the work also showed the potential of this method in the field of ESE. After Sargent and his co-workers introduced the state-task-network (STN) representation for batch scheduling in 1993 [21], and for synthesis of distillation systems in 1998 [22], Yeomans and Grossmann presented a general framework of superstructure optimization in 1999 [44]. In their paper, two fundamental representations of superstructure, STN and its complementary, state-equipment-network (SEN), were given. Then they used generalized disjunctive programming (GDP) to model the superstructure, and reformulated the GDP problem as an MILP problem, which could be steadily solved. This systematic approach was widely adopted in Grossmann's later research. Furthermore, the STN superstructure representation proposed in this paper, combined with Grossmann's systematic scheme for design problems [43], became a commonly used method in ESE research. This method is especially useful in energy generating systems, as the production process of energy products is naturally in such a form, including states from feedstocks, intermediates to finals, and the technologies linking these states. Typical application will be discussed in the following sections.

3.2. Development of Environmentally Conscious Process Design

Pollution minimization is also an important topic in the field of ESE. As people were more aware of the importance of environmental protection and thus relevant regulations became stricter, the costs of pollution treatment to comply with these regulations were increasing rapidly [45]. These costs were so high that pollution prevention seemed to be a better way than pollution disposal. As many researchers had shown, reduction in pollution can lead to cost savings, and thus bring an increase in profits [46]. This led to a new objective in design problems, i.e., to minimize the wastes generated in the process. Furthermore, since there was no appropriate way to convert the amount of pollution to numerical values that were comparable with economic costs, it was infeasible to include pollution and costs in one objective function. Thus, environmentally conscious process design problems were inevitably multi-objective optimization problems.

Multi-objective optimization (MOO) was at first an economic problem, whose concept was proposed by Pareto in 1896. Several decades later, it became a popular topic in engineering. Methods for solving these problems were well developed, such as the ε-constrained method, which was proposed by Haimes et al. in 1971 [47], and the parametric programming method. However, it was not until 1980s that MOO was implemented in chemical engineering [48]. Additionally, as is stated above, such implementation could be attributed to the awareness of pollution reduction. One of the first attempts to include environmental considerations in design problems by MOO is the work by Grossmann and his co-workers in 1982 [49]. In their paper, the traditional optimal design framework was combined with the newly introduced environmental factors. To be specific, the considered type of pollution was the toxicity of the chemicals, and the two objectives were maximizing net present value (NPV) and minimizing a toxicity index. The ε-constrained method was adopted to solve this MOO problem. However, they only took into account the toxicity of material flows in the system, while ignoring the environmental impacts that had already been made before these materials flew into the system. This led to an unreasonable conclusion that the production of all intermediates should be carried out by suppliers [45].

Such a problem can be avoided by the life cycle assessment, also known as life cycle analysis (LCA), which is a methodology considering the environmental impact of a product in every stage of its life cycle. The first paper using LCA as a tool for process design was published in 1995 by Stefanis et al. [50]. They presented a method for minimizing environmental impact (MEI), which took advantage of LCA to give a systematic quantification of the wastes generated. Contrary to conventional design problems that only considered energy generation processes, in this method, the system boundary was expanded to include energy generation processes as well as raw material extraction processes, so that the pollution produced in the latter would not be neglected. Moreover through a case study, they found that minimizing pollution within the expanded system boundary led to lower operation costs than those obtained by minimizing pollution within the conventional boundary, which showed that LCA was a potentially useful tool in optimal design. One year later, Stefanis et al. applied this method of MEI in solvent design and reaction path synthesis problems [51]. This time they posed the problem in MOO, with a performance indicator and several environmental indices as the set of objective functions. In 1997, they extended this method to the optimal design and scheduling of batch processes, and similarly, formulated the problem as a MOO problem with economic costs and environmental criteria as two objectives [52]. The ε-constrained method was used to solve such problems, resulting in Pareto curves based on which compromise solutions could be obtained. Motivated by these researches, in 1999, Azapagic and Clift reviewed the application of LCA in process optimization [53]. In this article, they gave a formal definition of LCA and showed that MOO coupled with LCA could provide a powerful tool for balancing economic and environmental performance of a process from cradle to crave. Since then, MOO and LCA had become two common methodologies used for environmental problems in the field of ESE. These years, some software has also been developed to perform MOO and LCA in environmentally conscious process design, such as SustainPro [54] and LCSoft [55], both developed by Gani and co-workers. Although such software is still novel, it is believed that they will be put into wide use in the near future, bringing convenience to the implementation of sustainable design.

3.3. Other Techniques

Sargent has also made great contributions in many other fields of PSE, such as optimization under uncertainty, solution of NAEs and DAEs, optimal control, and others. In this part, his work on the former problems, which has been developed by later researchers to create widely-known algorithms or software nowadays, is given a detailed discussion.

When planning for a long future period, it is likely that changes will happen in the environment of the system. For example, the situation of the market might change for the better. Revolutions might happen in existing technologies. Such uncertainties should be involved in optimization, leading to a new field—optimization under uncertainty. One preferable approach to perform optimization under uncertainty is stochastic programming. In addition, Sargent was one of the first to propose a method of this kind. In 1978, Grossmann and Sargent proposed an efficient method for NLP problems with uncertain parameters [19]. In their work, the uncertainty was expressed by a bounded range of values of the parameters, and the objective of their method was to design the plants that could meet the specifications over such bounds, while obtaining the optimum value of a weighted cost function. Their model was a two-stage stochastic programming model, and the requirement that the process should be feasible under all possible realizations of the uncertain parameters was essentially the idea of robust optimization. In 1982, Halemane and Grossmann gave a new mathematical formulation of Sargent and Grossmann's strategy so that the feasible operation under all possible situations could be rigorously ensured [56]. In 1985, Swaney and Grossmann considered another method to take into account uncertainty in their paper [57], in which an index of flexibility was proposed. Based on this work, in 1993, Straub and Grossmann proposed a method to maximize the stochastic flexibility [58]. Moreover, in this work, the uncertainty considered was in the form of probability distribution functions, rather than specific bounds. In 1995, Ierapetritou et al. proposed a combined multiperiod/stochastic programming formulation for optimization problems with uncertain parameters including both deterministic (expressed in bounded ranges) and stochastic (expressed in probability distribution functions) ones [59]. In the same year, Acevedo and Pistikopoulos gave several integration schemes for the approximation of the expected profit, which could be used in Ierapetritou et al.'s framework to find a trade-off between economic optimality and design flexibility [60]. These methods are widely applied to optimization problems under uncertainty.

Solution of systems of NAEs or DAEs is also a problem that frequently occurs in PSE research, and Sargent has made great contributions to this area. In terms of solution of NAEs, Sargent had presented a comprehensive article in 1980 [23], in which he discussed the existing solution methods of NAEs, the methods for ensuring global convergence, and the decomposition techniques for large problems. In addition, he also showed great interest in the way to facilitate the solution of NAEs through automatic algebraic manipulation. This idea was later further developed, giving rise to algorithms in the program developed by Sargent and his co-workers, SPEED-UP [3]. With regard to DAEs, Sargent and his co-workers were the first to point out the need for solving DAEs in PSE [61]. Since then Sargent and co-workers have made much effort to study available solution methods of DAEs [24] and to apply them to dynamic simulation of chemical processes [25]. Many of the algorithms they had developed were also incorporated in SPEED-UP. Around 1988, many researchers that had participated in the development of SPEED-UP started to design a new software, which is now one of the most frequently-used tools for process simulation, gPROMS [62]. This software is used in a large number of ESE studies, especially in modelling of energy-related devices. Typical applications include Kikkinides et al.'s work on hydrogen storage tanks [63], Kouramas and Pistikopoulos's work on wind turbines [64], and others. There are also many other useful tools for dynamic simulation. In 1988, Cameron and Gani developed an adaptive Runge–Kutta algorithm for solution of ordinary differential equation systems [65]. This algorithm was incorporated in an equation-oriented dynamic simulator, DYNSIM, developed by Sorensen and co-workers in 1990 [66]. Later Gani et al. applied this software to solve simulation problems in the design and analysis of chemical processes [67].

3.4. Typical Work in ESE

In the above paragraphs, the development history of ESE has been combed, especially on how researchers came to apply PSE methods, including superstructure, MIP, LCA, MOO, optimization under uncertainty, and others, to dealing with energy and environmental problems. In this part, typical work in ESE will be listed. The work to be mentioned is only a little portion of the numerous volumes of ESE research, but it is possible to see from it the wide application of those methods that was developed from Sargent's work or thoughts.

As mentioned above, energy problems and environmental problems are the two main questions in the field of ESE. These two kinds of problems are not separate problems. As many studies have already shown, the increase in energy efficiency has important impacts on the decrease in environmental pollution. As a result, the application of methods that can improve energy efficiency is paid much attention to.

One way to improve energy efficiency is to adopt co-generation. For processes that share the same intermediates, such as liquid fuel synthesis and combined cycle power generation, co-production can be made to achieve higher efficiency, which is the concept of polygeneration energy systems. In 2007, Liu et al. built a model for strategic decisions in the development progress of a polygeneration energy system [68]. The model was presented in a superstructure method, and the problem was first formulated as an MINLP one, while linearization techniques were adopted to convert it into an MILP one which could be solved more efficiently. The model was used to analyze the investment planning of polygeneration energy systems co-producing methanol and electricity in China between 2010 and 2035. In addition, it was shown that polygeneration technologies were indeed superior to stand-alone technologies. In 2009, based on the previous work, Liu et al. further studied the design optimization of polygeneration energy systems [69]. The problem was formulated as a superstructure-based non-convex MINLP problem. Moreover, the model was used for detailed analysis of a polygeneration plant for electricity and methanol. In 2010, Liu et al. presented a framework for optimal design of polygeneration energy systems under uncertainty [70]. They incorporated uncertainty in their previous model, leading to a two-stage stochastic programming program, which was then converted to a large-scale multiperiod MINLP problem through a decomposition-based strategy. A case study was conducted, in which uncertainty in the price of coal and the price and demand of methanol and electricity was considered. In 2011, Liu et al. studied the strategic planning of polygeneration energy systems in the supply chain level and conducted a case study for the UK [71].

In power generation, some energy will be inevitably discarded as waste heat. Combined heat and power (CHP) system is a system that co-generates electricity and heat so that a portion of waste heat can be used for heat supply, thus leading to higher energy efficiency. Further development of CHP systems led to combined cooling, heat and power (CCHP) systems, which use the waste energy not only for heat supply, but also as the driving power for refrigerators to produce cooling. Since the produced heat or cooling cannot be transported for a long distance, CHP or CCHP systems are usually distributed energy systems (DES), with the consumption of energy close to the production. In 2007, Arcuri et al. studied the optimal design of the CCHP system in a hospital complex by performing MILP optimization [72]. Their model was rather simple, in which only the energy provided by fuel was taken into account. In 2010, Liu et al. used the ESE framework to deal with the optimal design of DES in a supermarket [73]. In their model, fuel as well as other available energy sources were considered, including wind, solar and biomass energy. In 2013, Zhou et al. proposed an even more elaborate model, including six types of energy resources and twenty types of equipment [74]. Their model was used for optimal design of DES for a hypothetic hotel in Beijing, and it was shown that the system described by their model did better than centralized systems and traditional, simpler CCHP systems. In 2013, Kopanos et al. considered uncertainty in the scheduling of household CHP (or called micro-CHP) systems, and used parametric programming methods for solution [75].

The use of clean energy, such as hydrogen, is also given serious attention to. Since hydrogen was a novel kind of energy, its whole supply chain was to be designed. In 2005, Hugo et al. performed superstructure-based multi-objective MILP to study the optimal design of hydrogen infrastructure [76]. In every stage of the infrastructure, including production, transmission, storage and others, various technological options were considered. Greenhouse gas (GHG) emissions were evaluated using LCA, and MOO was solved to find the set of optimal trade-off solutions between investment and environmental pollution. In 2008, Li et al. extended this model for application to case studies in China [77]. Due to high transportation costs of hydrogen, they put emphasis on methanol pathway, which means methanol was to be produced and delivered as an alternative energy carrier, and hydrogen would be produced onsite from it. They found that methanol pathway had great potential for future Chinese hydrogen economy. In 2018, Ogumerem et al. developed a model similar to that of Hugo et al. [76], and deployed it for a case study in the state of Texas [78]. A creative point of their work was that they considered the possibility to collect and sell oxygen as a source of revenue, which was a by-product from electrolysis. They found that the processing of oxygen for selling could make electrolysis a more economically viable choice for hydrogen production.

Improvement in technology will inevitably lead to the need to construct new power plants and decommission old ones. The scheduling of such tasks has received wide attention. In 2012, Zhang et al. studied the planning of China's power sector with consideration of carbon dioxide (CO_2) emission [79]. In their research, renewable energy was paid much attention to, and plants with application of integrated gasification combined cycle plants (IGCC) or carbon dioxide capture and sequestration (CCS) were considered. CO_2 mitigation was considered by incorporating CO_2 emission costs in the objective function, and the problem was solved through single objective optimization. One year later, Zhang et al. further studied the problem under uncertainty using the levelized optimization method [80]. However, in Zhang et al.'s research, the power system of China was considered as a whole, in which energy transmission was neglected. However, in reality, the transmission losses could lead to very high costs, especially when serious mismatches happened among the distribution of power generation and power demand. In 2014, Koltsaklis et al. developed a spatial MILP model for optimal planning of Greek's power system [81]. The whole Greek was divided into five geographical zones, and energy transmission and resource transportation among zones were taken into account. In 2015, Cheng et al. also developed a multi-regional planning model for China's power system [82], in which China was modelled as 10 regions. Additionally, as was in Zhang et al.'s research [79,80], their research featured the consideration of renewable energy as well as application of IGCC and CCS. Guo et al. further considered the transmission capacity between regions [83]. They also further divided China into seventeen regions [84]. Furthermore, in previous research electricity supply and demand were balanced on a yearly basis, while Guo et al. sub-divided one year into several time blocks so that variations in the availability of various energy resources could be considered [85].

The operation or scheduling problems of existing plants can also be tackled using MIP optimization. However, when developing the model of the plant, it is necessary to determine the design variables from the plant's measurement data, and since given data are usually not satisfactory, data reconciliation techniques are usually adopted to improve their accuracy. One of the earliest works to use data reconciliation for modelling was conducted by Papalexandri et al. in 1996. They applied data reconciliation methods to the modeling of a steam production network, and performed multi-period MINLP optimization based on this model to get optimal operating schemes of the network under various demands. Such technologies are also widely studied and applied in ESE research. In 2014, Jiang et al. used the data reconciliation method for gross error detection in a real-life coal-fired power plant [86]. In 2016, Guo et al. proposed an inequality constrained nonlinear data reconciliation framework, applied it in real-life power plants, and found that their approach could lead to more accurate operational data and better estimates of parameters than the equality constrained approach [87]. In 2017, Li et al. [88] adopted Guo et al.'s data reconciliation method [87] to preprocess the operating data of a thermal power unit under various working conditions, and thus developed a mathematical model for the unit, after which operation optimization was performed.

4. Process Systems Engineering in China

The development of PSE in China can be generally divided into three periods. From 1973 to 1980, the discipline was introduced into China, and much fundamental research was conducted. In early and middle 1980s, people came to pay more attention to this field, and began to explore more fields such as process synthesis. Since the late 1980s, Chinese researchers have strengthened their communication with researchers all over the world. In addition, Sargent's idea of superstructure and NLP/MIP optimization has begun to be embodied in Chinese research. Some important events in the development of PSE in China are illustrated in Figure 2.

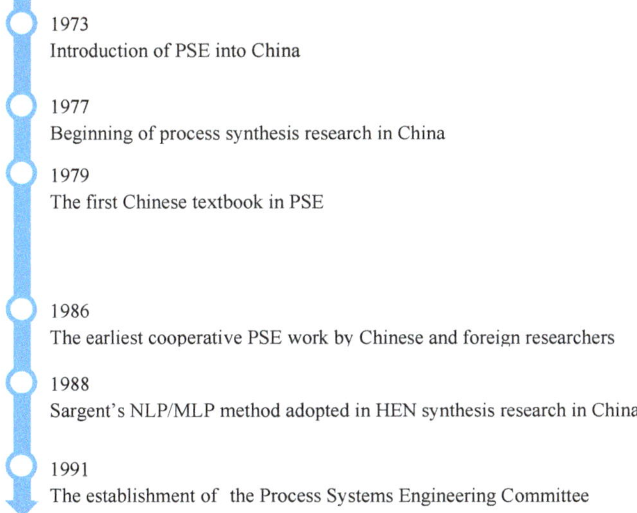

Figure 2. Important event in the development of process systems engineering (PSE) in China.

4.1. Before and In 1970s: Origin of PSE in China

From an international perspective, process systems engineering (PSE) can be traced back to 1959, when Williams delivered his lecture at the University of Texas with the title, "Systems Engineering for the Process Industries" [89], marking the beginning of the application of systems engineering to chemical engineering. In 1963, the exact term "process systems engineering" first appeared, as the title of the CEP Symposium Series Volume 59 [90]. The first textbook in PSE was "Strategy of Process Engineering" by Rudd and Watson in 1968 [91]. Next year, Japanese scientists Yagi and Nishimura published another book, "chemical process engineering" [92]. These two earliest textbooks had a significant impact on the spread of PSE, and their appearance could be regarded as the beginning of PSE as a discipline.

China was also amongst the earliest communities to recognize the PSE discipline. In 1973, the Research Institute of Petroleum Processing of Ministry of Fuel and Chemical Industries published the paper "Mathematical Simulation of Chemical Engineering Processes—A Brief Introduction of Process Systems Engineering Methodology" on the Conference on Computer Applications in Chemical Refining held in Lanzhou, China [93,94]. In the first paragraph of this paper, it was stated that, mathematical simulation of chemical processes was a rapidly developing method which had already shown great impacts on the industry, for which a new discipline had been formed, which was called "process systems engineering" in western countries. This article marked the point at which PSE was first introduced to China. Furthermore, the main focus of this paper is the concept, method and application of mathematical simulation to the analysis of chemical processes. Though problems in optimal design and optimal control were touched upon, they were not paid much attention to. In addition, the content about optimal design in this article was mostly from the two early textbooks in PSE [91,92], while Sargent's ideas were barely recognized. As a result, in the first few years after PSE was introduced in China, Chinese researchers were mainly conducting research in the field of process simulation, while little work in optimal design had been done, not to mention recognizing the importance of Sargent's work and thoughts in this field.

In reality, before 1973, a few companies had already been contacting PSE theories from abroad, compiling internal documentation. However, it was since 1973 that PSE gradually became well-known among Chinese universities and companies, in which researchers started to pay attention to the field of PSE. In the 1970s, much work had been done. Chinese researchers introduced advanced process flowsheeting systems from foreign countries and reviewed the development of PSE theories. For example, the research institute of Nanjing Chemical Industries Company introduced the GIPS synthetic ammonia simulation system from Denmark [93]. Li reviewed the methods for synthesis of heat exchanger networks [95]. Creative work had also been conducted in both application and theoretical areas. The designing institute of Lanzhou Chemical Industries Company invented the process flowsheeting system for petroleum gas compression separation, which is the first chemical process flowsheeting system independently developed by China [96]. Researchers from Shanghai Research Institute of Chemical Industry and East China Institute of Chemical Engineering (now named East China University of Science and Technology) cooperated to apply the theory of PSE to the simulation and analysis of combined production of synthetic ammonia and urea [97]. Chen and Chen conducted research in process decomposition, studying the criterion of optimal tearing [98]. The country was putting more and more emphasis on this discipline. In 1979, receiving the commission from Bureau of Science and Technology of Ministry of Chemical Industry, Tsinghua University held the first "chemical systems engineering course", and compiled the first Chinese textbook in PSE [99]. This decade witnessed the introduction of PSE in China and its application in Chinese industry. However, during this period, consensus understanding on this new discipline was not reached. In the 1978–1985 development layout for chemical engineering, PSE was placed as a subtitle under "chemical engineering dynamics and optimal process control" [93], which showed that the importance of PSE had still not been fully recognized in China.

Furthermore, in terms of Sargent's impact, although his research on optimal design was barely recognized, his work in other fields was gradually inspiring Chinese researchers. A typical example was Chen and Chen's research in process decomposition in 1979 [98]. In the field of system decomposition, one of the first research results is Sargent and Westerberg's tearing method based on dynamic programming, which was proposed in 1964 [6]. However, their method had the deficiency that it aimed at breaking the recycle streams with the minimum number of variables, but as some variables might have poorer convergence properties than others, such an approach didn't necessarily result in a minimum iteration time. Motivated by such shortcomings, in 1975, Upadhye and Grens developed a method for selecting the set of torn streams that could lead to a minimum number of iterations [100]. They found that multi-tearing a stream would lead to degradation of the convergence rate, so their method aimed to minimize the number of tearing that was made. Chen and Chen carefully studied this criterion and thought that it didn't take into account the convergence properties of the loops. They formulated a generalized expression of the eigen-polynomial of the iteration equations, and derived a new criterion for optimal tearing. They found that Upadhye and Gren's criterion was only a special case of their criterion. Later in 1983, Zhang et al. adopted their method to propose a new plate-to-plate method for calculating distillation processes [101].

4.2. In 1980s: Rapid Development of PSE in China

The 1980s saw the rapid development of PSE in China. Chinese universities began to set up courses on PSE, and pioneer institutes such as Tsinghua University and Tianjin University, founded their teaching and research group on PSE [102]. The first generation of Chinese PSE experts made great efforts to enlighten students on this subject. Famous textbooks, such as Yagi and Nishimura's *"Chemical Process Engineering"* [92] and Takamatsu's *"Chemical Process Systems Engineering"* [103], were translated into Chinese. Various textbooks compiled by Chinese researchers were also published, such as Zhang's *"Process Systems Engineering"* [104] and Yang's *"Practical Process Systems Engineering"* [105].

Before 1980, the research on PSE in China was mainly about the simulation of chemical processes. Since 1980, Chinese researchers had paid attention to new problems, such as process synthesis. As in western countries, the earliest studies in this field dealt with the synthesis problem of HEN. The first work in China was made in 1977 by researchers from Institute of Computing Technology of Chinese Academy of Sciences and Technical Laboratory of Design Institute of Beijing General Petrochemical Plant [106]. As mentioned above, there were generally two types of methods for the HEN synthesis, one using pure MP, another based on thermodynamics. Their work belongs to the former, in which they put the problem into an assignment problem, which was a special case of MILP, and solved it using Munkres assignment algorithm. In 1980s, Chen et al. conducted another pioneering study in the HEN synthesis [107]. They adopted a method based on thermodynamics, instead of MILP-based methods as in the previous work [106], which was because they thought that using MP to solve this problem required complicated computation and could be very different from reality. A worth mentioning point is that, since China put much emphasis on energy saving, Chen et al. formulated the problem as a multi-objective one, using energy consumption and equipment investment as two objectives, so that the weight on energy consumption could be adjusted according to China's energy policies. However, at that time, when foreign researchers studied this problem, they usually just added together the costs of energy and the investment on equipment to obtain a total cost which was to be minimized, which meant that they put equal weight on energy and other costs. Back to Chen et al.'s work, they used the linear combination method, i.e., adding the two objective functions together with respective weighting factors. However, this aggregate objective function was not to be minimized directly. According to thermodynamic analysis, they required that heat flows and cold flows should exchange their heat in a countercurrent way. So under any specific value of energy consumption, they designed the optimal structure of HEN based on this requirement, and calculated the corresponding value of the objective function. Then they changed the value of energy consumption, designed the structure and calculated the function value time and again, until they found the minimum objective function value. In their method, the optimization problem was not solved on a mathematical basis, but through making thermodynamic analyses in trial and error. Two years later, Chen et al. further studied this problem, and this time they required low sensibility of the HEN [108]. As in the earlier research, the problem was a multi-objective one, and requirements based on thermodynamics were put forward. Furthermore, to achieve low sensibility, they concluded some empirical requirements from several examples, and added them to the original thermodynamic ones. Then the design process could be done in a similar way. Their research was applied to improve the HEN structures of oil refining plants in China. Meanwhile, as problems were usually formulated as MOO ones in China, strategies for solving these problems were studied. Shen et al. discussed the MOO methods, which could be used in optimal scheduling of oil refining plants [109]. In this problem, the two objectives were maximizing profits and minimizing energy consumption. Motivated by various deficiencies in traditional methods for MOO, Shen et al. proposed a new method, in which they used two parameters to represent the difference between the energy consumption or the profit under any feasible solution and the minimum energy consumption or the maximum profit. The sum of the two parameters was the final objective function, and its value closer to zero, the system closer to the optimum. This method was later used in a computer software designed for optimal management of oil refining plants.

As mentioned above, during this period Sargent was also conducting much research in process synthesis. Grossmann and his work in 1978 [35] had already taken the first step into MINLP-based process synthesis, in which both the structure and the parameters could be decided through one pure MP problem, though computational techniques were still immature for solution. However, in China, such methods were regarded as impractical, and thus didn't receive much attention. Actually, it was not until Grossmann and his co-workers successfully solved process synthesis problems using MIP that Chinese researchers started to lay emphasis on MP-based design, which will be discussed in the following section.

4.3. Since Late-1980s: Sargent's Ideas Integrated into Development of PSE in China

Since 1980, a large number of international conferences had been held, strengthening the communication between researchers from different countries. In 1982, the first International Symposia on PSE was held in Kyoto, Japan. From then on it was held every three years. In addition, it was recorded that one of the pioneering institute in China PSE, Tsinghua University, began to attend this series of conferences in 1988, when the 3rd International Symposia on PSE was held. Furthermore, through a literature search, which might be not that sufficient, it seemed that the first cooperative work by Chinese and foreign researchers is that published in 1986 by Chen and Westerberg [110]. As a result, although it is hard to identify the exact point at which Chinese researchers began to have extensive contact with foreign researchers, the approximate time should be in middle or late 1980s. Back to Chen and Westerberg's paper, they studied the structural flexibility of heat integrated distillation processes. Flexibility is the property that the system be able to work properly when variations in parameters occur. As in the article by Grossmann et al. in 1987 [111], one way to introduce flexibility is to design the process for a number of different operating conditions, for which a pioneering work was done by Grossmann and Sargent in 1979 [37], or to design under uncertainty, which was explored by Grossmann and Sargent in 1978 [19]. Another way is to derive an index of flexibility, such as in Swaney and Grossmann's work in 1985 [57], and this idea served as the motivation for Chen and Westerberg's research. More importantly, Papoulias and Grossmann's success in using superstructure-based mixed-integer optimization in the design of utility systems [39] inspired Chinese researchers to pay more attention to MP methods for optimal design. In 1988, researchers in Tsinghua University combined such MP-based methods with the thermodynamic method they had developed before, and managed to apply it to the optimization of the operation of Jinxi Oil Refining Plant [99]. Under the above facts, it is reasonable to say that the integration of Sargent's idea of optimal design into China PSE research did occur in late 1980s.

With more and more successful application of PSE in industry, the importance of this discipline became widely recognized in China. In 1991, affiliated to the Systems Engineering Society of China, the Process Systems Engineering Committee was established, which greatly strengthened exchanges in China and cooperation between Chinese and foreign researchers [112]. Since the beginning of the 1990s decade, Sargent's work had affected Chinese researchers much, embodied in the fact that the methods derived from his ideas began to be widely adopted and developed. In process design, Sargent was one of the first to recognize the importance of optimization in process engineering, and he had conducted significant work on NLP and superstructure. This research was further developed by his descendants, giving rise to systematic process synthesis methods combining superstructure-based modelling with NLP/MIP optimization. Such technologies were widely applied in China. One traditional as well as popular problem in process design is the synthesis of HEN. Chinese researchers have made great efforts for more efficient and reliable solution of such problems. In 1995, Yin et al. proposed a superstructure-based MINLP model, through which multiple objectives could be optimized, such as utility cost, exchanger areas and number of units [113]. In 1999, Wang et al. [114] proposed a new superstructure-based MINLP model for synthesis of large-scale HEN without stream splitting, and adopted the improved genetic algorithm (GA) they developed earlier [115] for solution. In 2000, Li et al. formed a superstructure representation of HEN with stream splitting based on expert systems and presented the corresponding MINLP model [116]. In 2004, Wei et al. proposed a hybrid method of GA and simulated annealing algorithm (SA) to solve the MINLP problem in the HEN synthesis [117]. In 2008, Ma et al. [118] used the MINLP model for multi-period synthesis of multi-stream HEN and adopted the GA/SA algorithm developed by Wei et al. [117] for solution. In 2015, Peng et al. developed a two-level method for simultaneous HEN synthesis using SA mechanism [119].

Similar methods can also be found in research on design of chemical processes or their operating units. In 1995, Wang et al. studied the optimization of operation for multipurpose batch chemical plants with multiple production routes. In their work, a MILP model for production plan was proposed [120]. In 2011, Jiao et al. presented a multi-objective superstructure-based model for design of hydrogen network in refineries [121]. In 2012, Jin et al. proposed a double-level optimization method for superstructure-based synthesis of reactor network composed of continuous stirred-tank reactors and plug flow reactors [122]. In 2017, Cui et al. developed a superstructure of 4-column methanol distillation systems (MDSs), based on which they performed simultaneous MINLP optimization for the system as well as its HEN and work exchanger network (WEN) [123]. Qian et al. used superstructure and NLP to deal with the synthesis of a reactor system composed of rotating packed bed reactors and traditional packed bed reactors, and demonstrated their method through a case study on the desulfurization process for crack gas in refineries [124]. In 2018, Zhang et al. proposed a method combining superstructure-based modelling and the GDP optimization method, developed by Grossmann and co-workers [44], for design of reaction and separation processes [125]. In addition, they applied the method to two important chemical industrial processes—cyclohexane oxidation process and benzene chlorination process. Furthermore, the idea of superstructure and MIP is also embodied in other fields, such as biology, pharmacy and others. In 2000, Hou et al. performed NLP optimization for the leaching process of angelica, showing that a 20% increase in profit could be achieved [126]. The authors commented that PSE methodology was useful in herb leaching processes, encouraging the modernization of traditional Chinese medicine production. In 2015, Du et al. adopted superstructure-based MINLP optimization for optimal synthesis of reverse osmosis networks with split partial second pass design for seawater desalination [127]. In 2016, Yan et al. proposed a superstructure representation of biomass to biomethane system through digestion and formed a multi-objective MINLP problem, which was solved using the nondominated sorting genetic algorithm (NSGA-II) [128]. Additionally, they concluded that the superstructure optimization method they proposed could be a useful tool for improving the performance of biomethane production process.

Moreover, ESE was also developing fast. As is mentioned above, as soon as PSE was introduced into China, Chinese researchers began to use its methodology to solve problems considering the costs (or profits) as well as the energy consumption of the system. As a result, the origin of China ESE was almost at the same time of the origin of China PSE. Since 1980, much research had been done on the modelling of energy systems in the scope of plants, districts and the whole country, amongst which many of the results were used for making energy policies [129]. For example, Li et al. proposed a combined approach of thermodynamic analysis and MILP for optimal design of steam-power energy system in 1992 [130]. They set up the superstructure of system by thermodynamic rules, formulated the problem as MILP one, and solved it using the BB method. Since the concept of ESE was formally proposed in 2007 [26], Chinese researchers have made significant contributions in this field, conducting a great amount of pioneering work to apply the systematic methodology from PSE to the research of various energy systems. Typical work in polygeneration energy systems [68–71], CHP/CCHP systems [73,74], hydrogen infrastructure [77], power systems [79–85] and data reconciliation [86–88] has already been mentioned in Section 3.4.

5. Industrial Applications of PSE and ESE Methods

The PSE and ESE methods mentioned above have also been widely applied to the chemical industry and led to significant economic and environmental contributions. In this part, some representative industrial practices are presented, showing the feasibility and helpfulness of PSE and ESE theories in real-life industries.

As is mentioned above, Sargent's work has led to tools that are widely used in the industry. Based on Sargent's work on NLP/MIP optimization, researchers developed the NLP solver, MINOS [29], and the MINLP solver, DICOPT [42], both of which can now be founded in GAMS. In addition, Sargent and his co-workers in Imperial College developed the SPEED-UP software, in which many algorithms they had proposed were incorporated, such as the decomposition techniques and the solution methods for DAEs. Many of these developers later participated in the research and development of gPROMS [62]. In addition, they founded PSE Limited with gPROMS as a major product. gPROMS is undoubtedly a huge success, which is used for design and optimization tasks in a large number of universities and companies all over the world [131].

In addition to those commercialized packages, some theoretical methodologies, which were originally proposed in academic research, have been popularized and applied in real-life projects. In 1996, Siirola [132] pointed out that the systematic process synthesis technologies, such as the superstructure-based MIP approach developed by Sargent, Grossmann and others and the pinch analysis method proposed by Linnhoff and colleagues, had been put into industrial practice and achieved great success. Typically, the application of these technologies could lead to a 50% saving of energy and a 35% reduction in costs. In their paper in 2003, Dunn and El-Halwagi also presented many industrial applications of those process synthesis methods, including projects on energy conservation, optimization of utility systems, and others [133].

Applications of PSE and ESE methods in China have also achieved great success. In 2008, Gao et al. proposed a hybrid MOO method combining SQP and NSGA-II for the periodic operation of the naphtha pyrolysis process [134]. Their algorithm was incorporated in a new software for simulation and optimization of ethylene cracking furnace, which was developed by researchers from Tsinghua University, PetroChina Research Institute of Petrochemical Technology and Lanzhou Petrochemical Company. This software, EPSOS, was used in a 240 kt/a ethylene plant of the Lanzhou Petrochemical Company and resulted in a 1.08% increase in production [135]. According to Wang's paper in 2009 [136], petrochemical companies in China have been positively applying PSE technologies, including modern control, process synthesis and others, to the restructuring of the plants and facilities and have made great progress. From 2005 to 2009, the two leading petrochemical enterprises in China, Sinopec and PetroChina, had on average achieved a 12.5% reduction in the energy consumption of their oil refining processes. In addition, Sinopec's overall energy consumption had decreased by 15.4% in this period. In the field of ESE, the monitoring and optimization methods for power units via data reconciliation, developed by researchers from Tsinghua University [86–88], have been applied to several Chinese power plants, making contributions to their energy saving. Furthermore, much of the strategic research on development strategies of China's power sector has been incorporated into the National Energy Development Strategy by 2030, whose effects will be seen with the restructuring of China's energy system in the future.

6. Conclusions

In his research career, Professor Sargent had made great contributions to the field of PSE, and his work surely took a guiding role in the development of ESE and China PSE. Sargent is one of the first to recognize the importance of optimization in process design. His attempts to apply optimization in chemical engineering and to develop computational techniques for NLP problems, have led to a combined superstructure representation and MIP modelling framework for design problems, which is widely applied in ESE. With people becoming more conscious of environmental protection, pollution minimization was regarded as an important objective by more researchers, giving rise to much work on MOO and LCA, which are now common methods in ESE research. Sargent and co-workers' work in other fields, such as optimization under uncertainty and solution of DAEs and NAEs, have also led to useful algorithms and software that are frequently adopted in the field of ESE.

The beginning of the discipline of PSE in China was not much later than that in other countries. However, during the initial period, Sargent's idea of optimal design did not receive much attention in China. In early research on HEN synthesis problems, Chinese researchers preferred thermodynamic methods, while regarding MP methods as infeasible. However, Sargent's work in other aspects was gradually inspiring Chinese researchers, such as in system decomposition. It was in late 1980s that Sargent's ideas became integrated to China PSE research. Since then, methods developed under his leadership have been widely studied and applied in the industry. Chinese researchers also paid much attention to the field of ESE, resulting in advanced achievements.

Looking forward to the future, the pioneering concepts and methodological contributions of Sargent are still inspiring new research directions. Taking energy systems for instance, operation at a more intermittent mode, in a much wider range, and with more severe pollutants and greenhouse-gas emissions constraints, pose great challenges for design and operation of these systems. However, increasing availability of operational data, improving data quality, and higher computational capability, when integrated with mathematical programming, provide opportunities to tackle these challenges, giving rise to a merging research direction of integrated mathematical programming and data analytics. This fits well with the idea of using increasingly powerful mathematical and computational methods to tackle unprecedented problems with process design and operation, originally proposed by Sargent and developed by his co-workers and followers.

Author Contributions: Conceptualization, W.Q., P.L. and Z.L.; Methodology, P.L.; Resources: W.Q.; Writing-Original Draft Preparation, W.Q.; Writing-Review & Editing, P.L.; Supervision, P.L.

Funding: This research was funded by the National Key Research and Development Program of China (2018YFB0604301), National Natural Science Foundation of China (71690245), and the Phase III Collaboration between BP and Tsinghua University.

Conflicts of Interest: The authors declare no conflict of interest.

References

1. Sargent, R.W.H. Process Systems Engineering: A Retrospective View with Questions for the Future. *Comput. Chem. Eng.* **2005**, *29*, 1237–1241. [CrossRef]
2. Glavic, P. Thirty Years of International Symposia on Process Systems Engineering. *Curr. Opin. Chem. Eng.* **2012**, *1*, 421–429. [CrossRef]
3. Doherty, M.F.; Grossmann, I.E.; Pantelides, C.C. A Tribute to Professor Roger Sargent: Intellectual Leader of Process Systems Engineering. *AIChE J.* **2016**, *62*, 2951–2958. [CrossRef]
4. Sargent, R.W.H. Integrated Design and Optimization of Processes. *Chem. Eng. Prog.* **1967**, *63*, 71–78.

5. Sargent, R.W.H. Forecasts and Trends in Systems Engineering. In Proceedings of the Annual Meeting of Israel Institute of Chemical Engineers, Haifa, Israel, 29 August–2 September 1971.

6. Sargent, R.W.H.; Westerberg, A.W. SPEED-UP in Chemical Engineering Design. *Trans. Inst. Chem. Eng.* **1964**, *42*, T190–T197.

7. Sargent, R.W.H. The Decomposition of Systems of Procedures and Algebraic Equations. In *Numerical Analysis, Proceedings of the Biennial Conference, Dundee, UK*; Springer: Berlin/Heidelberg, Germany, 1978.

8. Leigh, M.J. A Computer Flowsheeting Programme Incorporating Algebraic Analysis of the Problem Structure. Ph.D. Thesis, University of London, London, UK, 1973.

9. Davidon, W.C. Variable Metric Method for Minimization. *SIAM J. Optim.* **1991**, *1*, 1–17. [CrossRef]

10. Fletcher, R.; Powell, M.J.D. A Rapidly Convergent Descent Method for Minimization. *Comput. J.* **1963**, *6*, 163–168. [CrossRef]

11. Rosen, J.B. The Gradient Projection Method for Nonlinear Programming. Part I. Linear Constraints. *J. Soc. Ind. Appl. Math.* **1960**, *8*, 181–217. [CrossRef]

12. Murtagh, B.A.; Sargent, R.W.H. A Constrained Minimization Method with Quadratic Convergence. In *Optimization*; Academic Press: London, UK, 1965; pp. 215–245.

13. Sargent, R.W.H.; Murtagh, B.A. Projection Methods for Non-linear Programming. *Math. Program.* **1973**, *4*, 245–268. [CrossRef]

14. Murtagh, B.A.; Sargent, R.W.H. Computational Experience with Quadratically Convergent Minimisation Methods. *Comput. J.* **1970**, *13*, 185–194. [CrossRef]

15. Sargent, R.W.H.; Sebastian, D.J. On the Convergence of Sequential Minimization Algorithms. *J. Optim. Theory Appl.* **1973**, *12*, 567–575. [CrossRef]

16. Sargent, R.W.H. Convergence Properties of Projection Methods for Nonlinear Programming. In Proceedings of the 8th International Symposium on Mathematical Programming, Stanford, CA, USA, 26–31 August 1973.

17. Sargent, R.W.H. An Efficient Implementation of the Lemke Algorithm and Its Extension to Deal with Upper and Lower Bounds. *Math. Program Stud.* **1978**, *7*, 36–54.

18. Sargent, R.W.H.; Ding, M. A New SQP Algorithm for Large-scale Nonlinear Programming. *SIAM J. Optim.* **2001**, *11*, 716–747. [CrossRef]

19. Grossmann, I.E.; Sargent, R.W.H. Optimum Design of Chemical-Plants with Uncertain Parameters. *AIChE J.* **1978**, *24*, 1021–1028. [CrossRef]

20. Sargent, R.W.H.; Gaminibandara, K. Optimum design of plate distillation columns. In *Optimization in Action*; Dixon, L.C.W., Ed.; Academic Press: London, UK, 1976; pp. 267–314.

21. Yee, T.F.; Grossmann, I.E.; Kravanja, Z. Simultaneous Optimization Models for Heat Integration. III. Process and Heat Exchanger Network Optimization. *Comput. Chem. Eng.* **1990**, *14*, 1185–1200. [CrossRef]

22. Sargent, R.W.H. A Functional Approach to Process Synthesis and Its Application to Distillation Systems. *Comput. Chem. Eng.* **1998**, *22*, 31–45. [CrossRef]

23. Sargent, R.W.H. A Review of Methods for Solving Nonlinear Algebraic Equations. In Proceedings of the Engineering Foundation Conference, Henniker, NH, USA, 3–8 August 1980.

24. Cameron, I.T. Numerical Solution of Differential-Algebraic Systems in Process Dynamics. Ph.D. Thesis, University of London, London, UK, 1981.

25. Gritsis, D.; Pantelides, C.C.; Sargent, R.W.H. The Dynamic Simulation of Transient Systems Described by Index Two Differential Algebraic Equations. In Proceedings of the PSE'88 The Third International Symposium on Process Systems Engineering, Sydney, Australia, 28 August–2 September 1988.

26. Pistikopoulos, E.N. Energy Systems Engineering—An Integrated Approach for the Energy Systems of the Future. In Proceedings of the European Congress of Chemical Engineering—6, Copenhagen, Denmark, 16–20 September 2007.

27. Liu, P.; Georgiadis, M.C.; Pistikopoulos, E.N. Advances in Energy Systems Engineering. *Ind. Eng. Chem. Res.* **2011**, *50*, 4915–4926. [CrossRef]

28. Sargent, R.W.H.; Murtahg, B.A. Design of Plate Distillation Columns for Multicomponent Mixtures. *Trans. Inst. Chem. Eng.* **1969**, *47*, T85–T94.

29. Murtagh, B.A.; Saunders, M.A. Large-Scale Linearly Constrained Optimization. *Math. Program.* **1978**, *14*, 41–72. [CrossRef]

30. Sargent, R.W.H.; Ming, D. *A Modified Sargent-Gaminibandara Superstructure for Synthesis of Distillation Sequences*; Report No. A94-23; Centre for Process Systems Engineering, Imperial College: London, UK, 1994.

31. El-Temtamy, S.A.; Hamid, I.; Gabr, E.M.; Sayed, A.E.-R. The Use of Pinch Technology to Reduce Utility Consumption in a Natural Gas Processing Plant. *Petrol. Sci. Technol.* **2010**, *28*, 1316–1330. [CrossRef]
32. Rudd, D.F. Synthesis of System Designs. I. Elementary Decomposition Theory. *AIChE J.* **1968**, *14*, 343–349. [CrossRef]
33. Gundersen, T.; Naess, L. The Synthesis of Cost Optimal Heat Exchanger Networks—An Industrial Review of the State of the Art. *Comput. Chem. Eng.* **1988**, *12*, 503–530. [CrossRef]
34. Sargent, R.W.H. Introduction: 25 Years of Progress in Process Systems Engineering. *Comput. Chem. Eng.* **2004**, *28*, 437–439. [CrossRef]
35. Grossmann, I.E.; Sargent, R.W.H. Optimum Design of Heat-Exchanger Networks. *Comput. Chem. Eng.* **1978**, *2*, 1–7. [CrossRef]
36. Vázquez-Román, R.; Inchaurregui-Méndez, J.A.; Ponce-Ortega, J.M.; Mannan, M.S. A Heat Exchanger Networks Synthesis Approach Based on Inherent Safety. *J. Chem. Eng. Res. Update* **2015**, *2*, 22–29. [CrossRef]
37. Grossmann, I.E.; Sargent, R.W.H. Optimum Design of Multipurpose Chemical Plants. *Ind. Eng. Chem. Proc. Des. Dev.* **1979**, *18*, 343–348. [CrossRef]
38. Grossmann, I.E.; Santibanez, J. Applications of Mixed-Integer Linear-Programming in Process Synthesis. *Comput. Chem. Eng.* **1980**, *4*, 205–214. [CrossRef]
39. Papoulias, S.A.; Grossmann, I.E. A Structural Optimization Approach in Process Synthesis. Part 1. Utility Systems. *Comput. Chem. Eng.* **1983**, *7*, 695–706. [CrossRef]
40. Duran, M.A.; Grossmann, I.E. An Outer-Approximation Algorithm for A Class of Mixed-Integer Nonlinear Programs. *Math. Program.* **1986**, *36*, 307–339. [CrossRef]
41. Kocis, G.R.; Grossmann, I.E. Relaxation Strategy for the Structural Optimization of Process Flowsheets. *Ind. Eng. Chem. Res.* **1987**, *26*, 1869–1880. [CrossRef]
42. Kocis, G.R.; Grossmann, I.E. Computational Experience with DICOPT Solving MINLP Problems in Process Systems Engineering. *Comput. Chem. Eng.* **1989**, *13*, 307–315. [CrossRef]
43. Grossmann, I.E. MINLP Optimization Algorithms and Strategies for Process Synthesis. In Proceedings of the FOCAPD'89, Snowmass Village, CO, USA, 9–14 July 1989.
44. Yeomans, H.; Grossmann, I.E. A Systematic Modeling Framework of Superstructure Optimization in Process Synthesis. *Comput. Chem. Eng.* **1999**, *23*, 709–731. [CrossRef]
45. Cano-Ruiz, J.A.; McRae, G.J. Environmentally Conscious Chemical Process Design. *Annu. Rev. Energy Environ.* **1998**, *23*, 499–536. [CrossRef]
46. Douglas, J.M. Process Synthesis for Waste Minimization. *Ind. Eng. Chem. Res.* **1992**, *31*, 238–243. [CrossRef]
47. Haimes, Y.Y.; Ladson, L.S.; Wismer, D.A. Bicriterion Formulation of Problems of Integrated System Identification and System Optimization. *IEEE Trans. Syst. Man Cybern.* **1971**, *SMC1*, 296–297.
48. Bhaskar, V. Application of Multi-Objective Optimization in Chemical Engineering. *Rev. Chem. Eng.* **2000**, *16*. [CrossRef]
49. Grossmann, I.E.; Drabbant, R.; Jain, R.K. Incorporating Toxicology in the Synthesis of Industrial Chemical Complexes. *Chem. Eng. Commun.* **1982**, *17*, 151–170. [CrossRef]
50. Stefanis, S.K.; Livingston, A.G.; Pistikopoulos, E.N. Minimizing the Environmental Impact of Process Plants—A Process Systems Methodology. *Comput. Chem. Eng.* **1995**, *19*, S39–S44. [CrossRef]
51. Stefanis, S.K.; Buxton, A.; Livingston, A.G.; Pistikopoulos, E.N. A Methodology for Environmental Impact Minimization: Solvent Design and Reaction Path Synthesis Issues. *Comput. Chem. Eng.* **1996**, *20*, S1419–S1424. [CrossRef]
52. Stefanis, S.K.; Livingston, A.G.; Pistikopoulos, E.N. Environmental Impact Considerations in the Optimal Design and Scheduling of Batch Processes. *Comput. Chem. Eng.* **1997**, *21*, 1073–1094. [CrossRef]
53. Azapagic, A.; Clift, R. The Application of Life Cycle Assessment to Process Optimisation. *Comput. Chem. Eng.* **1999**, *23*, 1509–1526. [CrossRef]
54. Carvalho, A.; Matos, H.A.; Gani, R. SustainPro—A Tool for Systematic Process Analysis, Generation and Evaluation of Sustainable Design Alternatives. *Comput. Chem. Eng.* **2013**, *50*, 8–27. [CrossRef]
55. Kalakul, S.; Malakul, P.; Siemanond, K.; Gani, R. Integration of Life Cycle Assessment Software with Tools for Economic and Sustainability Analyses and Process Simulation for Sustainable Process Design. *J. Clean. Prod.* **2014**, *71*, 98–109. [CrossRef]
56. Halemane, K.P.; Grossmann, I.E. Optimal Process Design under Uncertainty. *AIChE J.* **1983**, *29*, 425–433. [CrossRef]

57. Swaney, R.E.; Grossmann, I.E. An Index for Operational Flexibility in Chemical Process Design. Part I. Formulation and Theory. *AIChE J.* **1985**, *31*, 621–630. [CrossRef]
58. Straub, D.A.; Grossmann, I.E. Design Optimization of Stochastic Flexibility. *Comput. Chem. Eng.* **1993**, *17*, 339–354. [CrossRef]
59. Ierapetritou, M.G.; Acevedo, J.; Pistikopoulos, E.N. An Optimization Approach for Process Engineering Problems under Uncertainty. *Comput. Chem. Eng.* **1996**, *20*, 703–709. [CrossRef]
60. Acevedo, J.; Pistikopoulos, E.N. Computational Studies of Stochastic Optimization Algorithms for Process Synthesis under Uncertainty. *Comput. Chem. Eng.* **1996**, *20*, S1–S6. [CrossRef]
61. Mah, R.S.H.; Michaelson, S.; Sargent, R.W.H. Dynamic Behaviour of Multi-Component Multi-Stage Systems—Numerical Methods for the Solution. *Chem. Eng. Sci.* **1962**, *17*, 619–639. [CrossRef]
62. Pantelides, C.C.; Barton, P.I. Equation-Oriented Dynamic Simulation Current Status and Future Perspectives. *Comput. Chem. Eng.* **1993**, *17*, S263–S285. [CrossRef]
63. Kikkinides, E.S.; Georgiadis, M.C.; Stubos, A.K. On the Optimization of Hydrogen Storage in Metal Hydride Beds. *Int. J. Hydrog. Energy* **2006**, *31*, 737–751. [CrossRef]
64. Kouramas, K.; Pistikopoulos, E.N. Wind Turbines Modeling and Control. In *Process Systems Engineering: Volume 5: Energy Systems Engineering*; Georgiadis, M.C., Kikkinides, E.S., Pistikopoulos, E.N., Eds.; Wiley-VCH: Weinheim, Germany, 2011; pp. 195–214.
65. Cameron, I.T.; Gani, R. Adaptive Runge-Kutta Algorithms for Dynamic Simulation. *Comput. Chem. Eng.* **1988**, *7*, 705–717. [CrossRef]
66. Sorensen, E.L.; Johansen, H.; Gani, R.; Fredenslund, A. A Dynamic Simulator for Design and Analysis of Chemical Processes. In *Process Technology Proceedings*; Bussemaker, H.T., Iedema, P.D., Eds.; Elsivier: Amsterdam, The Netherlands, 1990; Volume 9, pp. 13–18.
67. Gani, R.; Sorensen, E.L.; Perregaard, J. Design and Analysis of Chemical Processes Through DYNSIM. *Ind. Eng. Chem. Res.* **1992**, *31*, 244–254. [CrossRef]
68. Liu, P.; Gerogiorgis, D.I.; Pistikopoulos, E.N. Modeling and Optimization of Polygeneration Energy Systems. *Catal. Today* **2007**, *127*, 347–359. [CrossRef]
69. Liu, P.; Pistikopoulos, E.N.; Li, Z. A Mixed-integer Optimization Approach for Polygeneration Energy Systems Design. *Comput. Chem. Eng.* **2009**, *33*, 759–768. [CrossRef]
70. Liu, P.; Pistikopoulos, E.N.; Li, Z. Decomposition Based Stochastic Programming Approach for Polygeneration Energy Systems Design under Uncertainty. *Ind. Eng. Chem. Res.* **2010**, *49*, 3295–3305. [CrossRef]
71. Liu, P.; Whitaker, A.; Pistikopoulos, E.N.; Li, Z.; Chen, Y. A Mixed-integer Programming Approach to Strategic Planning of Chemical Centres: A Case Study in the UK. *Comput. Chem. Eng.* **2011**, *35*, 1359–1373. [CrossRef]
72. Arcuri, P.; Florio, G.; Fragiacomo, P. A Mixed Integer Programming Model for Optimal Design of Trigeneration in a Hospital Complex. *Energy* **2007**, *32*, 1430–1447. [CrossRef]
73. Liu, P.; Pistikopoulos, E.N.; Li, Z. An Energy Systems Engineering Approach to the Optimal Design of Energy Systems in Commercial Buildings. *Energy Policy* **2010**, *38*, 4224–4231. [CrossRef]
74. Zhou, Z.; Zhang, J.; Liu, P.; Li, Z.; Georgiadisb, M.C.; Pistikopoulos, E.N. A Two-stage Stochastic Programming Model for the Optimal Design of Distributed Energy System. *Appl. Energy* **2014**, *103*, 135–144. [CrossRef]
75. Kopanos, G.M.; Georgiadis, M.; Pistikopoulos, E.N. Scheduling Energy Cogeneration Units under Energy Demand Uncertainty. *IFAC Proc. Vol.* **2013**, *46*, 1280–1285. [CrossRef]
76. Hugo, A.; Rutter, P.; Pistikopoulos, S.; Amorelli, A.; Zoia, G. Hydrogen Infrastructure Strategic Planning Using Multi-objective Optimization. *Int. J. Hydrog. Energy* **2005**, *30*, 1523–1534. [CrossRef]
77. Li, Z.; Gao, D.; Chang, L.; Liu, P.; Pistikopoulos, E.N. Hydrogen Infrastructure Design and Optimization: A Case Study of China. *Int. J. Hydrog. Energy* **2008**, *33*, 5275–5286. [CrossRef]
78. Ogumerem, G.S.; Kim, C.; Kesisoglou, I.; Diangelakis, N.A.; Pistikopoulos, E.N. A Multi-Objective Optimization for the Design and Operation of a Hydrogen Network for Transportation Fuel. *Chem. Eng. Res. Des.* **2018**, *131*, 279–292. [CrossRef]
79. Zhang, D.; Liu, P.; Ma, L.; Li, Z.; Ni, W. A Multi-Period Modelling and Optimization Approach to the Planning of China's Power Sector with Consideration of Carbon Dioxide Mitigation. *Comput. Chem. Eng.* **2012**, *37*, 227–247. [CrossRef]

80. Zhang, D.; Liu, P.; Ma, L.; Li, Z. A Multi-Period Optimization Model for Optimal Planning of China's Power Sector with Consideration of Carbon Mitigation—The Optimal Pathway under Uncertain Parametric Conditions. *Comput. Chem. Eng.* **2013**, *50*, 196–206. [CrossRef]

81. Koltsaklis, N.E.; Dagoumas, A.S.; Kopanos, G.M.; Pistikopoulos, E.N.; Georgiadis, M.C. A Spatial Multi-Period Long-Term Energy Planning Model: A Case Study of the Greek Power System. *Appl. Energy* **2014**, *115*, 456–482. [CrossRef]

82. Cheng, R.; Xu, Z.; Liu, P.; Wang, Z.; Li, Z.; Jones, I. A Multi-Region Optimization Planning Model for China's Power Sector. *Appl. Energy* **2015**, *137*, 413–426. [CrossRef]

83. Guo, Z.; Ma, L.; Liu, P.; Jones, I.; Li, Z. A Multi-Regional Modelling and Optimization Approach to China's Power Generation and Transmission Planning. *Energy* **2016**, *116*, 1348–1359. [CrossRef]

84. Guo, Z.; Ma, L.; Liu, P.; Jones, I.; Li, Z. A Long-Term Multi-Region Load-Dispatch Model Based on Grid Structures for the Optimal Planning of China's Power Sector. *Comput. Chem. Eng.* **2017**, *102*, 52–63. [CrossRef]

85. Guo, Z.; Cheng, R.; Xu, Z.; Liu, P.; Wang, Z.; Li, Z.; Jones, I.; Sun, Y. A Multi-Region Load Dispatch Model for the Long-Term Optimum Planning of China's Electricity Sector. *Appl. Energy* **2017**, *185*, 556–572. [CrossRef]

86. Jiang, X.; Liu, P.; Li, Z. Data Reconciliation and Gross Error Detection for Operational Data in Power Plants. *Energy* **2014**, *75*, 14–23. [CrossRef]

87. Guo, S.; Liu, P.; Li, Z. Inequality Constrained Nonlinear Data Reconciliation of a Steam Turbine Power Plant for Enhanced Parameter Estimation. *Energy* **2016**, *103*, 215–230. [CrossRef]

88. Li, T.; Liu, P.; Li, Z. Operation Data based Modelling and Optimization of Thermal Power Units under Full Working Conditions. *Comput. Aided Chem. Eng.* **2017**, *40*, 2455–2460.

89. Perkins, J. Education in Process Systems Engineering: Past, Present and Future. *Comput. Chem. Eng.* **2002**, *26*, 283–293. [CrossRef]

90. Stephanopoulos, G.; Reklaitis, G.V. Process Systems Engineering: From Solvay to Modern Bio- and Nanotechnology. A History of Development, Successes and Prospects for the Future. *Chem. Eng. Sci.* **2011**, *66*, 4272–4306. [CrossRef]

91. Rudd, D.F.; Watson, C.C. *Strategy of Process Engineering*; John Wiley and Sons: New York, NY, USA, 1968.

92. Yagi, S.; Nishimura, H. *Chemical Process Engineering*; Maruzen Co., Ltd.: Tokyo, Japan, 1969. (In Japanese)

93. Yang, Y. Development of Process Systems Engineering. *Chem. Ind Eng. Prog.* **1983**, *5*, 8–13. (In Chinese) [CrossRef]

94. Research Institute of Petroleum Processing of Ministry of Fuel and Chemical Industries. Mathematical Simulation of Chemical Engineering Processes—A Brief Introduction of Process Systems Engineering Methodology. *Petrochem. Technol.* **1973**, *6*, 559–581. (In Chinese) [CrossRef]

95. Li, W. Optimal Design of Heat Exchanger Networks—Review and Evaluation of Existing Methods. *Chem. Eng. (China)* **1979**, *5*, 76–96. (In Chinese)

96. Yang, Y. Chemical Engineering Simulation System. *Petrochem. Technol.* **1978**, *4*, 373–381. (In Chinese)

97. Tong, H.; Hu, G.; Tang, Y.; Li, W. Application of Process Flowsheeting in the Development of Combined Production of Synthetic Ammonia and Urea. *J. Chem. Fertil. Ind.* **1981**, *5*, 2–5. (In Chinese)

98. Chen, M.; Chen, L. Decomposition of Nets—The Criterion of Optimal Tearing. *J. East China Univ. Sci. Technol.* **1979**, *1–2*, 11–28. [CrossRef]

99. Chen, B.; Yang, Y. Progress in Process Systems Engineering. In Proceedings of the 6th Annual Meeting of Systems Engineering Society of China, Tianjin, China, 2–6 August 1990. (In Chinese).

100. Upadhye, R.S.; Grens, E.A. Selection of Decompositions for Chemical Process Simulation. *AIChE J.* **1975**, *21*, 136–143. [CrossRef]

101. Zhang, R. A New Plate-to-Plate Method for Calculating Distillation Processes (I) Performance Prediction of Multicomponent Distillation Processes. *J. Chem. Ind. Eng. (China)* **1983**, *4*, 303–316.

102. Yang, Y.; Cheng, S. PSE in China Over the Past 20 Years: Retrospective and Prospects (Part I). *Mod. Chem. Ind.* **2012**, *32*, 1–5. [CrossRef]

103. Takamatsu, T. *Chemical Process Systems Engineering*; Chemical Industry Press: Beijing, China, 1981. (In Chinese)

104. Zhang, N. *Process Systems Engineering*; Chemical Industry Press: Beijing, China, 1982. (In Chinese)

105. Yang, Y. *Practical Process Systems Engineering*; Chemical Industry Press: Beijing, China, 1989. (In Chinese)

106. Institute of Computing Technology of Chinese Academy of Sciences; Technical Laboratory of Design Institute of Beijing General Petrochemical Plant. Optimal Synthesis of Heat Exchanger Network. *Pet. Process. Petrochem.* **1977**, *6*, 15–31. (In Chinese)

107. Chen, B.; Li, Y.; Shen, Z.; He, X.; Zhang, N. Optimal Synthesis of Heat Exchanger Network. *Pet. Process. Petrochem.* **1980**, *12*, 20–26. (In Chinese)

108. Chen, B.; Ouyang, P.; He, X.; Li, Y. Optimal Design of Heat Exchanger Network with Low Sensitivity. *Pet. Process. Petrochem.* **1982**, *2*, 18–24, 60.

109. Shen, J.; He, X.; Chen, B. Discussion on Multi-Objective Decision-Making Methods for Optimal Scheduling of Refineries. *Pet. Process. Petrochem.* **1985**, *11*, 29–44. (In Chinese)

110. Bingzhen, C.; Westerberg, A.W. Structural Flexibility for Heat Integrated Distillation Columns—I. Analysis. *Chem. Eng. Sci.* **1986**, *41*, 355–363. [CrossRef]

111. Grossmann, I.E.; Halemane, K.P.; Swaney, R.E. Optimization Strategies for Flexible Chemical Processes. *Comput. Chem. Eng.* **1983**, *7*, 439–462. [CrossRef]

112. Cheng, S.; Yang, Y. Developing History, Status and Prospect of Process Systems Engineering. *J. Tianjin Univ.* **2007**, *40*, 321–328.

113. Yin, H.; Yuan, Y.; Wang, X.; Shi, G. Multi-Target Simultaneous Optimization for Non-Isothermal Mixing Heat Exchanger Network Synthesis. *J. Dalian Univ. Technol.* **1995**, *5*, 639–643.

114. Wang, K.F.; Qian, Y.; Huang, Q.M.; Yuan, Y.; Yao, P.J. New Model and New Algorithm for Optimal Synthesis of Large Scale Heat Exchanger Networks without Stream Splitting. *Comput. Chem. Eng.* **1999**, *23*, S149–S152. [CrossRef]

115. Wang, K.F.; Qian, Y.; Yuan, Y.; Yao, P.J. Synthesis and Optimization of Heat Integrated Distillation Systems Using an Improved Genetic Algorithm. *Comput. Chem. Eng.* **1998**, *23*, 125–136. [CrossRef]

116. Li, Z.H.; Hua, B. Modeling and Optimizing for Heat Exchanger Networks Synthesis Based on Expert System and Exergo-Economic Objective Function. *Comput. Chem. Eng.* **2000**, *24*, 1223–1228. [CrossRef]

117. Wei, G.; Yao, P.; Luo, X.; Roetzel, W. Study on Multi-Stream Heat Exchanger Network Synthesis with Parallel Genetic/Simulated Annealing Algorithm. *Chin. J. Chem. Eng.* **2004**, *1*, 66–77.

118. Ma, X.; Yao, P.; Luo, X.; Wilfried, R. Synthesis of Multi-Stream Heat Exchanger Network for Multi-Period Operation with Genetic/Simulated Annealing Algorithms. *Appl. Thermal. Eng.* **2008**, *28*, 809–823. [CrossRef]

119. Peng, F.; Cui, G. Efficient Simultaneous Synthesis for Heat Exchanger Network with Simulated Annealing Algorithm. *Appl. Thermal. Eng.* **2015**, *78*, 136–149. [CrossRef]

120. Wang, Y.P.; Zheng, G.W.; Xu, Y.E. Study on Optimization of Operation for Multipurpose Batch Chemical Plants with Multiple Production Routes. *Comput. Chem. Eng.* **1995**, *19*, S101–S106. [CrossRef]

121. Jiao, Y.; Su, H.; Liao, Z.; Hou, W. Modeling and Multi-Objective Optimization of Refinery Hydrogen Network. *Chin. J. Chem. Eng.* **2011**, *19*, 990–998. [CrossRef]

122. Jin, S.; Li, X.; Tao, S. Globally Optimal Reactor Network Synthesis via the Combination of Linear Programming and Stochastic Optimization Approach. *Chem. Eng. Res. Des.* **2012**, *90*, 808–813. [CrossRef]

123. Cui, C.; Li, X.; Sui, H.; Sun, J. Optimization of Coal-Based Methanol Distillation Scheme Using Process Superstructure Method to Maximize Energy Efficiency. *Energy* **2017**, *119*, 110–120. [CrossRef]

124. Qian, Z.; Chen, Q.; Grossmann, I.E. Optimal Synthesis of Rotating Packed Bed Reactor. *Comput. Chem. Eng.* **2017**, *105*, 152–160. [CrossRef]

125. Zhang, X.; Song, Z.; Zhou, T. Rigorous Design of Reaction-Separation Processes Using Disjunctive Programming Models. *Comput. Chem. Eng.* **2018**, *111*, 16–26. [CrossRef]

126. Hou, K.F.; Zheng, Q.; Li, Y.R.; Shen, J.Z.; Hu, S.Y. Modeling and Optimization of Herb Leaching Processes. *Comput. Chem. Eng.* **2000**, *24*, 1343–1348. [CrossRef]

127. Du, Y.; Xie, L.; Liu, Y.; Zhang, S.; Xu, Y. Optimization of Reverse Osmosis Networks with Split Partial Second Pass Design. *Desalination* **2015**, *365*, 365–380. [CrossRef]

128. Yan, N.; Ren, B.; Wu, B.; Bao, D.; Zhang, X.; Wang, J. Multi-Objective Optimization of Biomass to Biomethane System. *Green Energy Environ.* **2016**, *2*, 58–67. [CrossRef]

129. Liu, B. Research Direction of China's Energy Systems Engineering. *Syst. Eng. Theory Pract.* **1984**, *2*, 3–7. (In Chinese)

130. Li, Y.; Zhu, M.; Chen, B. Energy System Integration of Chemical Processes. *J. Chem. Ind. Eng. (China)* **1992**, *7*, 75–82.

131. Grossmann, I.E.; Harjunkoski, I. Process Systems Engineering: Academic and Industrial Perspectives. *Comput. Chem. Eng.* **2019**, *126*, 474–484. [CrossRef]
132. Siirola, J.J. Industrial Applications of Chemical Process Synthesis. *Adv. Chem. Eng.* **1996**, *4*, 1–62. [CrossRef]
133. Dunn, R.F.; El-Halwagi, M.M. Process Integration Technology Review: Background and Applications in the Chemical Process Industry. *J. Chem. Technol. Biotechnol.* **2003**, *78*, 1011–1021. [CrossRef]
134. Gao, X.; Chen, B.; He, X.; Qiu, T.; Li, J.; Wang, C.; Zhang, L. Multi-Objective Optimization for the Periodic Operation of the Naphtha Pyrolysis Process Using a New Parallel Hybrid Algorithm Combining NSGA-II with SQP. *Comput. Chem. Eng.* **2008**, *32*, 2801–2811. [CrossRef]
135. Yu, J.; Zhang, L.; Li, J. Development Status and its Prospect of PetroChina's Ethylene Industry. *Petrochem. Technol. Appl.* **2010**, *3*, 258–263.
136. Wang, J. Energy Saving and Emission Reduction in Petrochemical Industry as well as Low-Carbon Development—A Trial Discussion on PSE's Application in Petrochemical Industry. *Pet. Petrochem. Today* **2010**, *9*, 1–4.

 *processes*

Article

A Hybrid Multi-Objective Optimization Framework for Preliminary Process Design Based on Health, Safety and Environmental Impact

Shin Yee Teh [1], Kian Boon Chua [1], Boon Hooi Hong [1], Alex J. W. Ling [1], Viknesh Andiappan [2], Dominic C. Y. Foo [1], Mimi H. Hassim [3] and Denny K. S. Ng [1,2,*]

[1] Department of Chemical and Environmental Engineering/Centre of Excellence for Green Technologies, The University of Nottingham Malaysia, Broga Road, Semenyih 43500, Malaysia; shinyee30@gmail.com (S.Y.T.); kianboon19@gmail.com (K.B.C.); b0onhui@hotmail.com (B.H.H.); alex1215.aiesec@gmail.com (A.J.W.L.); Dominic.Foo@nottingham.edu.my (D.C.Y.F.)
[2] School of Engineering and Physical Sciences, Heriot-Watt University Malaysia, Putrajaya 62200, Malaysia; v.murugappan@hw.ac.uk
[3] School of Chemical and Energy Engineering/Centre of Hydrogen Energy, Universiti Teknologi Malaysia, Johor Bahru 81310, Malaysia; mimi@cheme.utm.my
* Correspondence: Denny.Ng@hw.ac.uk; Tel.: +603-8894-3784

Received: 25 February 2019; Accepted: 25 March 2019; Published: 8 April 2019

Abstract: Due to increasingly stringent legal requirements and escalating environmental control costs, chemical industries have paid close attention to sustainable development without compromising their economic performance. Thus, chemical industries are in need of systematic tools to conduct sustainability assessments of their process/plant design. In order to avoid making costly retrofits at later stages, assessments during the preliminary design stage should be performed. In this paper, a systematic framework is presented for chemical processes at the preliminary design stage. Gross profit, Health Quotient Index (HQI), Inherent Safety Index (ISI) and the Waste Reduction (WAR) algorithm are used to assess the economic performance, health, safety and environmental impact of the process, respectively. The fuzzy optimization approach is used to analyse the trade-off among the four aspects simultaneously, as they often conflict with each other. Deviation between the solution obtained from mathematical optimization model and process simulator is determined to ensure the validity of the model. To demonstrate the proposed framework, a case study on 1, 4-butanediol production is presented.

Keywords: input-output model; fuzzy optimization; process synthesis; preliminary stage design

1. Introduction

Process design is a core element in the field of chemical engineering. It can be considered a centre point, bringing together all chemical engineering components as a whole. Process design is associated with creating processes or improving existing processes. An integral part of process design is *process synthesis*. Process synthesis is defined as "the discrete decision-making activities of conjecturing (1) which of the many available component parts one should use, and (2) how they should be interconnected to structure the optimal solution to a given design problem." [1]. The field of process synthesis has seen significant developments since its inception in the 1960s led by the late Roger W. H. Sargent [2]. Most notably, key pioneering contributions have been an integral part of establishing what process synthesis is and what it entails. For instance, Nishida et al. [3] provided an important overview of developments within the boundary of process synthesis and defined process synthesis as "an act of determining the optimal interconnection of processing units as well as the optimal type and design of the units within a process system". Meanwhile, the *Onion model* [4] was proposed as a systematic overview and guide to process synthesis thinking. The Onion model emphasized that a reactor is designed first followed by separation and recycle streams, heat recovery systems and utility systems [4]. Foo and Ng [5] then extended the *Onion model* further by incorporating material recovery and treatment systems (see Figure 1). On the other hand, Douglas' model is another well-accepted decision hierarchy approach that was proposed for process synthesis in the late 80s [6].

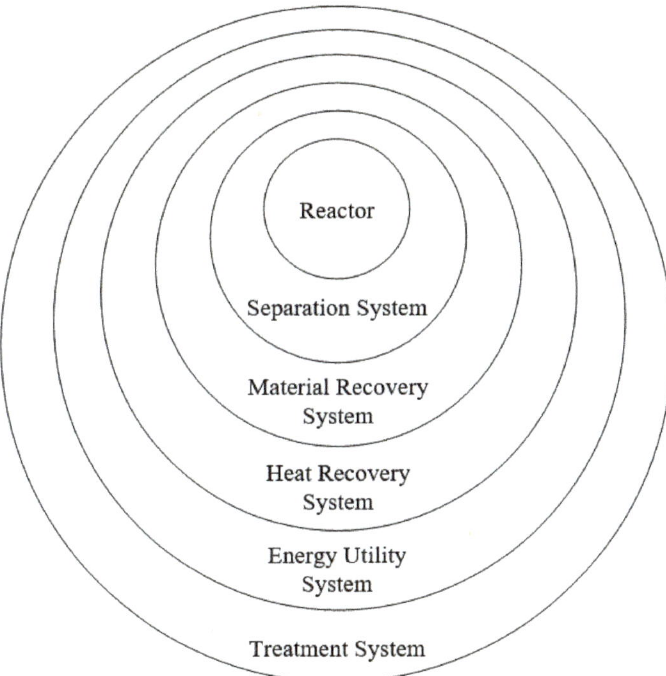

Figure 1. The extended Onion Model [5]. Reproduced with permission from Foo, D.C.Y.; Ng, D.K.S., Process Integration for Cleaner Process Design; In Handbook of Process Integration: Minimisation of Energy and Water Use, Waste and Emissions; published by Elsevier Science, 2013.

To date, there has been a vast number of developed process synthesis methodologies. Often, these methodologies pay close attention to the technical and economic performance of a process design. Meanwhile, other aspects such as safety, occupational health and environmental impacts are typically considered at the later/final stages of design [7,8]. However, several papers argue that these aspects should be considered at the preliminary stage of process synthesis as the cost of process improvement and operational risks can be significantly lowered compared to at the later stages (detailed design) [9,10]. In this respect, several alternative methods for evaluating occupational health, safety and environmental impacts have been presented.

In the area of safety evaluations, Edwards and Lawrence [11] developed the earliest method for assessing inherent safety in a given process. This method is known as *Prototype Index of Inherent Safety* (PIIS). PIIS ranks the inherent safety level of alternative chemical process routes based on main reactions and parameters such as pressure, temperature, yield, heat of reaction, inventory, flammability, toxicity and explosiveness. Despite considering the mentioned parameters, PIIS focuses solely on the main reaction of a process and not the other parts of the process. In view of this, Heikkila [12] proposed an alternative approach called the *Inherent Safety Index* (ISI). ISI considers the same factors as in the PIIS along with additional ones such as corrosions, side reactions, and inventory for both inside and outside the battery limits, types of equipment and process structure. Later, Palaniappan et al. [13]

added a few supplementary indexes such as worst reaction index, overall chemical index and total chemical index to the previously proposed ISI to form *i*-Safe.

Aside from safety, various methods have been developed to evaluate occupational health. For example, Hassim and Edwards [14] developed Process Route Healthiness Index (PRHI) to rank the process alternatives based on the potential of working activities and process conditions that may harm workers. However, PRHI is not suitable for assessment in the preliminary design stage, as it requires complete process information (e.g., points for manual handling etc.). In this respect, Hassim et al. developed the Inherent Occupational Health Index (IOHI) [10], which is suitable for application at the initial stage of the process research and development (R&D). This work was then extended to a more detailed assessment called the Health Quotient Index (HQI), which uses detailed process information available in the preliminary design stage [15]. HQI is able to rank the process alternatives based on health risk values from the fugitive emissions and to calculate the risk of a given process.

Apart from occupational health, there are many well-established approaches reported in the literature on environmental impact assessments. The Environmental Hazard Index (EHI) [16] and Atmospheric Hazard Index (AHI) [17] are among the early methods that assess inherent environmental performance of chemical process routes. Later, Gunasekera and Edwards [18] integrated EHI with AHI into a new method which is known as Inherent Environmental Toxicity Hazard (IETH). IETH can estimate the inherent environmental friendliness of a chemical plant in all media including air, soil and aquatic due to total loss of contaminants. Cabezas et al. [19] then introduced the generalized Waste Reduction (WAR) algorithm. This algorithm determines the total environmental impacts based on the Potential Environmental Index (PEI) balance, which assigned environmental impact values to different pollutants. Young and Cabezas [20] extended PEI balance to integrate the energy consumption of the chemical process into the environmental evaluation. Similarly, Andiappan et al. [21] presented the incremental environmental burden assessment (IEBA), which used the concept of economic potential assessment method developed by Seider et al. [22] to determine the new environmental burden for each process.

It is essential to note that the abovementioned evaluation methods focus only on assessing the performance of a process design based on just one aspect (e.g., safety or occupational health or environmental impact) and not addressing them simultaneously. In this respect, there have been attempts to develop process synthesis approaches that address safety, occupational health and environmental impacts simultaneously. For instance, Azapagic et al. [8] and Othman et al. [23] presented sustainable assessment and design selection approaches that consider economic, environmental, and social aspects. Besides, Al-Sharrah et al. [24] presented a multi-objective optimization model which considers environmental impact, economic performance and operational risk simultaneously for the petrochemical industry. Liew et al. [25] developed a systematic approach to the screening of sustainable chemical reaction pathways at the research and development (R&D) stage. Based on the proposed approach, fuzzy optimization is used to trade-off economic performance, health, safety and environmental impact. Similarly, Ng et al. [26] presented a multi-objective process synthesis approach which considered economic performance, health, safety and environmental impact for biorefineries. Following this, a visualization tool called the *Piper diagram* was also proposed for considering economic, safety and environmental aspects simultaneously [27]. However, due to its graphical nature, the Piper diagram is unable to consider process optimization.

It is clear that the abovementioned work presents approaches to evaluate safety, occupational health and environmental impacts at the preliminary stage of design. However, these approaches focus on preliminary screening without considering operating conditions in a given process. The choice of process operating conditions has a process direct impact on screening decisions and this should be given closer attention. To address this, process simulators can be used to analyse operating conditions for a given process. Process simulators can be used to represent chemical processes in terms of mathematical models and solve them to attain insights on their performance [28]. Process simulation is an essential and complementary task of process synthesis, as it predicts how a process design would behave under defined operating conditions [29,30]. Despite the benefits of process simulators, they suffer from various limitations. For instance, simulators are limited to considering a single objective (e.g., product yield or production rate) at a given run. This limitation disables simulators from finding a trade-off between multiple aspects such as process economics, occupational health, safety and environmental impacts for a given process design. In order to consider multiple aspects simultaneously, multi-objective optimization is required.

As such, this paper presents a systematic framework which combines the benefits from both process simulation and multi-objective optimization to address economic, environmental, health and safety aspects simultaneously at the preliminary design stage. This paper is structured as follows: First, a formal problem statement is given in Section 2. A systematic framework for preliminary process design is presented in Section 3; To illustrate the proposed framework, a case study on the screening of 1,4-butanediol production processes is presented in Section 4; and the work is finally concluded in Section 5.

2. Problem Statement

The problem addressed in this work is stated as follows; given a set of alternative processes k to produce chemical product (in a given output stream) p. Each alternative process k has a set of unit operation j with operating capacities, $x_{j,k}$ and a set of (output) stream p to or from unit operation j represented by matrix $a_{p,j,k}$. Alongside this, each alternative process k differs in economic performance, health, safety and environment impacts. In this work, four objectives are considered for performing a preliminary evaluation and screening of alternative process k. The economic performance is determined based on the gross profit of alternative process k (GP). The health impacts of process alternative k are evaluated using Health Quotient Index (HQI). Meanwhile, the safety and environmental impact of each alternative process k are determined via Inherent Safety Index (ISI) and Potential Environmental Index (WAR) respectively.

This work combines the advantages of process simulation, input-output modelling (IOM) and fuzzy optimization to determine the trade-off between the GP, HQI, ISI and WAR objectives. In fuzzy optimization, the degree of satisfaction (λ_k) is introduced to quantify the degree of satisfaction for four objectives in each alternative process k. Following this, the alternative process k with the highest λ_k will be selected. A detailed description of the proposed framework to address the stated problem is presented in the following section.

3. Framework for Preliminary Process Design

Figure 2 presents the proposed framework for preliminary process design, which considers economic performance, health, safety and environment impacts simultaneously. As shown, the proposed framework begins by identifying the product i that is expected to be produced. Next, alternative process k which produce product i is determined. In order to analyse the performance of each alternative process k, process simulation tools (e.g., Aspen HYSYS, SuperPro Designer, PRO/II, etc.) can be used to simulate the process. Based on the simulation results, process data such as mass flow rates and energy requirements for each process unit j can be determined. These data are then used to develop an input-output model for each process. Input-output model was presented by Leontief [31,32] to analyse the relationship among the raw materials requirement, goods production and the exchange of materials within different economic sectors. Tan et al. [33] extended the usage of input-output model in life cycle assessment and ecological footprint analysis. Aviso et al. [34] integrated the approach with optimization framework to analyse the eco-industrial supply chain under a water footprint constraint. Recently, the input-output model has been used to describe the material and energy balance of processes in a system [35]. Based on the approach [35], the optimal operational adjustment in multi-functional energy systems in response to process inoperability can be determined. Kasivisvanathan et al. [36] also used the input-output model with robust optimization for process synthesis and design of multi-functional energy systems with uncertainties. Most recently,

Foong et al. [37] extended the use of the input-output model for sustainable oil palm plantation development. A detailed description on the procedure to develop an input-output model is presented with an example in Andiappan et al. [38]. The input-output model for each alternative process k, given as in Equation (1)

$$\sum_{j=1}^{J} a_{p,j,k} x_{j,k} = y_{p,k} \; \forall p \, \forall k \tag{1}$$

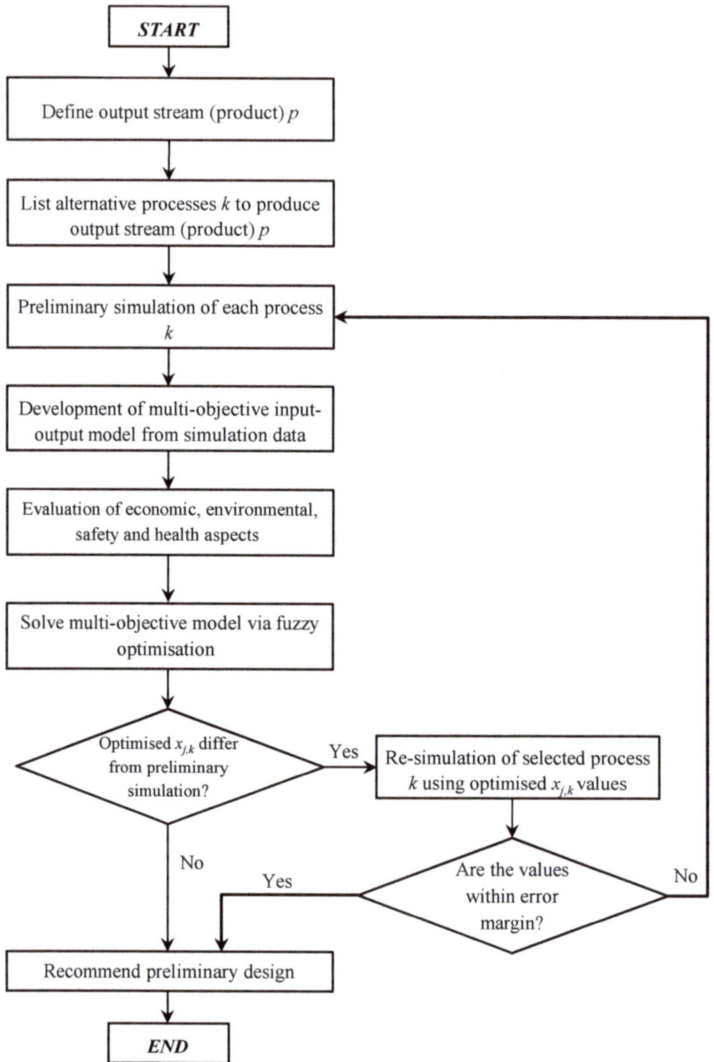

Figure 2. Systematic design framework for preliminary process design.

In Equation (1), $a_{p,j,k}$ represents the matrix of input or output mass flow rate of stream p to/from unit operation j in process k. $x_{j,k}$ represents the capacity or size of the unit operation j in process k and $y_{p,k}$ is the net output of stream p in process k. A positive value of $y_{p,k}$ indicates that it is an output stream which is either product or effluent stream whereas a negative value indicates that it is an input stream and a value of zero denotes that it is an intermediate stream.

Meanwhile, the lower and upper limits of the capacity of unit operation j in process k are presented in Equation (2):

$$x_{j,k}^{L} \leq x_{j,k} \leq x_{j,k}^{U} \forall j \forall k \tag{2}$$

In addition to Equations (1) and (2), the economic, environmental, health and safety impact for each process k is determined. Firstly, in this work, economic performance for each process k is measured via gross profit as shown in Equation (3):

$$GP_k = \left(\sum_{p=1}^{P} y_{p,k}^{Prod} C_{p,k}^{Prod} + \sum_{p=1}^{P} y_{p,k}^{Raw} C_{p,k}^{Raw} \right) AOT_k \forall k \tag{3}$$

where $y_{p,k}^{Prod}$ and $y_{p,k}^{Raw}$ are the mass flow rate of the products and raw materials in process k respectively; $C_{p,k}^{Prod}$ and $C_{p,k}^{Raw}$ are the product price and cost of raw materials in process k respectively; AOT_k is the annual operating time for process k. Note that $y_{p,k}^{Raw}$ is obtained from the input-output model as a negative value. This is because it is an input stream into the system, as stated previously in Equation (1).

On the other hand, the environmental impact of each process k is also evaluated. For this work, the WAR algorithm is adapted to evaluate the environmental impact because it can determine the average possible impact of a chemical process on the environment, based on environmental impact values of different pollutants and their respective mass flows. According to the WAR algorithm, the total environmental impact generated by process k, WAR_k can be expressed as in Equation (4) as follows.

$$WAR_k = i_k^{(cp),\ out} + i_k^{(cp),\ in} + i_k^{(ep),\ out} \forall k \tag{4}$$

where $i_k^{(cp),in}$ and $i_k^{(cp),out}$ are the input and output rates of impact in the process k, $i_k^{(ep),out}$ is the power consumption for process k. The value of $i_k^{(cp),out}$, $i_k^{(cp),in}$ and $i_k^{(ep),out}$ are determined through Equations (5)–(9) respectively:

$$i_k^{(cp),out} = \sum_{i=1}^{I} y_{i,k}^{Out} \sum_{a=1}^{A} \sum_{j=1}^{J} w_{a,j,k} PEI_a \forall k \tag{5}$$

$$i_k^{(cp),in} = \sum_{i=1}^{I} y_{i,k}^{In} \sum_{a=1}^{A} \sum_{j=1}^{J} w_{a,j,k} PEI_a \forall k \tag{6}$$

$$i_k^{(ep),out} = \sum_{i=1}^{I} -y_{i,k}^{Elec} PEI^{Elec} \forall k \tag{7}$$

$$PEI_a = \sum_{l=1}^{L} \alpha_l PEI_{a,l} \forall k \tag{8}$$

$$PEI^{Elec} = \sum_{l=1}^{L} \alpha_l PEI_l^{Elec} \forall k \tag{9}$$

where $y_{i,k}^{Elec}$ is the power consumption of process k; PEI_a and PEI^{Elec} are the score of the potential environmental impact of chemical component a and electricity respectively; α_l is the weighting factor of the impact at category l; $PEI_{a,l}$ and PEI_l^{Elec} are the potential environmental impact score of chemical component a and electricity at each of the category l respectively. Note that a total of eight categories are considered for PEI, that is, human toxicity potential by exposure, both dermal and inhalation (HTPE), human toxicity potential by ingestion (HTPI), aquatic toxicity potential (ATP), terrestrial toxicity potential (TTP), ozone depletion potential (ODP), global warming potential (GWP), acidification potential (AP) and photochemical oxidation potential (PCOP).

In order to evaluate the health impact of process k, the Hazard Quotient Index (HQI) is selected for this work. This is because HQI can be used for simple process flow diagrams (PFDs) containing limited information, namely process drawings and process descriptions. Such method is suitable for the case of preliminary process design optimization as it allows for the comparison of alternative processes by ranking them based on the risk value based on minimal available information. The calculation of HQI consists of four parts, i.e., estimation of fugitive emissions, air volumetric flow rate, airborne chemical concentration and the health risk (HQI) value. Fugitive emission of each of the chemical components a in process k, $m_{a,k}$ can be calculated via Equation (10):

$$m_{a,k} = \sum_{j=1}^{J} \sum_{p=1}^{P} x_{j,k}^{0.5} w_{a,j,k} FE_{p,j,k} \forall a \forall k \tag{10}$$

where $x_{j,k}$, $w_{a,j,k}$ and $FE_{p,j,k}$ are the capacity of the unit operations j, weight composition of chemical component a and the estimated fugitive emission rates in stream p to or from unit operation j respectively. The pre-calculated fugitive emission rates database for the unit operation stream can be obtained from Hassim et al. [15]. The air volumetric flow rate is determined by the following Equations (11)–(14).

$$A_k^T = \sum_{j=1}^{J} A_{j,k} \forall k \tag{11}$$

$$s_k = \left(A_k^T\right)^{\frac{1}{2}} \forall k \tag{12}$$

$$A_k^c = h_k s_k \forall k \tag{13}$$

$$Q_k = v A_k^c \forall k \tag{14}$$

where A_k^T and $A_{j,k}$ are the total process floor area and the floor area of each unit operations j in process k respectively; s_k is the side length of the process k; A_k^c is the cross-sectional area of the process; h_k is the average height of the main unit operations in process k; Q_k is the air volumetric flow rate; v is the wind speed. The average concentration of the chemical components in the air at the downwind edge of the plot area, $C_{a,k}$ can be determined using Equation (15):

$$C_{a,k} = \frac{m_{a,k}}{Q_k} \forall a \forall k \tag{15}$$

HQI of component a in process k can be calculated using Equations (16) and (17):

$$HQI_{a,k} = \frac{C_{a,k}}{C_a^{EL}} \forall a \forall k \tag{16}$$

$$HQI_k = \sum_{a=1}^{A} HQI_{a,k} \forall k \tag{17}$$

where $HQI_{a,k}$ is the HQI of each chemical components a; C_a^{EL} is a constant which represents the threshold limit of the chemical component a; HQI_k is the total HQI of process k.

The safety impact of a process is assessed via Inherent Safety Index (ISI). ISI is selected because it is a simple scoring method that can be incorporated into an optimization framework. Moreover, ISI considers all parts of the process and equipment, unlike the PIIS method. The total ISI score for process k is represented by ISI_k. ISI_k is calculated by summing the Chemical ISI, I_k^{CI} and Process ISI, I_k^{PI} as shown in Equation (18):

$$ISI_k = I_k^{CI} + I_k^{PI} \forall k \tag{18}$$

The sub-index for Chemical ISI, I_k^{CI} is expressed as below:

$$I_k^{CI} = I_k^{RM,\ max} + I_k^{RS,\ max} + I_k^{INT,\ max} + I_k^{FL,\ max} + I_k^{EX,\ max} + I_k^{TOX,\ max} + I_k^{COR,\ max} \forall k \tag{19}$$

where $I_k^{RM,\ max}$, $I_k^{RS,\ max}$, $I_k^{INT,\ max}$, $I_k^{FL,\ max}$, $I_k^{EX,\ max}$, $I_k^{TOX,\ max}$ and $I_k^{COR,\ max}$ are the sub-index for the factor of the heat of main reaction, heat of side reaction, chemical interaction, flammability, explosiveness, toxic exposure and corrosiveness of the chemical components present in the process k respectively. The sub-index for process ISI, I_k^{PI} is given as:

$$I_k^{PI} = I_k^{I} + I_k^{T,\ max} + I_k^{P,\ max} + I_k^{EQ,\ max} + I_k^{ST,\ max} \forall k \tag{20}$$

where I_k^{I}, $I_k^{T,\ max}$, $I_k^{P,\ max}$, $I_k^{EQ,\ max}$ and $I_k^{ST,\ max}$ are the sub-index for the factor of inventory, process temperature, process pressure, equipment safety and safe process structure, respectively. The calculations for both I_k^{CI} and I_k^{PI} are performed on the basis of the worst-case scenario. Among all chemical components present in the process, the scores of the chemical with the most severe hazard in terms of flammability, explosiveness and toxicity are used in the I_k^{CI} calculation. Besides, the highest temperatures of the main and side reaction and the worst possible chemical substance interaction in the process are considered. Meanwhile, the maximum expected values for inventory, process temperature, and pressure and the worst process structure are taken into account for the calculation of I_k^{PI}. The sub-index for inventory is influenced by the total output of the process, y_k^{Out} which is calculated using Equation (21):

$$y_k^{Out} = \sum_{p=1}^{P} y_{p,k}^{Out} \forall k \tag{21}$$

where $y_{p,k}^{Out}$ is the mass flow rate of the output stream for product and effluent streams. The score of the sub-index for inventory is assigned using Equations (22) and (23) [39]. In Equation (22), $L_{k,r}$ and $U_{k,r}$ are the lower and upper bounds of a given criteria r. The criteria r can often be in the form of a range. For instance, $r = 1$ may refer to the criteria where mass flowrate is between the range of $300 - 500$ kg/hr. In this context, 300 kg/hr would be $L_{k,r}$ and 300 kg/hr would be $U_{k,r}$. If $y_{p,k}^{Out}$ falls between this range $r = 1$ or meets this criteria, Equation (22) will assign a specific score $S_{k,r}$, which is activated by binary variable $b_{k,r}$.

$$(L_{k,r} - U_{k,r}) \times b_{k,r} < y_k^{Out} - U_{k,r} < (L_{k,r+1} - U_{k,r}) \times (1 - b_{k,r}) \forall r \forall k \tag{22}$$

$$I_k^{I} = \sum_{r=1}^{R} b_{k,r} \times S_{k,r} \forall k \tag{23}$$

Although several objectives are considered simultaneously in this work, it is important to note that these objectives conflict with each other. To address such conflict, this work employs a multi-objective approach known as fuzzy optimization. Fuzzy optimization is a simple multi-objective optimization approach that was founded upon the fuzzy decision-making approach introduced by Bellman and Zadeh [40]. Zimmermann [41] then extended the fuzzy approach to deal with linear and non-linear programming problems that contain multiple objectives. In this approach, a continuous interdependence variable, λ, which is also known as the degree of satisfaction, is introduced. Every fuzzy constraint will be satisfied partially at least to λ. Thus, the multi-objective functions in the optimization can be integrated into a single objective function within the optimization framework.

Fuzzy optimization has been widely used to determine the optimum process alternative that considers all aspects simultaneously based on the pre-defined limits. The fuzzy optimization model is expressed as:

$$\lambda = \sum_{k=1}^{K} \lambda_k b_k \tag{24}$$

$$\sum_{k=1}^{K} b_k = 1 \tag{25}$$

where integer variable, b_k is used to indicate the existence (or absence) of λ_k of process k. Note that the formulation in Equation (24) results in the model being non-linear. The optimization objective of λ is maximized, subject to the predefined upper and lower bounds. All flexible targets (GP, HQI, ISI and WAR) are predefined as fuzzy goals which are given by a linear membership function bounded by the upper (GP^U, HQI^U, ISI^U, WAR^U) and lower limits (GP^L, HQI^L, ISI^L, WAR^L). λ of each processes k is determined using Equations (26)–(29):

$$\frac{GP_k - GP^L}{GP^U - GP^L} \geq \lambda_k \qquad \forall k \tag{26}$$

$$\frac{WAR^U - WAR_k}{WAR^U - WAR^L} \geq \lambda_k \qquad \forall k \tag{27}$$

$$\frac{ISI^U - ISI_k}{ISI^U - ISI^L} \geq \lambda_k \qquad \forall k \tag{28}$$

$$\frac{HQI^U - HQI_k}{HQI^U - HQI^L} \geq \lambda_k \qquad \forall k \tag{29}$$

Following this, the model is solved via fuzzy optimization to select a more sustainable process configuration based on the aforementioned four objectives. Based on the optimized solution, the operating capacities, $x_{j,k}$ of the selected process configuration are then examined. If the operating capacities do not differ from the values used in the preliminary simulation, then the selected process k can be recommended for the next process design phase. However, if the operating capacities do differ from the preliminary simulation, the selected process k would be re-simulated in accordance to the optimized operating capacities. Once the selected process k is re-simulated, the operating capacities are checked. In the event where the operating capacities in the simulation exceed an allowable error margin (i.e., $\geq 10\%$), the preliminary simulation must be revisited for troubleshooting. For cases where the operating capacities are within the allowed error margin, the re-simulated conditions for process k can be recommended for the next process design phase. Note that the error margin can be changed based on several factors such as type of industry, experience and knowledge of the decision-maker, and type of equipment used.

4. Case Study

To illustrate the proposed framework, a case study on 1,4-butanediol ($C_4H_{10}O_2$) production process is presented. 1,4-butanediol is a colorless and non-corrosive solution which is widely used as a solvent in the industry to manufacture elastic fibers, technical plastics and polyurethanes [42]. Based on literature review, Reppe [42] and Davy [43,44] processes are the two common and well-established alternative processes available to produce 1,4-butanediol. Table 1 shows the reaction pathway for both processes.

Table 1. Reaction pathway for the production of 1,4-Butanediol ($C_4H_{10}O_2$).

Path	Reaction	Chemical Reaction
Reppe Process	Hydrogenation of Acetylene and Formaldehyde	$C_2H_2 + 2\,CH_2O \rightarrow C_4H_6O_2$ $2\,H_2 + C_4H_6O_2 \rightarrow C_4H_{10}O_2$
Davy Process	Hydrogenation of Dimethyl Maleate	$C_6H_8O_4 + 5H_2 \rightarrow C_4H_{10}O_2 + 2\,CH_3OH$

As shown in Table 1, the raw materials needed for Reppe process are formaldehyde (C_2H_2), acetylene (CH_2O) and hydrogen (H_2). Formaldehyde and acetylene are fed into two batch reactors which operate at 5 bar, 80 °C and are arranged in parallel. Formaldehyde is reacted with acetylene in the presence of copper, bismuth and silicon dioxide supporting catalyst to produce 1,4-butynediol as an intermediate product with a conversion rate of 60%. Aqueous sodium hydroxide and hydrochloric acid are used to maintain the pH of the medium in the reactor within the range of pH 6 to 8. Based on the reaction, process synthesis is started by establishing the remaining unit operations in the process. For example, the product from the batch reactors is sent to a buffer tank and is stored for approximately 1–30 h before feeding to a distillation column for further separation. After separation, the bottom product stream of the distillation column is rich in 1,4-butynediol and is fed to a trickling and hydrogenation reactor for further reaction to produce 1,4-butanediol as the main product with a conversion rate of 90%. The operating condition of this reactor is 160 °C and 150 bar. Figure 3 shows the process flow diagram for the Reppe process.

For the Davy process, the raw materials are dimethyl maleate (DMM), $C_6H_8O_4$ and hydrogen, H_2. DMM is completely vaporised with H_2 before feeding into the first fixed bed reactor. The fixed bed reactor operates at 15 bar, 150 °C and consists of two bed of catalysts which are palladium on alumina and copper-zinc oxide. In the reactor, DMM is fully converted into dimethyl succinate (DMS) through hydrogenation process in the presence of palladium on alumina as catalyst in the first reaction stage. In the second reaction stage, DMS is mainly converted into gammabutyrolactone (GBL) with trace amounts of tetrahydrofuran (THF) and 1,4-butanediol in the presence of copper-zinc oxide as catalyst. The conversion rate of the second reaction is 97%. Based on these two reactions, process synthesis is conducted to list all the other required unit operations. For instance, the vapour-phase product from the reactor is condensed at 30 °C and is fed into a gas-liquid separator to remove the unreacted H_2. The unreacted H_2 is then being recycled to the fixed bed reactor for further reaction. Then, the GBL rich stream from the first reactor is fed into the second hydrogenation reactor for further reaction to produce 1,4-butanediol with trace amounts of methanol and THF as by-products. The operating condition of the second reactor is 210 °C and 75 bar. The catalyst used in the second hydrogenation reactor is copper-zinc oxide and the conversion of GBL to 1,4-butanediol is around 95%. The PFD for the Davy process is shown in Figure 4. The annual production rates for both Reppe and Davy process are assumed to be 60,000 tonnes of 1,4-butanediol, operating at 8000 h/y. The acceptable range of fluctuation on the amount of the product output given by the decision maker is within ±30%. The average height of the main unit operations in the process is assumed to be below 7 m [45]. Wind speed is assumed to be 4 m/s, which is a typical value for outdoor facilities, since local average wind speed is not available [45].

Based on the information mentioned above, process simulation can be performed to analyse the performance for both Reppe and Davy processes. In this work, both processes are simulated with the aid of commercial process simulation software, Aspen HYSYS version 8.8 (Aspentech, Bedford, MA, USA, 2014) [46] (see simulation flowsheets in Figures S1,S2 and simulation files in Supplementary Materials respectively). Settings and parameters used for both simulations are summarized in Tables S1 and S2 (Supplementary Materials) respectively.

Based on the simulation result, data such as input and output mass flow rates and energy consumption are then extracted. The extracted data was then used to develop the input-output models for both processes (see Tables S3 and S4 in Supplementary Materials).

In the following step, the input-output models are evaluated based on economic, environmental, health and safety aspects. For the environmental aspects, the weighting factor of the impact at category l, α_l is assumed as one. This indicates that the impact for each category is equally important. Meanwhile, Table 2 shows the information obtained for both Reppe and Davy processes prior to the optimisation step. As shown in Table 2, the Davy process has higher initial GP and WAR; however, with a lower score of HQI and ISI, compared with the Reppe process. The latter has a lower initial GP mainly due to its higher cost of raw materials. Its higher HQI score is mainly due to the presence of harmful chemical components in the process, such as formaldehyde and 1,4-butynediol. Note that the threshold limit (TLV) of formaldehyde and 1,4-butanediol are very low, that is, 1.228 mg/m^3 and 0.5 mg/m^3 respectively. Besides, the higher ISI score for the Reppe process is mainly due to its highly exothermic side reaction involving formaldehyde, acetylene and hydrogen to form 2-propanol. Note that the Reppe process has a lower score of WAR (i.e., it is safer) because it converts the hazardous raw material (formaldehyde) to a product (1,4-butanediol) with a lower environmental impact. As such, the selection between these two processes is complex, particularly when four different aspects are considered simultaneously.

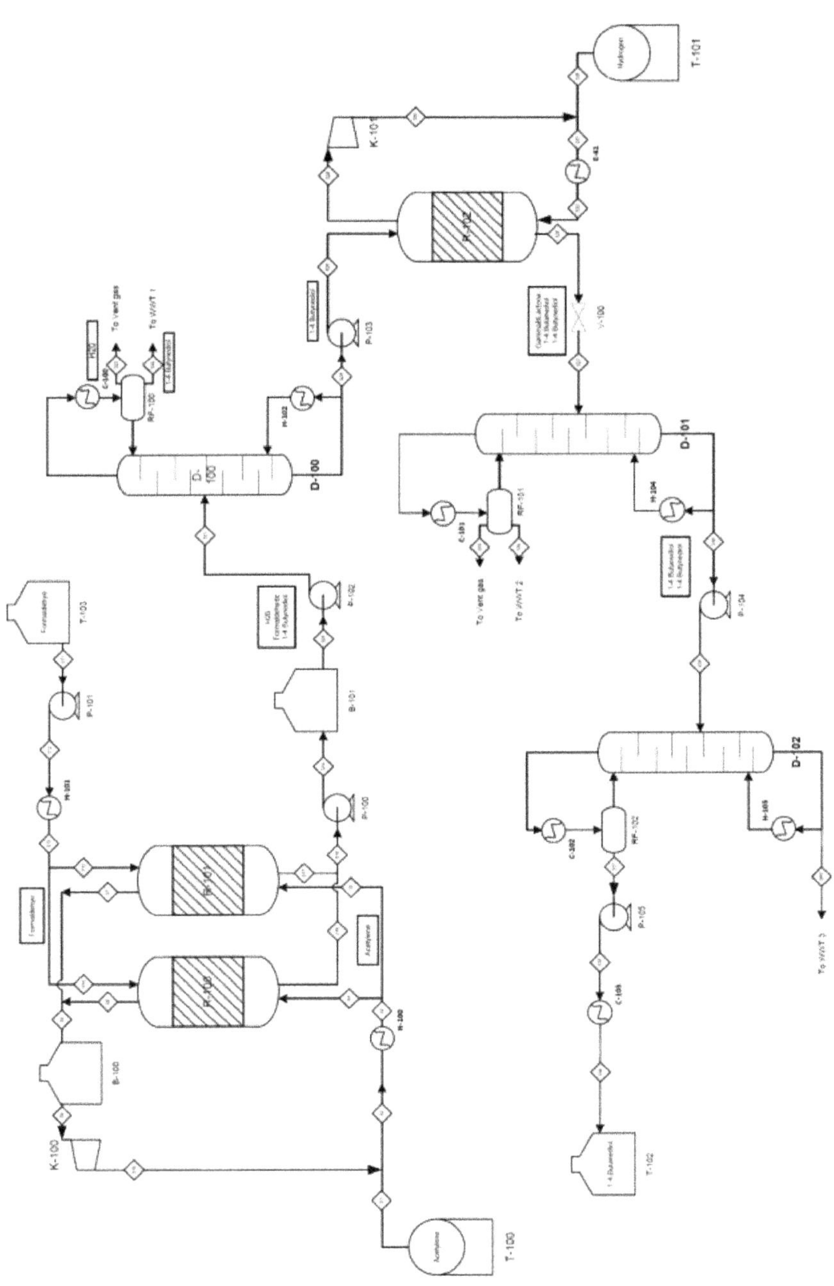

Figure 3. Process flow diagram of the Reppe Process.

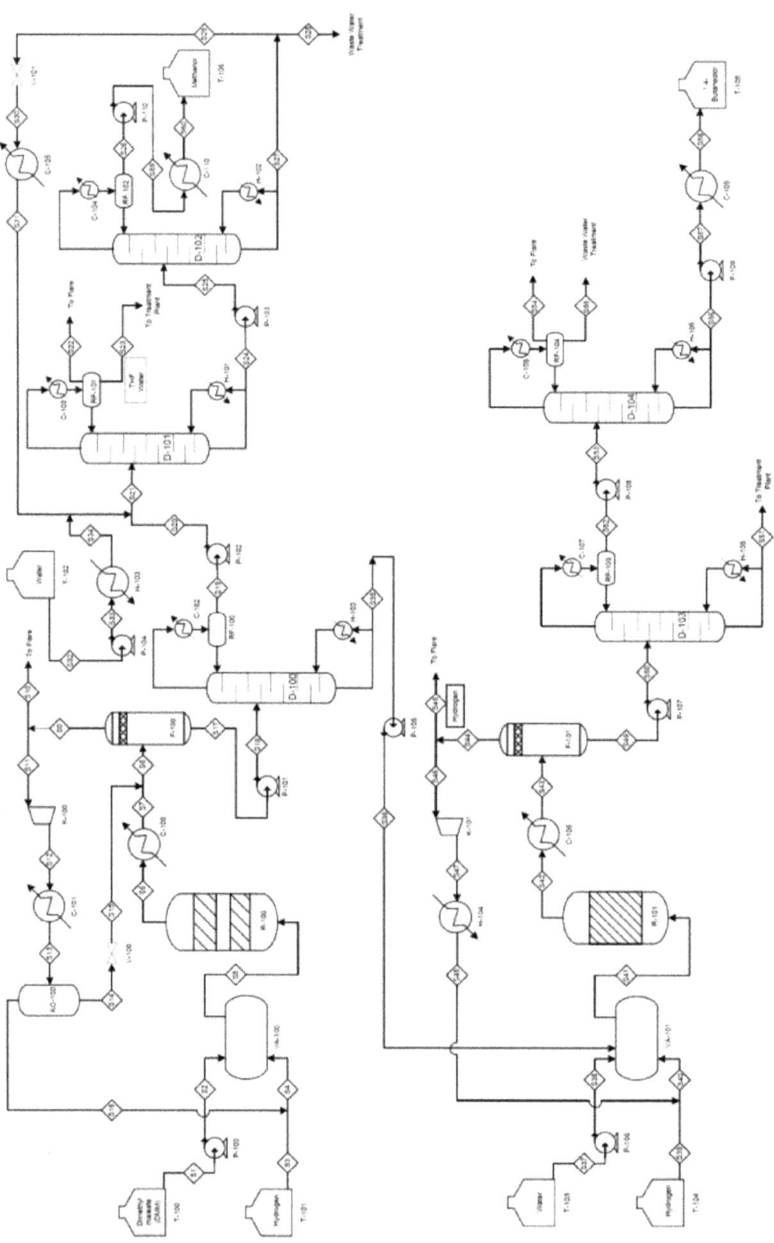

Figure 4. Process flow diagram of the Davy Process.

Table 2. Data obtained for Reppe and Davy processes before optimisation.

Aspects	Reppe Process	Davy Process
GP_k (10^6 USD/y)	45.633	123.408
HQI_k	0.484	0.025
ISI_k	36	28
WAR_k	2.091	33.679

In order to determine the optimum process with consideration of all aspects simultaneously, fuzzy optimisation is used. Based on Equations (26)–(29), it is noted that lower and upper limits for GP, HQI, ISI and WAR are required. To obtain these values, GP_k is first maximized to determine the upper limit GP^U. The corresponding values for HQI_k, ISI_k and WAR_k are used as upper limits in HQI^U, ISI^U and WAR^U respectively. Following this, WAR_k is minimized to determine WAR^L. The corresponding GP_k, ISI_k and WAR_k values were taken as lower limits GP^L, HQI^L and ISI^L respectively. The obtained values for GP^U, GP^L, HQI^U, HQI^L, ISI^U, ISI^L, WAR^U and WAR^L are 185.112×10^6 USD/y, 22.817×10^6 USD/y, 0.593, 0.017, 36, 27, 50.519 and 1.045 respectively. The model is then solved by maximizing λ in Equation (24) with constraints in Equations (1)–(23) and (26)–(29). The model was solved using LINGO version 14.0 (Lindo Systems, Chicago, IL, USA, 2015) on a Lenovo P700 with 8 GB RAM and Intel®Core™ i7, 2.60 GHz Processor. The mixed integer non-linear programming (MINLP) model consists of 187 variables, 6 integers and 341 constraints.

Based on the results obtained, the Davy process was selected as it has a higher value of λ, that is, 0.467, as compared to that of the Reppe process (0.111). The production of 1,4-butanediol after optimisation is scaled down from 7.9 ton/h to 7.64 ton/h, which is within the acceptable range for the amount of product output. Table 3 shows the comparison of the four aspects before and after optimisation for the Davy process.

Table 3. Comparison of the four aspects for the Davy process before and after optimisation step.

Aspects	Before	After
Capacity of process, $x_{j,k}$	1	0.967
GP_k (10^6 USD/y)	123.408	119.293
HQI_k	0.0246	0.0242
ISI_k	28	28
WAR_k	33.679	32.556

After optimisation, it is observed that the operating capacity of the process has been scaled down by 3.3% from the target production to obtain a trade-off among economic, health, safety and environmental aspects. The reason for this is that, in order to achieve the trade-off solution shown in Table 3, the operating capacity for the Davy process has to be scaled down by 3.3%. From here, the new scaled down operating capacity of the Davy process is re-simulated to study the deviations. The net output of the process streams obtained from the developed model has a slight difference to the solution generated from process simulator with a deviation of less than 10%. This is mainly due to the different models used in generating the solution. The solution obtained by the mathematical model is linear, while that used by the process simulator is in rigorous mode. Since the deviation obtained is small, the selected Davy process with its optimized operating capacity can be recommended for the next stage of design.

5. Conclusions

In conclusion, this work presented a systematic framework to screen and select a sustainable chemical process at the preliminary design stage. The presented framework consists of three main tasks. First, process simulation was carried out to analyse the performance of each alternative process. Based on the simulation, process data such as mass flow rates and energy requirements for each process unit were determined. These data were then used in the next task, which is to develop an input-output model for each process. Each input-output model was formulated to address four conflicting objectives. These four objectives, namely economic performance, health, safety and environment aspects were assessed using gross profit (GP), the Health Quotient Index (HQI), the Inherent Safety Index (ISI) and the WAR algorithm respectively. Following this, fuzzy optimization was used in the third task to optimize the abovementioned aspects simultaneously and to select the most sustainable process. A sustainable process in this context refers to the process with the highest degree of satisfaction (λ) among all four aspects considered. The proposed framework was illustrated with a case study on 1,4-butanediol production, where the proposed framework was used to determine the most sustainable process to produce 1,4-butanediol. Two processes, namely the Reppe and Davy processes were considered. Each process was simulated using Aspen HYSYS Version 8.8 and subsequently underwent input-output modelling. The developed input-output models were then optimized using fuzzy optimization. Results from the case study indicate that the Davy process is more sustainable compared to the Reppe process. In particular, the Davy process had the highest λ value. Furthermore, the results suggest that the operating capacity of the Davy process should be scaled down by 3.3% to meet the trade-off scores determined by the λ value obtained. Based on this recommendation, the Davy process was re-simulated. It was found that there was a small percentage of deviations between the mathematical and process simulation models. This proved that the developed model is feasible for determining a more sustainable process configuration. For future work, a selection of different unit operations may be incorporated into the model to determine their impact on flowsheet selection.

Supplementary Materials: The following are available online at http://www.mdpi.com/2227-9717/7/4/200/s1, Figure S1: Simulation Flowsheet of Davy Process via Aspen HYSYS version 8.8; Figure S2: Simulation Flowsheet of Reppe Process via Aspen HYSYS version 8.8; Table S1: Simulation Settings for Davy Process; Table S2: Simulation Settings for Reppe Process; Table S3: Input-Output Table for Reppe Process; Table S4: Input-Output Table for Davy Process; Simulation File for Davy Process (Filename: Davy Process); Simulation File for Reppe Process (Filename: Reppe Process).

Author Contributions: Conceptualization, D.C.Y.F. and D.K.S.N.; methodology, S.Y.T., K.B.C., D.C.Y.F., D.K.S.N. and M.H.H.; software, S.Y.T., K.B.C., B.H.H., A.J.W.L. and V.A.; validation, V.A. and D.K.S.N.; formal analysis, S.Y.T., K.B.C. and B.H.H.; investigation, S.Y.T., K.B.C., B.H.H.; resources, D.K.S.N.; data curation, B.H.H., A.J.W.L.; writing—original draft preparation, S.Y.T., V.A.; writing—review and editing, V.A., D.C.Y.F., M.H.H. and D.K.S.N.; visualization, S.Y.T.; supervision, V.A., D.C.Y.F., M.H.H. and D.K.S.N.

Funding: This research was funded by Ministry of Higher Education, Malaysia through the LRGS Grant (LRGS/2013/UKM-UNMC/PT/05).

Acknowledgments: The authors would like to acknowledge the financial support provided from the Ministry of Higher Education, Malaysia through the LRGS Grant (LRGS/2013/UKM-UNMC/PT/05).

Conflicts of Interest: The authors declare no conflict of interest.

Nomenclature

Index/Subscript

p	Index for stream of input or output
j	Index for unit operations
k	Index for alternative process
a	Index for chemical components
l	Index for environmental impact category
r	Index for criteria with a given range

Parameters

$a_{p,j,k}$	Matrix of input or output mass flow rate for stream p to or from unit operation j of process k
$x^L_{j,k}$	Lower limit of the capacity for unit operation j in process k
$x^U_{j,k}$	Upper limit of the capacity for unit operation j in process k
AOT_k	Annual operating time for process k
$C^{Prod}_{p,k}$	Product price of output p (products) in process k
$C^{Raw}_{p,k}$	Cost of input p (raw materials) in process k

EP^L	Lower limit for gross profit
EP^U	Upper limit for gross profit
WAR^L	Lower limit for environmental impact
WAR^U	Upper limit for environmental impact
ISI^L	Lower limit for inherent safety index
ISI^U	Upper limit for inherent safety index
HQI^L	Lower limit for health quotient index
HQI^U	Upper limit for health quotient index
PEI_a	Score of potential environmental impact for chemical component a
$PEI_{a,l}$	Score of potential environmental impact for chemical component a at each of the category l
PEI^{Elec}	Score of potential environmental impact for electricity
α_l	Weighting factor of the impact at category l
$FE_{i,j,k}$	Estimated fugitive emission rates in stream i to or from unit operation j in process k
A_k^T	Total process floor area in process k
$A_{j,k}$	The floor area of each unit operations j in process k
s_k	Side length of the process k
A_k^c	Cross-sectional area of process k
h_k	Average height of the main unit operations in process k
Q_k	Air volumetric flow rate in process k
V	Wind speed
$C_{a,k}$	Average concentration of chemical components a in the air at downwind edge of the plot area
C_a^{EL}	Threshold limit of the chemical component a
$L_{k,r}$	Lower bound for a given criteria r in process k
$U_{k,r}$	Upper bound for a given criteria r in process k
$S_{k,r}$	Score of sub-index for inventory in process k based on a given criteria r

Variables

$x_{j,k}$	Capacity or size of the unit operation j in process k
$y_{p,k}$	Net input or output of stream p in process k
$y_{p,k}^{Out}$	Net output of stream p in process k
$y_{p,k}^{In}$	Net input of stream p in process k
$y_{p,k}^{Elec}$	Power consumption of process k
$w_{a,j,k}$	Weight composition of chemical component a from unit j in process k
$m_{a,k}$	Fugitive emission of each of the chemical components a in process k
GP_k	Gross profit for process k
WAR_k	Total environmental impact generated by process k
$i_k^{(cp),out}$	Output rates of impact in the process k
$i_k^{(cp),in}$	Input rates of impact in the process k
$i_k^{(ep),out}$	Impact of power consumption for process k
I_k^{CI}	Chemical ISI score of process k
I_k^{PI}	Process ISI score of process k
$I_k^{RM,\,max}$	Sub-index for the factor of the heat of main reaction in process k
$I_k^{RS,\,max}$	Sub-index for the factor of the heat of side reaction in process k
$I_k^{INT,\,max}$	Sub-index for the factor of chemical interaction in process k
$I_k^{FL,\,max}$	Sub-index for the factor of flammability in process k
$I_k^{EX,\,max}$	Sub-index for the factor of explosiveness in process k
$I_k^{TOX,\,max}$	Sub-index for the factor of toxic exposure in process k
$I_k^{COR,\,max}$	Sub-index for the factor of corrosiveness of the chemical components in process k
I_k^{I}	Sub-index for the factor of inventory in process k
$I_k^{T,\,max}$	Sub-index for the factor of process temperature in process k
$I_k^{P,\,max}$	Sub-index for the factor of process pressure in process k
$I_k^{EQ,\,max}$	Sub-index for the factor of equipment safety in process k
$I_k^{ST,\,max}$	Sub-index for the factor of safe process structure in process k
ISI_k	Total ISI score of process k
$HQI_{a,k}$	HQI of each chemical components a in process k

HQIk	Total HQI score of process k
λ_k	Degree of satisfaction for multiple objectives in each alternative process k
b_k	Existence of alternative process k

References

1. Westerberg, A.W. *Recent Developments in Chemical Process and Plant Design*; Liu, Y.A., Henry, A., McGee, J., Epperly, W.R., Eds.; John Wiley and Sons: New York, NY, USA, 1987.
2. Stephanopoulos, G.; Reklaitis, G.V. Process systems engineering: From Solvay to modern bio- and nanotechnology. A history of development, successes and prospects for the future. *Chem. Eng. Sci.* **2011**, *66*, 4272–4306. [CrossRef]
3. Nishida, N.; Stephanopoulos, G.; Westerberg, A.W. A review of process synthesis. *AIChE J.* **1981**, *27*, 321–351. [CrossRef]
4. Linnhoff, B.; Townsend, D.W.; Boland, D.; Hewitt, G.F.; Thomas, B.E.A.; Guy, A.R.; Marsland, R.H. *User Guide on Process Integration for the Efficient Use of Energy*; Institution of Chemical Engineers: Rugby, UK, 1982.
5. Foo, D.C.Y.; Ng, D.K.S. Process Integration for Cleaner Process Design. In *Handbook of Process Integration: Minimisation of Energy and Water Use, Waste and Emissions*; Series in Energy No. 61; Klemeš, J.J., Ed.; Woodhead Publishing: Sawston, UK, 2013.
6. Douglas, J.M. *Conceptual Design of Chemical Processes*; McGraw-Hill: New York, NY, USA, 1988.
7. Azapagic, A.; Millington, A.; Collett, A. A methodology for integrating sustainability considerations into process design. *Chem. Eng. Res. Des.* **2006**, *84*, 439–452. [CrossRef]
8. Azapagic, A.; Millington, A.; Collett, A. Process design for sustainability: The case of vinyl chloride monomer. In *Sustainable Development in Practice: Case Studies for Engineers and Scientists*; John Wiley & Sons, Ltd.: Hoboken, NJ, USA, 2004; pp. 201–249.
9. Gentile, M.; Rogers, W.; Mannan, M. Development of a fuzzy logic-based inherent safety index. *Process Saf. Environ. Prot.* **2003**, *81*, 444–456. [CrossRef]
10. Hassim, M.; Hurme, M. Inherent occupational health assessment during preliminary design stage. *J. Loss Prev. Process Ind.* **2010**, *23*, 476–482. [CrossRef]
11. Edwards, D.; Lawrence, D. Assessing the inherent safety of chemical process routes: Is there a relation between plant costs and inherent safety? *Process Saf. Environ. Prot.* **1993**, *71*, 252–258.
12. Heikkila, A. *Inherent Safety in Process Plant Design*; Technical Research Centre of Finland: Espoo, Finland, 1999.
13. Palaniappan, C.; Srinivasan, R.; Tan, R. Expert system for the design of inherently safer processes. 1. Route selection stage. *Ind. Eng. Chem. Res.* **2002**, *41*, 6698–6710. [CrossRef]
14. Hassim, M.; Edwards, D. Development of a methodology for assessing inherent occupational health hazards. *Process Saf. Environ. Prot.* **2006**, *84*, 378–390. [CrossRef]
15. Hassim, M.H.; Perez, A.L.; Hurme, M. Estimation of chemical concentration due to fugitive emissions during chemical process design. *Process Saf. Environ. Prot.* **2010**, *88*, 173–184. [CrossRef]
16. Cave, S.; Edwards, D. Chemical process route selection based on assessment of inherent environmental hazard. *Comput. Chem. Eng.* **1997**, *21*, 965–970. [CrossRef]
17. Gunasekera, M.; Edwards, D. Estimating the environmental impact of catastrophic chemical releases to the atmosphere: An index method for ranking alternative chemical process routes. *Process Saf. Environ. Prot.* **2003**, *81*, 463–474. [CrossRef]
18. Gunasekera, M.; Edwards, D. Chemical process route selection based upon the potential toxic impact on the aquatic, terrestrial and atmospheric environments. *J. Loss Prev. Process Ind.* **2006**, *19*, 60–69. [CrossRef]
19. Cabezas, H.; Bare, J.; Mallick, S. Pollution prevention with chemical process simulators: The generalized waste reduction (WAR) algorithm-full version. *Comput. Chem. Eng.* **1999**, *23*, 623–634. [CrossRef]
20. Young, D.; Cabezas, H. Designing sustainable processes with simulation: The waste reduction (WAR) algorithm. *Comput. Chem. Eng.* **1999**, *23*, 1477–1491. [CrossRef]
21. Andiappan, V.; Ko, A.S.Y.; Lau, V.W.S.; Ng, L.Y.; Ng, R.T.L.; Chemmangattuvalappil, N.G.; Ng, D.K.S. Synthesis of sustainable integrated biorefinery via reaction pathway synthesis: Economic, incremental enviromental burden and energy assessment with multiobjective optimization. *AIChE J.* **2015**, *61*, 132–146. [CrossRef]

22. Seider, W.D.; Seader, J.D.; Lewin, D.R. *Product and Process Design Principles: Synthesis, Analysis and Evaluation*, 2nd ed.; John Wiley & Sons, Ltd.: Hoboken, NJ, USA, 2004.
23. Othman, M.; Repke, J.; Wozny, G.; Huang, Y. A modular approach to sustainability assessment and decision support in chemical process design. *Ind. Eng. Chem. Res.* **2010**, *49*, 7870–7881. [CrossRef]
24. Al-Sharrah, G.; Elkamel, A.; Almanssoor, A. Sustainability indicators for decisionmaking and optimisation in the process industry: The case of the petrochemical industry. *Chem. Eng. Sci.* **2010**, *65*, 1452–1461. [CrossRef]
25. Liew, W.; Hashim, M.; Ng, D.; Hurme, M. Fuzzy Optimization for Screening of Sustainable Chemical Reaction Pathways. *Chem. Eng. Trans.* **2012**, *29*, 529–534.
26. Ng, R.; Hassim, M.; Ng, D. Process synthesis and optimization of a sustainable integrated biorefinery via fuzzy optimisation. *AIChE J.* **2013**, *59*, 4212–4227. [CrossRef]
27. Teng, W.C.; Fong, K.L.; Shenkar, D.; Wilson, J.A.; Foo, D.C.Y. Piper Diagram–A Novel Visualisation Tool for Process Design. *Chem. Eng. Res. Des.* **2016**, *112*, 132–145. [CrossRef]
28. Motard, R.L.; Shacham, M.; Rosen, E.M. Steady state chemical process simulation. *AIChE J.* **1975**, *21*, 417–436. [CrossRef]
29. Dimian, A.C.; Bildea, C.S.; Kiss, A.A. *Integrated Design and Simulation of Chemical Processes*, 2nd ed.; Elsevier: Amsterdam, The Netherlands, 2014.
30. Foo, D.C.Y.; Chemmangattuvalappil, N.; Ng, D.K.S.; Elyas, R.; Chen, C.-L.; Elms, R.D.; Lee, H.-Y.; Chien, I.-L.; Chong, S.; Chong, C.H. *Chemical Engineering Process Simulation*; Elsevier: Amsterdam, The Netherlands, 2017; ISBN 978-0-12-803782-9.
31. Leontief, W. Quantitative Input and Output Relations in the Economic Systems of the United States. *Rev. Econ. Stat.* **1936**, *18*, 105–125. [CrossRef]
32. Leontief, W.W. *The Structure of the AMERICAN Economy*; Oxford University Press: New York, NY, USA, 1951.
33. Tan, R.; Culaba, A.; Aviso, K. A fuzzy linear programming extension of the general matrix-based life cycle model. *J. Clean. Prod.* **2008**, *16*, 1358–1367. [CrossRef]
34. Aviso, K.; Tan, R.; Culaba, A.; Cruz, J., Jr. Fuzzy input–Output model for optimizing eco-industrial supply chains under water footprint constraints. *J. Clean. Prod.* **2011**, *19*, 187–196. [CrossRef]
35. Kasivisvanathan, H.; Barilea, I.D.U.; Ng, D.K.S.; Tan, R.R. Optimal Operational Adjustment in Multi-functional Energy Systems in Response to Process Inoperability. *Appl. Energy* **2013**, *102*, 492–500. [CrossRef]
36. Kasivisvanathan, H.; Ubando, A.T.; Ng, D.K.S.; Tan, R.R. Robust Optimization for Process Synthesis and Design of Multifunctional Energy Systems with Uncertainties. *Ind. Eng. Chem. Res.* **2014**, *53*, 3196–3209. [CrossRef]
37. Foong, S.Z.Y.; Goh, C.K.M.; Supramaniam, C.V.; Ng, D.K.S. Input–Output optimisation model for sustainable oil palm plantation development. *Sustain. Prod. Consum.* **2019**, *17*, 31–46. [CrossRef]
38. Andiappan, V.; Ng, D.S.; Tan, R.R. Design operability and retrofit analysis (DORA) framework for energy systems. *Energy* **2017**, *134*, 1038–1052. [CrossRef]
39. El-Halwagi, M.M. *Process Integration*; Academic Press: Cambridge, MA, USA, 2006.
40. Bellman, R.E.; Zadeh, L.A. Decision making in a fuzzy environment. *Manag. Sci.* **1970**, *17*, 141–164. [CrossRef]
41. Zimmermann, H. Fuzzy programming and liner programming with several objective functions. *Fuzzy Sets Syst.* **1978**, *1*, 45–55. [CrossRef]
42. Haas, T.; Jaeger, B.; Weber, R.; Mitchell, S.; King, C. New diol processes: 1,3- propanediol and 1,4-butanediol. *Appl. Catal. A Gen.* **2005**, *1*, 83–88. [CrossRef]
43. Schlander, J.; Turek, T. Gas-phase hydrogenolysis of dimethyl maleate to 1,4- butanediol and -butyrolactone over copper/zinc oxide catalysts. *Ind. Eng. Chem. Res.* **1999**, *4*, 1264–1270. [CrossRef]
44. Bertola, A. Process for the production of Tetrahydrofuran, Gammabutyrolactone and Butanediol. U.S. Patent 6248906B1, 19 June 2001.
45. Mecklenburgh, J.C. *Process Plant Layout*; Halstead Press: New York, NY, USA, 1985.
46. Aspentech. *Getting Started in Aspen HYSYS*; Aspen Technology, Inc.: Bedford, MA, USA, 2014.

 *processes*

Article

Symmetry Detection for Quadratic Optimization Using Binary Layered Graphs

Georgia Kouyialis †, Xiaoyu Wang and Ruth Misener *

Department of Computing, Imperial College London, London SW7 2AZ, UK;
g.kouyialis14@imperial.ac.uk (G.K.); xiao.wang10@imperial.ac.uk (X.W.)
* Correspondence: r.misener@imperial.ac.uk; Tel.: +44-207-594-8315
† Current address: Schlumberger Cambridge Research.

Received: 16 May 2019; Accepted: 5 November 2019; Published: 9 November 2019

Abstract: Symmetry in mathematical optimization may create multiple, equivalent solutions. In nonconvex optimization, symmetry can negatively affect algorithm performance, e.g., of branch-and-bound when symmetry induces many equivalent branches. This paper develops detection methods for symmetry groups in quadratically-constrained quadratic optimization problems. Representing the optimization problem with adjacency matrices, we use graph theory to transform the adjacency matrices into binary layered graphs. We enter the binary layered graphs into the software package nauty that generates important symmetric properties of the original problem. Symmetry pattern knowledge motivates a discretization pattern that we use to reduce computation time for an approximation of the point packing problem. This paper highlights the importance of detecting and classifying symmetry and shows that knowledge of this symmetry enables quick approximation of a highly symmetric optimization problem.

Keywords: symmetry; quadratic optimization; quadratically-constrained quadratic optimization

1. Introduction

When the optimization variables can be permuted without changing the structure of the underlying optimization problem, we say that the *formulation group* of an optimization problem is *symmetric* [1,2]. For motivation, consider the circle packing problem illustrated in Figure 1 [3]. Given an integer $n > 0$, the circle packing problem asks: what is the largest radius r for which n non-overlapping circles can be placed in the unit square? Costa et al. [3] show that the formulation group, i.e., a subgroup of symmetry group generated by permuting variables and constraints, is isomorphic to a symmetry group created by permuting the variable indices and switching the two coordinates in a unit square ($C_2 \times S_n$). Solution methods for nonconvex optimization problems lacking symmetry-aware formulations and/or solution procedures may end up exploring all of these equivalent solutions. In other words, symmetry may cause classical optimization methods such as branch-and-bound to explore many unnecessary subtrees.

More generally, a number of authors have considered a range of symmetry detection methods, e.g., for constraint programming [4], integer programming [1,5–8], and mixed-integer nonlinear optimization [2]. These automatic symmetry detection methods can then be used to mitigate the computational difficulties caused by symmetries, e.g., with symmetry-breaking constraints [9–11], objective perturbation [12], specialized branching strategies [13,14], cutting planes [15,16], and extended formulations [17]. The recent computational comparison of Pfetsch and Rehn [18] indicates that these state-of-the-art symmetry handling methods expedite the solution process for the MIPLIB 2010 instances and additionally enable more instances to be solved in a time limit.

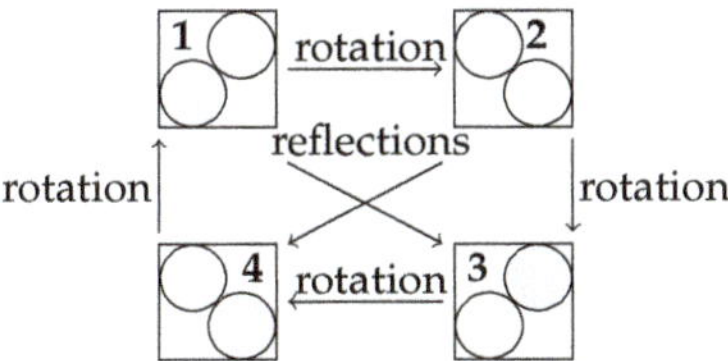

Figure 1. Given an integer $n > 0$, the circle packing problem asks: what is the largest radius r for which n non-overlapping circles can be placed in the unit square? Already for $n = 2$, there are four equivalent solutions [3]; these solutions are related to one another via rotations and reflections.

Researchers have also developed symmetry-handling methods for specific applications including covering design [19], circle packing [3,20], scheduling [21], transmission switching [22], unit commitment [23–26], and heat exchanger network synthesis [27,28]. As a concrete example of the type of contributions researchers have made, consider a job shop scheduling problem that minimizes makespan on two identical machines. Good scheduling formulations and/or solution procedures, e.g., Maravelias and Grossmann [29], Maravelias [30], and Mistry et al. [31], will implicitly exclude two of the three equivalent solutions diagrammed in Figure 2.

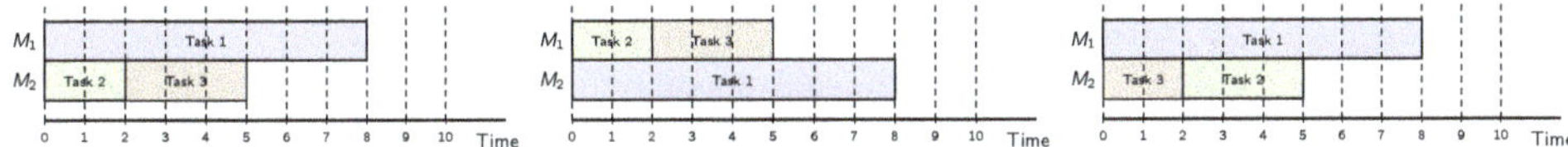

Figure 2. To observe symmetries that may arise in scheduling, consider a job shop scheduling problem that minimizes makespan on two identical machines. Lacking symmetry-aware formulations and/or solution procedures, a solution procedure may end up exploring all three of these equivalent solutions.

This paper develops detection methods for symmetry groups in quadratically-constrained quadratic optimization problems. Representing the optimization problem with adjacency matrices, we use graph theory to transform the adjacency matrices into binary layered graphs. We enter the binary layered graphs into the software package nauty that generates important symmetric properties of the original problem. Symmetry pattern knowledge motivates a discretization pattern that we use to reduce computation time for an approximation of the point packing problem.

2. Formulation Symmetry for Quadratically-Constrained Quadratic Optimization Problems

Consider the quadratically-constrained quadratic optimization problem (QCQP):

$$\min_{x \in \mathbb{R}^n} \quad f_0(x)$$
$$\text{s.t.} \quad f_k(x) \leq 0 \qquad \forall k = 1, \ldots, m \qquad \text{(QCQP)}$$
$$x_i \in \left[x_i^L, x_i^U \right] \quad \forall i = 1, \ldots, n,$$

where:

$$f_k(x) = \sum_{i=1}^{n} \sum_{j=1}^{n} \alpha_{ij}^k x_i x_j + \sum_{i=1}^{n} \alpha_{i0}^k x_i + \alpha_{00}^k \ \forall k = 0. \ldots, m, \qquad (1)$$

with finite variable bounds $x_i^L, x_i^U \in \mathbb{R}$, $\forall i$ and coefficients $\alpha_{ij}^k \in \mathbb{R}$ for $i, j \in \{0, \ldots, n\}, k \in \{0, \ldots, m\}$.

To represent symmetry in the QCQP formulation, consider S_n, the symmetric group of order n formed by the $n!$ possible permutation operations. The *formulation group* of QCQP , or $\mathcal{G}_{QCQP}$, is the set of variable index permutations that preserve the objective and constraint structure [2]. For a variable index permutation $\pi \in S_n$, we seek the constraint index permutations $\sigma \in S_n$ that maintain both the objective value and the constraint structure on the feasible domain $dom(f)$ where $f = [f_1, \ldots, f_m]$. More formally:

Definition 1 (Formulation group of QCQP).

$$\mathcal{G}_{QCQP} = \{\pi \in S_n \mid \forall x \in dom(f_0)\, f_0(\pi x) = f_0(x) \wedge \forall x \in dom(f)\, \exists \sigma \in S_m\, (\sigma f(\pi x) = f(x))\}.$$

Because $dom(f)$ may be nonconvex and difficult to compute, this paper considers a $\mathcal{G}_{QCQP}$ restriction that enforces symmetry on the entire box bounds, i.e., we assume that $dom(f) = [x^L, x^U]$ for the purpose of computing $\mathcal{G}_{QCQP}$. The next subsections represent formulation symmetry in two ways: (i) expression graphs in Section 2.1 and (ii) tensors in Section 2.2. Representing formulation symmetry using expression graphs is due to Liberti [2] and the tensor representation is new to this paper.

2.1. Symmetry Detection with Expression Graphs

One option to compare two functions is to compare their expression trees, i.e., a directed acyclic graph representation of each function that incorporates the relevant operations, constants, and variables [2]. These expression tree models were first developed for mixed integer nonlinear optimization (MINLP) by Smith and Pantelides [32] and are common in most global MINLP solvers [33–39] and other MINLP-related software [40–42]. Figure 3 illustrates a simple example of an expression tree for $3x_1 + 2x_4^2 + 2x_2x_3$. A tree comparison algorithm may recursively compare two trees to determine equivalence [2]. More advanced implementations may detect equivalent but differently-formulated expressions, e.g., $(x_1 + x_2)^2$ versus $x_1^2 + 2x_1x_2 + x_2^2$.

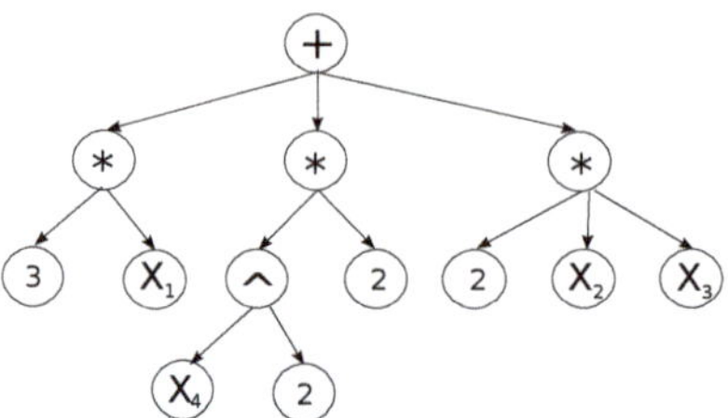

Figure 3. Example of an expression tree for $3x_1 + 2x_4^2 + 2x_2x_3$.

With a directed acyclic graph representation, Liberti [2] computes the formulation symmetry group using the graph isomorphism problem, i.e., a problem that can be solved using off-the-shelf software nauty [43]. Liberti [2] also proves how to map the automorphism group of a directed acyclic graph to the formulation group of the original MINLP.

2.2. Symmetry Detection with Tensors

As an alternative to the expression tree representation, Figure 4 illustrates that QCQP can be represented as a tensor: $A_{QCQP} \in \mathbb{R}^{(n+1) \times (n+1) \times (m+2n)}$. Each of the two dimensions $(n + 1)$ corresponds to a constant term and the variables. Each two-dimensional slice of the tensor corresponds to the constant, linear, and quadratic terms in a constraint. The first m slices correspond to Equation (1) and have entries a_{ij}^k. The next $2n$ slices correspond to the box constraints, i.e., $x_i \geq x_i^L$ and $x_i \leq x_i^U, \forall i$.

The formulation group of this representation is:

$$\mathcal{G}_{QCQP,T} = \{\pi \in S_n \mid \forall x \in dom(f_0)\, f_0(\pi x) = f_0(x) \wedge \forall x \in dom(f)\, \exists \sigma \in S_m\, (A_{QCQP}(\pi, \pi, \sigma) = A_{QCQP})\}.$$

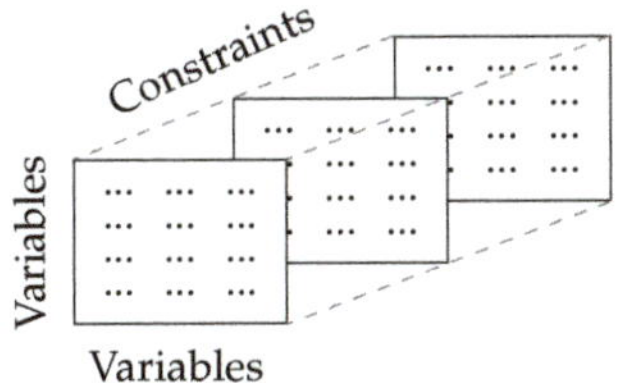

Figure 4. Tensor representation of the symmetry.

2.2.1. Sparse Tensor Representation

For a given tensor A_{QCQP}, consider a sparse representation, illustrated in Figure 5, that reduces the memory required to store the tensor. Instead of storing the entire tensor, we store arrays of length s where s is the number of nonzero entries in QCQP. The first array, $M = (M_1, \ldots, M_s)$ stores all nonzero entries α_{ij}^k of QCQP. The next three arrays, $I = (I_1, \ldots, I_s), J = (J_1, \ldots, J_s), K = (K_1, \ldots, K_s)$ represent the indices corresponding to the nonzero α_{ij}^k entries. The maximum size of s is $(n+1)^2(m+2n)$, but, in practice, most arrays will be significantly shorter.

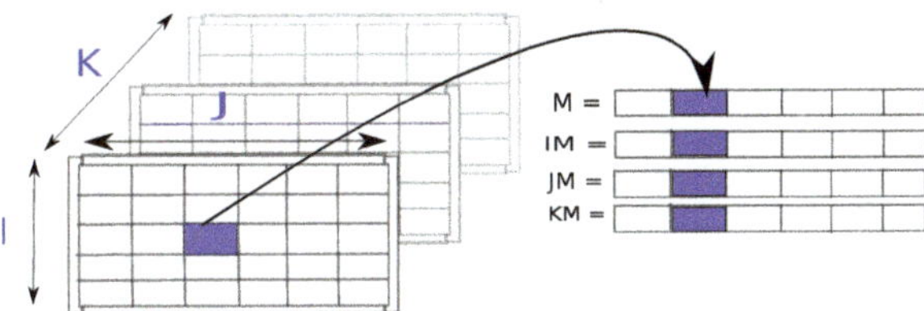

Figure 5. Sparse tensor representation models A_{QCQP} as 4 arrays with the nonzero entry α_{ij}^k in M and arrays I, J, K holding the index.

2.2.2. Converting Matrices to Edge-Labeled, Vertex-Colored Graphs

We convert the sparse tensor representation of A_{QCQP} into an edge-labeled, vertex-colored graph. Given the edge-labeled, vertex-colored graph, generating graph automorphisms to the original problem symmetries is well-known [1,5,44,45]. To construct the edge-labeled, vertex-colored graph, consider a graph $G = (V, E, c)$ corresponding to an instance M, I, J, K. The function $c : E \mapsto r$, for $r \in \{0, \ldots, \ell - 1\}$ is an edge coloring where $\ell \in \mathbb{Z}^+$ is the number of different coefficients in M. Each unique element in M is stored in a vector $U \in R^\ell$. We also partition (color) the vertex set into four subsets: a set V_F representing the objective function, V_C nodes for the constraints, a constant node V_S, and V_R variable nodes. The automorphism definition prevents vertices from being mapped onto a vertex of a different color, so these colors prevent, for example variables becoming constraints. The equivalence relation is [2]:

$$\forall u, v \in V_P\, u \sim v \implies (u, v \in V_F \wedge \ell(u) = \ell(v)) \vee (u, v \in V_C \wedge \ell(u) = \ell(v))$$
$$\vee (u, v \in V_S \wedge \ell(u) = \ell(v)) \vee (u, v \in V_R \wedge \ell(u) = \ell(v)).$$

Figure 6 illustrates the edge-labeled, vertex-colored graph. Initially, the edge set is empty $E = \varnothing$. For $i = \{0, \ldots, s\}$ where $s = |M|$, add an edge $v_{I_i}^{(r)}$ to $v_{K_i}^{(r)}$, i.e., from a vertex in the set that represents the constant element / variables to a vertex in the set of the objective function / constraints, with the relevant color. The graph construction incorporates edges between variable nodes V_R for the quadratic bilinear terms. For $i = \{0, \ldots, s\}$:

- If $I_i = J_i$, i.e., a quadratic term, then $E = E \cup \{\{(v_{I_i}, v_{K_i})^r\} \cap \{(v_{I_i})^r\}\}$.
- else for bilinear term, $I_i \neq J_i$, then $E = E \cup \{\{(v_{I_i}, v_{K_i})^r\} \cap \{(v_{J_i}, v_{K_i})^r\} \cap \{(v_{I_i}, v_{J_i})^r\}\}$.

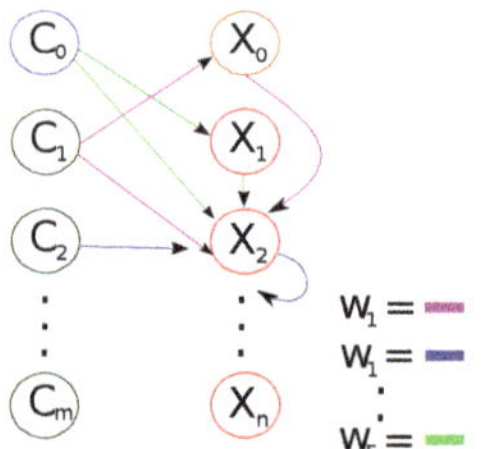

Figure 6. The tensor A_{QCQP} as an edge-labeled, vertex-colored graph.

3. Formulation Symmetry Detection via Binary Layered Graphs

The software nauty [43], which detects symmetry, accepts vertex-colored graphs but does not accept the Section 2.2.2 edge-labeled, vertex-colored graphs. Thus, we associate edge colors with layers in a graph and transform the edge-labeled, vertex-colored graph into a vertex-colored graph. Since the transformation from an edge-labeled, vertex-colored graph to a vertex-colored graph is isomorphic [43], the transformation does not lose anything. Using the resulting *binary layered graphs*, we generate the automorphism group and find symmetry in the original QCQP.

To convert a graph $G = (V, E, c)$ with ℓ colors into an ℓ - *layered graph* [43], we replace each vertex $v_j \in V$ with a fixed connected graph of ℓ vertices $v_j^{(0)}, \ldots, v_j^{(\ell-1)}$. If an edge $(v_j, v_{j'})$ has color r, add an edge from $v_j^{(r)}$ to $v_{j'}^{(r)}$. Finally, we partition the vertices by the superscripts, $V_r = \{v_0^{(r)}, \ldots, v_{n-1}^{(r)}\}$. Alernatively, a binary representation avoids too many layers in G when the number of colors is large.

Definition 2 (Binary Layered Graph). *Let $\ell \in \mathbb{Z}^+$ be the number of edge labels of G. A binary layered graph is a vertex-colored graph where the number of layers $L = \lceil \log_2 (\ell + 1) \rceil$ matches a binary representation.*

We assign a unique positive integer $\mu(z)$ to each $z \in U$ and map edge labels $\mu(z)$ to a binary representation that switches on/off parameters c_t to represent the edge colors as layers. If $c_t = 1$, add a new edge from v_i^t to v_j^t for every $c_t \in \{c_1, \ldots, c_{L-1}\}$:

$$\mu(z) = 2^{L-1} \cdot c_{L-1}(z) + 2^{L-2} \cdot c_{L-2}(z) + \cdots + 2^0 \cdot c_0(z), \text{ for } c_t \in \{0, 1\}, t = \{0, \ldots, L-1\}. \quad (2)$$

Figure 7 illustrates the resulting binary labeled graph with its $L = \lceil \log_2 (\ell + 1) \rceil + 2$ layers. There are vertices for the objective function and each constraint and layers of copies of these constraints (connected with vertical edges). The horizontal edges encode the problem coefficients. On the upper part of Figure 7, there are vertices for a constant element and each variable and a layer of variable copies (connected with vertical edges). Here, the horizontal edges and loops distinguish the linear and bilinear terms. Algorithms 1 and 2 summarize computing the vertex and edge sets, respectively.

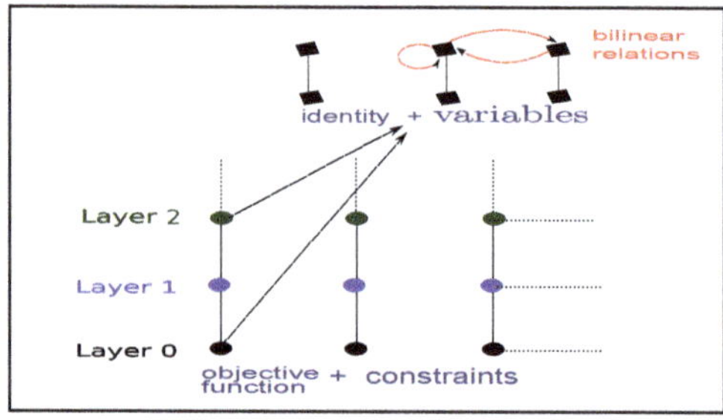

Figure 7. Binary layered graph representation of QCQP using the tensor representation A_{QCQP}.

Algorithm 1 Algorithm constructing the vertex set

```
 1: procedure G=(V, E)
 2:     V ← ∅, V_s ⊂ V ← ∅ ∀ s, E ← ∅
 3:     L ∈ ℤ, L = ⌈log₂(|U| + 1)⌉ + 2                    ▷ Define L ∈ ℤ⁺ the number of layers
 4:     for s = 0 → L − 3 do                             ▷ Partition of vertices representing the constraints
 5:         for k = 0 → m do
 6:             V_s ← V_s ∪ {v_k^(s)}                     ▷ Copies of vertices representing the constraints
 7:         end for
 8:         V ← V ∪ V_s
 9:     end for
10:     for s = L − 2, L − 1 do                          ▷ Partition of vertices representing the variables
11:         for i = 0 → n do
12:             V_s ← V_s ∪ {v_i^(s)}                     ▷ Copies of vertices representing the variables
13:         end for
14:         V ← V ∪ V_s
15:     end for
16:     return G
17: end procedure
```

Algorithm 2 Algorithm constructing the edge set

```
 1: procedure G=(V, E)
 2:     V ← V
 3:     E ← ∅
 4:     for s = 0 → L − 4 do                                 ▷ Vertical edges between copies of vertices
 5:         for k = 0 … m do                                 ▷ Copies of vertices representing the constraints
 6:             E = E ∪ E ∪ (v_k^(s), v_k^(s+1))
 7:         end for
 8:     end for
 9:                                                          ▷ Vertical edges between copies of vertices
10:     for j = 0 … n do                                     ▷ Copies of vertices representing the variables
11:         E = E ∪ (v_j^(L−2), v_j^(L−1))
12:     end for
13:     for F = 0, … N − 1 do                                ▷ Bilinear terms
14:         if IM(F) = JM(F) then                            ▷ Add a loop
15:             E = E ∪ (v_{IM(F)}^(L−1), v_{IM(F)}^(L−1))
16:         else
17:             if IM(F) < JM(F) then                        ▷ Add an edge
18:                 E = E ∪ (v_{IM(F)}^(L−1), v_{JM(F)}^(L−1))
19:             else
20:                 E = E ∪ ∅
21:             end if
22:         end if
23:     end for
24:     for F = 0 … N − 1 do
25:         for k = 0 … m do
26:             if KM(F) = k then
27:                 if IM(F) = 0 then
28:                     E = E ∪ (v_k^(s), v_{JM(F)}^(L−2))
29:                 else
30:                     E = E ∪ (v_{IM(F)}^(L−1), v_{JM(F)}^(L−1))
31:                 end if
32:             end if
33:         end for
34:     end for
35:     return G
36: end procedure
```

---

4. Numerical Discussion and Comparison to the State-of-the-Art

The following example incorporates the algorithms proposed in this paper. We construct the binary labeled graph and then enter it into nauty through the dreadnaut command line interface:

$$\max_{x_1,x_2,x_3,x_4 \in [0,1]} 3x_1 + 3x_4 + 2x_2x_3 \qquad (c_0),$$

$$x_2 + x_1{}^2 + 1 \leq 0 \qquad (c_1),$$
$$x_3 + x_4^2 + 1 \leq 0 \qquad (c_2),$$
$$x_2 + x_3 + 1 \leq 0 \qquad (c_3).$$

The optimization problem has sparse matrix representation: $M = (3,3,2,1,1,1,1,1,1,1,1,1)$, $I = (0,0,2,0,0,1,0,0,4,0,0,0)$, $J = (1,4,3,0,2,1,0,3,4,0,2,3)$, $K = (0,0,0,1,1,1,2,2,2,3,3,3)$, vector of unique elements $U = [1,2,3]$, and $L = \lceil \log_2 4 \rceil = 2$ layers.

Equation (2) computes the binary representation of each unique element, e.g., $3 = 2^1 + 2^0$ indicates that there is an edge between vertices on layer zero and another edge between the same vertices on layer 1. The graph consists of four layers and $|V| = 18$, one associated with a constant element and one with the objective function and the rest for the problem variables and constraints. The left-hand side of Figure 8 illustrates the graph representation. Nauty generates permutations: $\pi = (1,2)(5,6)(9,12)(10,11)(14,17)(15,16)$. To see how these Nauty-generated permutations usefully explain the symmetry properties of the entire problem, observe: (i) Permutations $(1,2)(5,6)$, as shown in Figure 8, permute the constraints c_1, c_2 and (ii) Permutations $(9,12)(10,11)$ are associated with the variables x_1, x_4 and x_2, x_3 with $(14,17)(15,16)$ their copies. These permutations therefore allow us to automatically calculate the formulation group $\mathcal{G} = (x_1x_4)(x_2x_3)$.

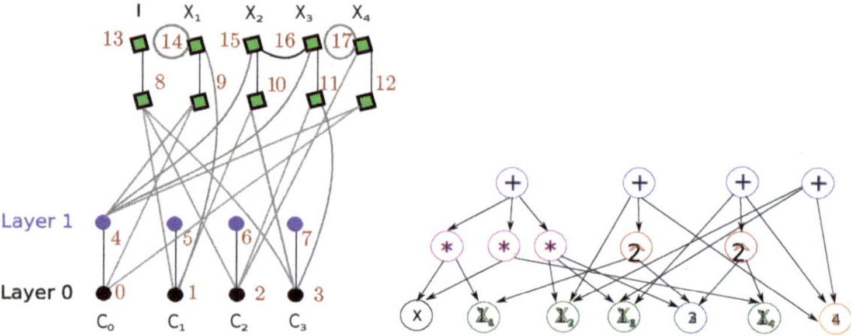

Figure 8. Illustration of the example problem using the binary labeled graph representation (**left**) and the directed acyclic graph representation (**right**).

The right-hand side of Figure 8 uses Section 2.1 to develop a directed acyclic graph representation for the same problem. The graph colors represent the vertex partitioning that enables node exchanges. In this case, the directed acyclic graph representation uses a smaller number of vertices and edges than the tensor-based representation. However, the representations generate the same formulation group.

Comparison. To evaluate the trade-offs between the tensor and directed acyclic graph representations, observe that both methods will search for the same formulation group symmetries. However, the tensor representation may be especially useful when working with problems with many differently-valued coefficients, i.e., the logarithmic number of layers may reduce the number of nodes. The function assigning integer values to the problem coefficients lets us work not only with 0–1 coefficients, but also with any other value.

5. Exploiting Symmetry in the Point Packing Problem

Once symmetry has been detected, we can use our knowledge of the symmetry to mitigate the computational difficulties caused by symmetry. Here, we focus on solving the point packing problem [46–49]. The point packing problem concerns packing n points to a unit square. The aim is to maximize the in-between distance between any two points:

$$\max \quad \theta$$

$$\text{subject to} \quad (x_i - x_j)^2 + (y_i - y_j)^2 \geq \theta \quad 1 \leq i < j \leq n,$$

$$0 \leq x_i \leq 1, 0 \leq y_i \leq 1 \quad\quad 1 \leq i \leq n,$$

where θ denotes the minimum distance between any two points. To approximate this problem, consider a grid approach that approximates the optimal solution by adding grid lines to the unit square and forcing the points to be placed only to the vertices generated by these grid lines. Note that the point packing problem has significant applications, e.g., in placing mobile phone towers.

For n points, there will be at most a $k* \times k*$ grid, where $k*$ is the smallest number whose square is the least integer that is greater than n, i.e., $(k* - 1)^2 < n$ and $k*^2 \geq n$. In other words, we add at most $2k*$ grid lines. On this $k* \times k*$ grid, points will occupy most of the vertexes and the unit length of spacing has been maximized. This approach, unfortunately, has the potential of missing the optimal solution. Consider fitting six points to the square. The most fitting grid is 3×3 and the optimal θ we achieve is 0.5. However, the optimal solution of 0.6009 is achieved by the arrangement shown in Figure 9, which is not available on a 3×3 grid. Although an approximation, $k*$ is still a useful pruning tool, e.g., points need to be at least $\frac{1}{k*-1}$ away from any other point.

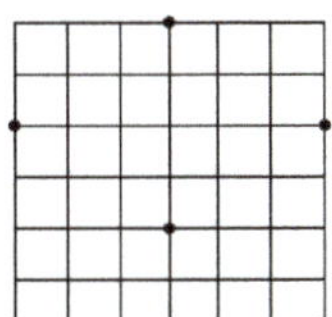

Figure 9. Optimal arrangement of six points.

Exhaustive Search and 2D Symmetry Removal

First, consider Algorithm 3, an exhaustive search method. To break the symmetries, start by calculating how many grids can possibly have points and the upper limit on the number of points on any occupied grid lines. Once calculated, the complete set of all possible combinations on the x-axis is calculated. For example, on a 5×5 grid, one possible x-coordinate setup is $(2, 0, 1, 0, 2)$. If we label the five horizontal grid lines from 1 to 5, this setup means that there are two points on grid 1, no points on grid 2, one point on grid 3, no points on grid 4 and two points on grid 5. To remove x-axis symmetries, we consider reflections as a duplication, i.e., only one of $(1, 0, 2, 0, 2)$ and $(2, 0, 2, 0, 1)$ are considered.

Algorithm 3 Algorithm 1—Exhaustive Search

Input: number of points n, number of grids k, number of occupied grids m.
Output: The optimal solution d^*

1: **Step1** Calculate the upper bound of number of points on each occupied grid u
2: **Step2** Generate complete set of combinations of x-coordinate
3: **Step3** Symmetries removal on x-coordinates
4: **Step4** Generate all y-coordinate sets based on binomial coefficient
5: **Step5** Y-coordinates pruning

Algorithm 3 considers the unique x-axis combinations. The y-axis, i.e., the horizontal grid lines, should preserve the same characteristics. Thus, in the improved Algorithm 4, we generate the set O of possible point arrangement such as $(2, 0, 1, 0, 2)$, and this is applied to both horizontal and vertical grid lines; in other words, we now consider the product $O \times O$ to give the exact coordinates of all n points.

Algorithm 4 Algorithm 2—2D Symmetry Removal

    Input: number of points n, number of grids k, number of occupied grids m.
    Output: The optimal solution d^*

1: **Step1** Calculate the upper bound of number of points on each occupied grid u
2: **Step2** Generate complete set of combinations of x-coordinate
3: **Step3** Symmetries removal on x-coordinates
4: **Step4** Use the x-coordinate set to determine y-coordinate
5: **Step5** Y-coordinates pruning

Example 1. *Consider for five points on a 5×5 grid where we have O as $\{O_1 = (1, 0, 2, 0, 2), O_2 = (1, 2, 0, 0, 2), O_3 = (1, 0, 0, 2, 2), O_4 = (2, 1, 0, 0, 2), O_5 = (2, 0, 1, 0, 2)\}$. The finalized set of full coordinates would contain 25 elements where the product of O with itself is taken. One particular point setup generated by this approach is for example, $O_3 \times O_4$. If we call the vertical grids as V_1 to V_5, respectively, and the horizontal grids as H_1 to H_5, respectively, $O_3 \times O_4$ would mean that we have the following point locations:*

1. *One point on V_1, 2 points on V_4 and 2 points on V_5. This is from O_3.*
2. *Two points on H_1, 1 point on H_2 and 2 points on H_5. This is from O_4.*

For $O_3 \times O_4$, we would have five complete point setups, as shown in Figure 10. In the last setup, we have also added the labels for grids to match what we defined earlier. This diagram contains all possible arrangements of points under this particular orbit partitioning.

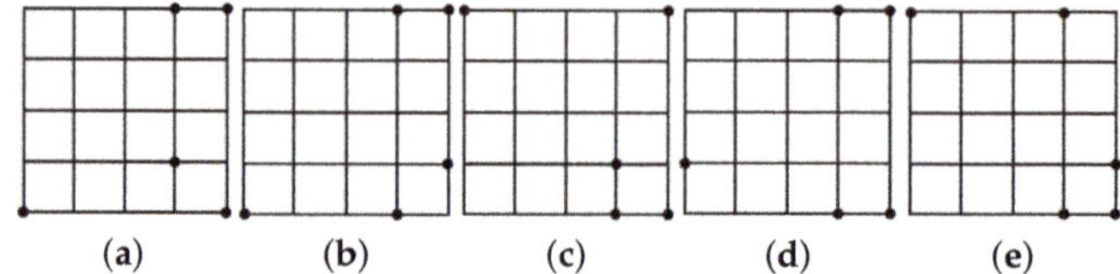

Figure 10. Five point setups for $O_3 \times O_4$. (**a**) point setup1; (**b**) point setup2; (**c**) point setup3; (**d**) point setup4; (**e**) point setup5.

To improve further, we can implement pruning mechanisms such as using the bound $\frac{1}{k^*-1}$ to prune the non-optimal setups. In this particular example, we would have pruned all five setups as none of them satisfy the 0.5 bound from the most fitting grid. This means that $O_3 \times O_4$ is not the optimal point setup for five points on a 5×5 grid.

6. Results and Comparisons

Both algorithms have been implemented and run on the same devices (HP EliteDesk 800 G2 TWR Intel Core i7-6700 3.4 GHz) to provide effective comparisons. Figure 11 shows the experimental results of 2D Symmetry Removal for packing 6 points. We eliminated the result from Exhaustive Search as it is clear that 2D Symmetry Removal outperforms Exhaustive Search in terms of run-time. We notice that m, the number of occupied grids (in the diagram, this corresponds to the occupied vertical grids), has an impact on the run-time. To a reasonable extent, the larger the m, the more choices we have regarding where we place the points so it in general takes longer time to compute. Although our strategies effectively convert the point packing QCQP into a mixed-integer linear optimization problems, we could have alternatively designed a branch-and-bound algorithm that is symmetry aware.

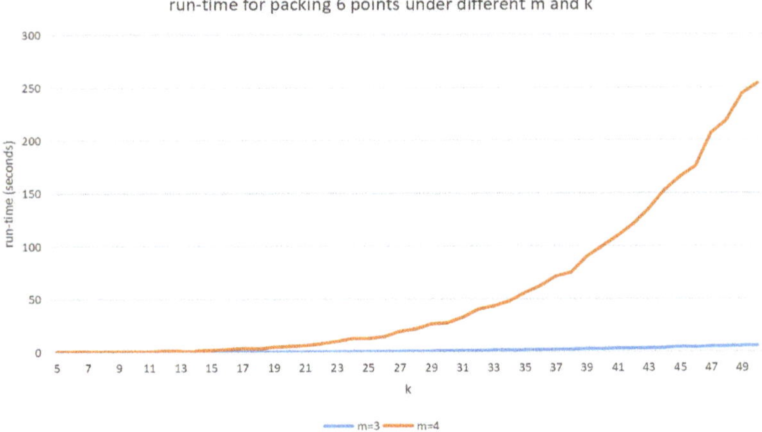

Figure 11. Run-time for six points on different grids with different number of occupied grids. The line $m = 4$ is above the line $m = 3$.

7. Conclusions

This paper has explored alternative representations for finding symmetry in formulation groups of a quadratically-constrained optimization problem. We also show that, after knowing the symmetry, we can design significantly better methods to solve the optimization problems. The contributions in this paper are relevant to industrial problems that contain a point packing element [50–52].

Author Contributions: Conceptualization, G.K. and R.M.; methodology, G.K.; validation, X.W.; writing, G.K., X.W., and R.M.; supervision, R.M.

Funding: This work was funded by an Engineering & Physical Sciences Research Council (ESPRC) Research Fellowship to R.M. (Grant No. EP/P016871/1) and an EPSRC DTP to G.K.

Conflicts of Interest: The authors declare no conflict of interest. The funders had no role in the design of the study; in the collection, analyses, or interpretation of data; in the writing of the manuscript, or in the decision to publish the results.

Abbreviations

The following abbreviations are used in this manuscript:

MINLP   Mixed-integer nonlinear optimization
QCQP   Quadratically-constrained quadratic optimization

References

1. Margot, F. Symmetry in Integer Linear Programming. In *50 Years of Integer Programming 1958–2008: From the Early Years to the State-of-the-Art*; Springer: Berlin/Heidelberg, Germany, 2010; pp. 647–686.
2. Liberti, L. Reformulations in mathematical programming: Automatic symmetry detection and exploitation. *Math. Program.* **2012**, *131*, 273–304. [CrossRef]
3. Costa, A.; Hansen, P.; Liberti, L. On the impact of symmetry-breaking constraints on spatial Branch-and-Bound for circle packing in a square. *Discret. Appl. Math.* **2013**, *161*, 96–106. [CrossRef]
4. Puget, J.F. Automatic Detection of Variable and Value Symmetries. In Proceedings of the 11th International Conference on Principles and Practice of Constraint Programming—CP 2005, Sitges, Spain, 1–5 October 2005; van Beek, P., Ed.; Springer: Berlin/Heidelberg, Germany, 2005; pp. 475–489.
5. Salvagnin, D. A Dominance Procedure for Integer Programming. Master's Thesis, University of Padua, Padua, Italy, 2005.

6.   Berthold, T.; Pfetsch, M. Detecting Orbitopal Symmetries. In Proceedings of the Annual International Conference of the German Operations Research Society (GOR), Augsburg, Germany, 3–5 September 2008; Springer: Berlin/Heidelberg, Germany, 2009; pp. 433–438.

7.   Bremner, D.; Dutour Sikirić, M.; Pasechnik, D.V.; Rehn, T.; Schürmann, A. Computing symmetry groups of polyhedra. *LMS J. Comput. Math.* **2014**, *17*, 565–581. [CrossRef]

8.   Knueven, B.; Ostrowski, J.; Pokutta, S. Detecting almost symmetries of graphs. *Math. Program. Comput.* **2018**, *10*, 143–185. [CrossRef]

9.   Sherali, H.D.; Smith., J.C. Improving Discrete Model Representations via Symmetry Considerations. *Manag. Sci.* **2001**, *47*, 1396–1407. [CrossRef]

10. Liberti, L. Automatic Generation of Symmetry-Breaking Constraints. In *Combinatorial Optimization and Applications*; Yang, B., Du, D.Z., Wang, C.A., Eds.; Springer: Berlin/Heidelberg, Germany, 2008; pp. 328–338.

11. Liberti, L.; Ostrowski, J. Stabilizer-based symmetry breaking constraints for mathematical programs. *J. Glob. Optim.* **2014**, *60*, 183–194. [CrossRef]

12. Ghoniem, A.; Sherali, H.D. Defeating symmetry in combinatorial optimization via objective perturbations and hierarchical constraints. *IIE Trans.* **2011**, *43*, 575–588. [CrossRef]

13. Ostrowski, J.; Linderoth, J.; Rossi, F.; Smriglio, S. Constraint Orbital Branching. In Proceedings of the 13th International Conference on Integer Programming and Combinatorial Optimization IPCO, Bertinoro, Italy, 26–28 May 2008; pp. 225–239.

14. Ostrowski, J.; Linderoth, J.; Rossi, F.; Smiriglio, S. Orbital branching. *Math. Program.* **2011**, *126*, 147–178. [CrossRef]

15. Margot, F. Pruning by isomorphism in branch-and-cut. *Math. Program.* **2002**, *94*, 71–90. [CrossRef]

16. Kaibel, V.; Peinhardt, M.; Pfetsch, M.E. Orbitopal fixing. *Discret. Optim.* **2011**, *8*, 595–610. [CrossRef]

17. Faenza, Y.; Kaibel, V. Extended Formulations for Packing and Partitioning Orbitopes. *Math. Oper. Res.* **2009**, *34*, 686–697. [CrossRef]

18. Pfetsch, M.E.; Rehn, T. A computational comparison of symmetry handling methods for mixed integer programs. *Math. Program. Comput.* **2019**, *11*, 37–93. [CrossRef]

19. Margot, F. Small covering designs by branch-and-cut. *Math. Program.* **2003**, *94*, 207–220. [CrossRef]

20. Costa, A.; Liberti, L.; Hansen, P. Formulation symmetries in circle packing. *Electron. Notes Discret. Math.* **2010**, *36*, 1303–1310. [CrossRef]

21. Ostrowski, J.; Vannelli, A.; Anjos, M.F. *Symmetry in Scheduling Problems*; Cahier du GERAD G-2010-69; GERAD: Montreal, QC, Canada, 2010.

22. Ostrowski, J.; Wang, J.; Liu, C. Exploiting Symmetry in Transmission Lines for Transmission Switching. *IEEE Trans. Power Syst.* **2012**, *27*, 1708–1709. [CrossRef]

23. Ostrowski, J.; Wang, J. Network reduction in the Transmission-Constrained Unit Commitment problem. *Comput. Ind. Eng.* **2012**, *63*, 702–707. [CrossRef]

24. Ostrowski, J.; Anjos, M.F.; Vannelli, A. Modified orbital branching for structured symmetry with an application to unit commitment. *Math. Program.* **2015**, *150*, 99–129. [CrossRef]

25. Lima, R.M.; Novais, A.Q. Symmetry breaking in MILP formulations for Unit Commitment problems. *Comput. Chem. Eng.* **2016**, *85*, 162–176. [CrossRef]

26. Knueven, B.; Ostrowski, J.; Wang, J. The Ramping Polytope and Cut Generation for the Unit Commitment Problem. *INFORMS J. Comput.* **2018**, *30*, 739–749. [CrossRef]

27. Kouyialis, G.; Misener, R. Detecting Symmetry in Designing Heat Exchanger Networks. In Proceedings of the International Conference of Foundations of Computer-Aided Process Operations-FOCAPO/CPC, Tucson, AZ, USA, 8–12 January 2017.

28. Letsios, D.; Kouyialis, G.; Misener, R. Heuristics with performance guarantees for the minimum number of matches problem in heat recovery network design. *Comput. Chem. Eng.* **2018**, *113*, 57–85. [CrossRef]

29. Maravelias, C.T.; Grossmann, I.E. A hybrid MILP/CP decomposition approach for the continuous time scheduling of multipurpose batch plants. *Comput. Chem. Eng.* **2004**, *28*, 1921–1949. [CrossRef]

30. Maravelias, C.T. Mixed-Time Representation for State-Task Network Models. *Ind. Eng. Chem. Res.* **2005**, *44*, 9129–9145. [CrossRef]

31. Mistry, M.; Callia D'Iddio, A.; Huth, M.; Misener, R. Satisfiability modulo theories for process systems engineering. *Comput. Chem. Eng.* **2018**, *113*, 98–114. [CrossRef]

32. Smith, E.M.B.; Pantelides, C.C. A symbolic reformulation/spatial branch-and-bound algorithm for the global optimisation of nonconvex MINLPs. *Comput. Chem. Eng.* **1999**, *23*, 457–478. [CrossRef]

33. Tawarmalani, M.; Sahinidis, N.V. A polyhedral branch-and-cut approach to global optimization. *Math. Program.* **2005**, *103*, 225–249. [CrossRef]

34. Belotti, P.; Lee, J.; Liberti, L.; Margot, F.; Wachter, A. Branching and Bounds Tightening techniques for Non-convex MINLP. *Optim. Methods Softw.* **2009**, *24*, 597–634. [CrossRef]

35. Youdong, L.; Linus, S. The global solver in the LINDO API. *Optim. Methods Softw.* **2009**, *24*, 657–668.

36. Misener, R.; Floudas, C.A. ANTIGONE: Algorithms for coNTinuous/Integer Global Optimization of Nonlinear Equations. *J. Glob. Optim.* **2014**, *59*, 503–526. [CrossRef]

37. Vigerske, S. Decomposition in Multistage Stochastic Programming and a Constraint Integer Programming Approach to Mixed-Integer Nonlinear Programming. Ph.D. Thesis, Humboldt-Universität zu Berlin, Berlin, Germany, 2013.

38. Mahajan, A.; Leyffer, S.; Linderoth, J.; Luedtke, J.; Munson, T. Minotaur: A mixed-integer nonlinear optimization toolkit. *Optim. Online* **2017**, *6275*.

39. Boukouvala, F.; Misener, R.; Floudas, C.A. Global optimization advances in Mixed-Integer Nonlinear Programming, MINLP, and Constrained Derivative-Free Optimization, CDFO. *Eur. J. Oper. Res.* **2016**, *252*, 701–727. [CrossRef]

40. Fourer, R.; Maheshwari, C.; Neumaier, A.; Orban, D.; Schichl, H. Convexity and concavity detection in computational graphs: Tree walks for convexity assessment. *INFORMS J. Comput.* **2010**, *22*, 26–43. [CrossRef]

41. Hart, W.E.; Laird, C.; Watson, J.; Woodruff, D.L. Pyomo: Modeling and solving mathematical programs in python. *Math. Program. Comput.* **2011**, *3*, 219–260. [CrossRef]

42. Ceccon, F.; Siirola, J.D.; Misener, R. SUSPECT: MINLP special structure detector for Pyomo. *Optim. Lett.* **2019**. [CrossRef]

43. McKay, B.D.; Piperno, A. Practical graph isomorphism, II. *J. Symb. Comput.* **2014**, *60*, 94–112. [CrossRef]

44. Ramani, A.; Aloul, F.; Markov, I.; Sakallah, K.A. Breaking instance-independent symmetries in exact graph coloring. *J. Artif. Intell. Res.* **2006**, *1*, 324–329. [CrossRef]

45. Ramani, A.; Markov, I.L. Automatically Exploiting Symmetries in Constraint Programming. In *Recent Advances in Constraints*; Faltings, B.V., Petcu, A., Fages, F., Rossi, F., Eds.; Springer: Berlin/Heidelberg, Germany, 2005; pp. 98–112.

46. Anstreicher, K.M. Semidefinite programming versus the reformulationlinearization technique for nonconvex quadratically constrained quadratic programming. *J. Glob. Optim.* **2009**, *43*, 471–484. [CrossRef]

47. Misener, R.; Floudas, C.A. Global optimization of mixed-integer quadratically-constrained quadratic programs (MIQCQP) through piecewise-linear and edge-concave relaxations. *Math. Program.* **2012**, *136*, 155–182. [CrossRef]

48. Misener, R.; Floudas, C.A. GloMIQO: Global mixed-integer quadratic optimizer. *J. Glob. Optim.* **2013**, *57*, 3–50. [CrossRef]

49. Furini, F.; Traversi, E.; Belotti, P.; Frangioni, A.; Gleixner, A.; Gould, N.; Liberti, L.; Lodi, A.; Misener, R.; Mittelmann, H. QPLIB: A library of quadratic programming instances. *Math. Program. Comput.* **2019**, *11*, 237–265. [CrossRef]

50. Jones, D.R. A fully general, exact algorithm for nesting irregular shapes. *J. Glob. Optim.* **2014**, *59*, 367–404. [CrossRef]

51. Misener, R.; Smadbeck, J.B.; Floudas, C.A. Dynamically generated cutting planes for mixed-integer quadratically constrained quadratic programs and their incorporation into GloMIQO 2. *Optim. Methods Softw.* **2015**, *30*, 215–249. [CrossRef]

52. Wang, A.; Hanselman, C.L.; Gounaris, C.E. A customized branch-and-bound approach for irregular shape nesting. *J. Glob. Optim.* **2018**, *71*, 935–955. [CrossRef]

 *processes*

Article

Modeling and Analysis of Coal-Based Lurgi Gasification for LNG and Methanol Coproduction Process

Jingfang Gu [1], Siyu Yang [1,*] and Antonis Kokossis [2,*]

[1] School of Chemical Engineering, South China University of Technology, Guangzhou 510641, China; ceswan@mail.scut.edu.cn

[2] School of Chemical Engineering, National Technical University of Athens, GR-15780 Athens, Greece

* Correspondence: cesyyang@scut.edu.cn (S.Y.); akokossis@mail.ntua.gr (A.K.); Tel.: +86-20-8711-2056 (S.Y.); +30-2-10-772-4275 (A.K.)

Received: 23 May 2019; Accepted: 30 September 2019; Published: 2 October 2019

Abstract: A coal-based coproduction process of liquefied natural gas (LNG) and methanol (CTLNG-M) is developed and key units are simulated in this paper. The goal is to find improvements of the low-earning coal to synthesis natural gas (CTSNG) process using the same raw material but producing a low-margin, single synthesis natural gas (SNG) product. In the CTLNG-M process, there are two innovative aspects. Firstly, the process can co-generate high value-added products of LNG and methanol, in which CH_4 is separated from the syngas to obtain liquefied natural gas (LNG) through a cryogenic separation unit, while the remaining lean-methane syngas is then used for methanol synthesis. Secondly, CO_2 separated from the acid gas removal unit is partially reused for methanol synthesis reaction, which consequently increases the carbon element utilization efficiency and reduces the CO_2 emission. In this paper, the process is designed with the output products of 642,000 tons/a LNG and 1,367,800 tons/a methanol. The simulation results show that the CTLNG-M process can obtain a carbon utilization efficiency of 39.6%, bringing about a reduction of CO_2 emission by 130,000 tons/a compared to the CTSNG process. However, the energy consumption of the new process is increased by 9.3% after detailed analysis of energy consumption. The results indicate that although electricity consumption is higher than that of the conventional CTSNG process, the new CTLNG-M process is still economically feasible. In terms of the economic benefits, the investment is remarkably decreased by 17.8% and an increase in internal rate of return (IRR) by 6% is also achieved, contrasting to the standalone CTSNG process. It is; therefore, considered as a feasible scheme for the efficient utilization of coal by Lurgi gasification technology and production planning for existing CTSNG plants.

Keywords: coproduction; Lurgi syngas; cryogenic separation; methanol synthesis; LNG

---

1. Introduction

China is the main source of global energy growth as well as the largest energy consumer in the past 20 years [1]. In 2016, China's natural gas production was 148.7 billion Nm^3, with a yearly rate increase of 8.5%. Meanwhile, the total imports are 92 billion m^3, with an annual growth rate of 27.6%. However, the total yearly gas consumption is 237.3 billion m^3, which is 15.3% higher than that of 2015 [2]. If keeping with the same growth rates, the natural gas will be in insufficient supply in the near future. To alleviate such an energy shortage, the Chinese government encourages the build and operate (B&O) development of coal to conduct synthetic natural gas (CTSNG) projects. Therefore, many CTSNG projects have been launched recently and are being run successfully all over the country, as shown in Table 1.

Table 1. B&O CTSNG projects in China (2018).

Project	Status	Company	Capacity (10^9 Nm^3/a)	Location
Keqi Coal-based Gas Project Phase I	Operating	Datang International Power Generation Company	1.33	Chifeng, Inner Mongolia
Xinjiang Kingho SNG Project Phase I	Operating	Xinjiang Kingho Energy Group Co., Ltd.	1.38	Yili, Xinjiang
Huineng Ordos SNG projects Phase I	Operating	Inner Mongolia Huineng Coal Chemical Industry Co., Ltd.	4	Ordos, Inner Mongolia
Yili Xintian SNG Project	Operating	Yili Xintian Coal Chemical Co., Ltd.	2	Yili, Xinjiang
Keqi Coal-based Gas Project Phase II/III	Building	Datang International Power Generation Company	2.67	Chifeng, Inner Mongolia
Datang Fuxin SNG Project	Building	Datang International Power Generation Company	4	Fuxin, Liaoning
Xinjiang Kingho's SNG Project Phase II	Building	Xinjiang Kingho Energy Group Co., Ltd.	4.13	Yili, Xinjiang
Huineng Ordos SNG Projects Phase II	Building	Inner Mongolia Huineng Coal Chemical Industry Co., Ltd.	1.6	Ordos, Inner Mongolia
Xinjiang Zhundong SNG Demonstration Project	Building	Suxin Energy Hefeng Co., Ltd.	4	Changji, Xinjiang
Beijing Enterprises JT Ordos SNG Project	Building	Inner Mongolia Beijing Enterprises JT Energy Development Co., Ltd.	4	Ordos, Inner Mongolia

However, CTSNG projects face some challenges. Firstly, the market price of synthetic natural gas (SNG) products is not based on its cost structure, nor according to the guidance from a market mechanism. Departments of China offer a rather lower price to the public, and priority is given to civil use, transportation field, etc. Thus, lower economic returns are common to all CTSNG projects. The price of natural gas has fallen sharply since November 2015 in the country after the National Development and Reform Commission issued a report about price adjustment [2]. In Figure 1, it shows that, during 2015 to 2018, the price of SNG is reduced from 2.75 to 1.82 CNY/Nm^3. After 2018, the price further decreased to 1.78 CNY/Nm^3. In such case, the market price of SNG is 0.97 CNY/ Nm^3 lower than that of 2015. Taking Keqi Coal-Based Gas Project Phase I for an example, in 2017, it produced 1.03 billion Nm^3 SNG products, yet with a big deficit of 650 million CNY. It can be seen that if natural gas price keeps fluctuating at a low-price level, these CTSNG projects are likely to face severe losses.

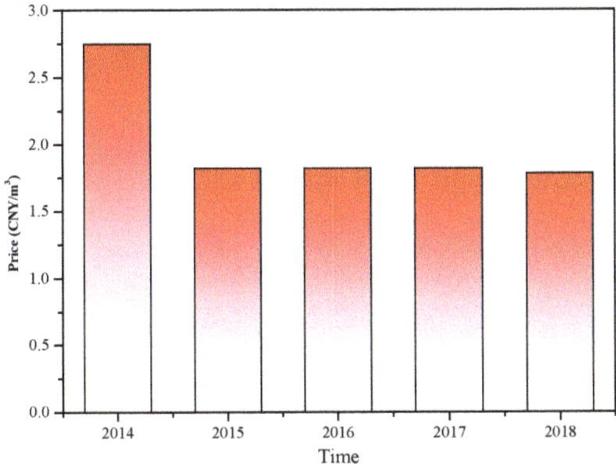

Figure 1. China's synthesis natural gas (SNG) market price recording (2014–2018).

Secondly, the SNG product is mainly supplied for civil use like urban heating in the winter. Thus, there is a peak–valley difference of natural gas demand between winter and summer. In most of the northern cities of China, the top demand of natural gas in the heating period is from November to March of the following year. In contrast, the demand in the non-heating period remains at a lower level from April to October. According to statistics from Ji [3], the consumption in the heating period is up to 10 times of that in the non-heating period. Taking the winter of 2018 as an example, the gap between supply and demand of natural gas is about 24 billion m^3. Since natural gas cannot be stored for a long time, coal-based gas projects are facing production cuts during the non-heating period, which brings huge economic losses.

Further, in many CTSNG projects, a Lurgi gasification technology has been employed to produce qualitative coal-based syngas for synthesis reaction. Major units in the process can be seen in Figure 2, including coal gasification, water–gas shift, acid gas removal, and methanation synthesis units [4,5].

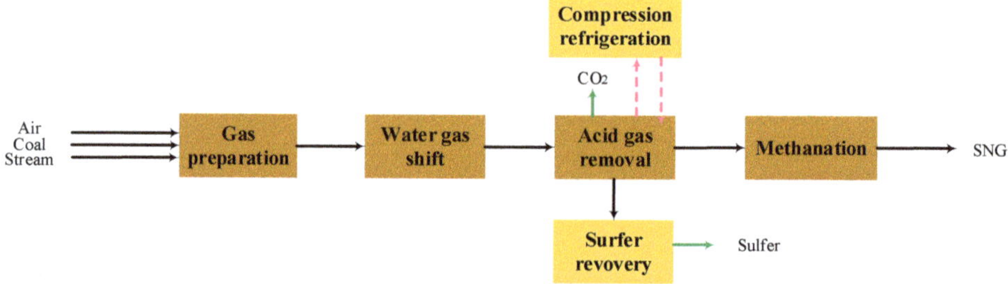

Figure 2. Flowchart of the CTSNG process.

However, the hydrogen:carbon ratio of crude syngas from Lurgi gasifier is about 2.7 [6]. According to the requirement of synthesis gas reaction, it is necessary to use water–gas shift technology to increase that ratio to 3.1 for methanation. However, as the Equation (1) of water–gas shift reaction shows, CO_2 emission is inevitably increased in that process [7,8].

$$CO + H_2O(vapor) = CO_2 + H_2, \Delta H = 41.19 \text{ kJ/mol} \tag{1}$$

The coproduction process alternative is a practical way to address this challenge. As known, traditional CTSNG processes have a single gaseous coal-based natural gas product. However, it is possible that the same raw materials can be converted to various designed products under the coproduction process structure. Till now, there are different studies devoted to comprehensive processing of coal syngas. These studies prove that coproduction systems can improve the resource utilization and energy efficiencies. Some works are also being proven by demonstration projects.

Yi et al. (2017) studied the modelling and optimization theories of coproduction systems. In their studies, the coproduction process can be very flexible for its integration of technologies and raw material distribution. Besides, it is pointed out that systematic design can improve the process performance like better carbon conversion ratio and improved energy saving size [9–11].

Hao et al. (2015) proposed a coproduction process of methanol and electricity with coal and coke oven gas as raw materials. The new system is compared with the process based on CH_4/CO_2 dry reforming technology, in terms of exergy efficiency, exergy cost, and CO_2 emissions. Through the new system, the exergy efficiency can be increased by 7.8%. Besides, the exergy cost can be reduced by 0.88 USD/GJ and the CO_2 emission can be reduced by 0.023 kg/MJ [12–14]. Han et al. (2010) introduced a methanol production and integrated gasification combined cycle power generation system using coal and natural gas as fuel. The syngas derived from natural gas and coal is firstly used for methanol synthesis. The unreacted syngas is used in the power plant as fuel. Comparing with the single production system, the coproduction system can save about 10% of fossil fuels [15]. Tu et al. (2015)

found that a methanol and electricity coproduction system can obtain the best benefit when the recycle ratio of unreacted gas is assigned with the value between 2.17 and 4.44, with relative energy saving rate and unit energy production approaching an optimum [16]. Huang et al. (2018) introduced a low-energy CO_2 capture process after the water–gas shift unit in a poly-generation process. A part of the unreacted syngas is used to generate power. Energy consumption for CO_2 capture is 0.7 GJ/t-CO_2, bringing a 40.6% reduction compared to that of the coal-to-methanol process [17].

In addition, Bai et al. (2015) studied a poly-generation system of generating methanol and power with the solar thermal gasification of the biomass. The syngas from the biomass gasification is used to produce the methanol via a synthesis reactor. The un-reacted gas is used for the power generation via a combined cycle power unit. The thermodynamic and economic performances of the system are investigated. A portion of the concentrated solar thermal energy can be chemically stored into the syngas. The highest energy efficiency of the poly-generation system is approximately 56.09%, which can achieve the stable utilization of the solar energy and the mitigation of CO_2 emission [18].

Many researchers from outside China are also interested in this field. You et al. (2011) studied the optimal distribution of raw materials in different production routes to maximize the benefit of the coproduction process. A superstructure optimization model is formulated as a mixed-integer nonlinear program to determine the optimal process design, and the proposed framework is applied to a comprehensive superstructure of an integrated shale gas for chemical processing, which involves steam cracking of ethane, propane, *n*-butane, and *i*-butane [19,20].

The above studies are mainly based on thermodynamics to reach a higher energy utilization, achieve a reduction on energy consumption, and realize the optimization of reaction conditions, like gas recycle ratio, operating temperature and pressure, etc. However, studies are less focused on matching products proposal and syngas component ratio, like $(H_2 - CO_2)/(CO + CO_2)$ ratio, which is specified for chemical synthesis.

Considering all difficulties that existing B&O CTSNG projects are facing, this paper studies a coproduction process with LNG and methanol (CTLNG-M). The CTLNG-M process is developed based on a rational distribution study on hydrogen and carbon elements in the processing, which reduce CO_2 emission by converting more carbon to chemicals and increase unit product income for a high valued liquefied natural gas (LNG) product. Section 2 gives the description of the new process on what measures have been taken. Section 3 gives the detailed modeling and simulation with respect to key parameters of added units in the CTLNG-M process. In Section 4, a discussion about the carbon utilization efficiency, energy efficiency, energy consumption, and economic performance of the CTLNG-M process is given.

2. LNG and Methanol Coproduction Process

The syngas from a Lurgi gasifier contains 12% to 18% methane [21]. Because of this high composition of methane, Lurgi gasification technology is usually used in CTSNG projects [22,23]. However, from another point of view, LNG products can also be obtained by separating methane from the syngas through an added cryogenic separation technology. LNG is a relatively high value-added product form of coal-based natural gas, whose price can reflect the supply and demand mechanism. In Figure 3, it can be seen that LNG prices have shown an upward trend from 3206 CNY/tons to 5373 CNY/tons in heating period time (January in each year) since 2017, with an average price of 3122 CNY/tons and a highest price up to 5613 CNY/tons. In addition, LNG can be transported and stored in a more flexible way [24].

Thus, use of a separation unit to remove methane from syngas out of the Lurgi gasifier is taken into consideration, as the remaining syngas can be used for methanol synthesis. In this paper, two units are added, the cryogenic separation unit is placed before a methanol synthesis unit. Thereby, the content of effective gas in methanol synthesis reaction is increased, and the production efficiency also improved.

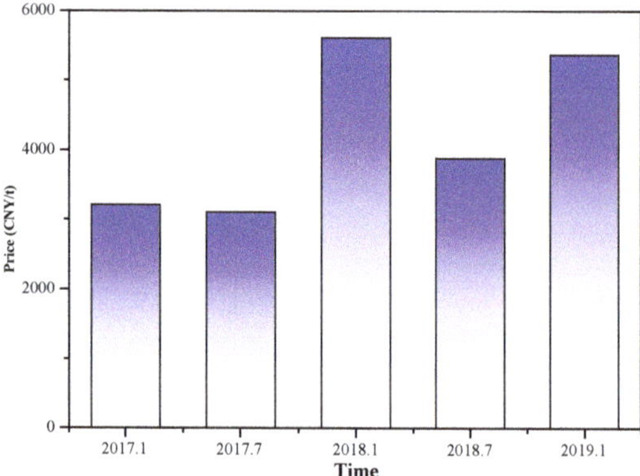

Figure 3. China's liquefied natural gas (LNG) market price recording (January 2017–January 2019).

This paper proposes a coproduction process for matching the product distribution of Lurgi gasification technology. The CTLNG-M process is highlighted through a schematic diagram as shown in Figure 4. In this process, the syngas from the Lurgi gasifier is separated to get LNG product by cryogenic separation unit. The remaining syngas has the H:C ratio close to 2.4. An additional carbon source is needed to decrease that ratio near to 2.1 before methanol synthesis. In that case, the additional carbon resource can be provided by CO_2 extracted from the acid gas removal unit. Thereby, the carbon emission of the system is also reduced.

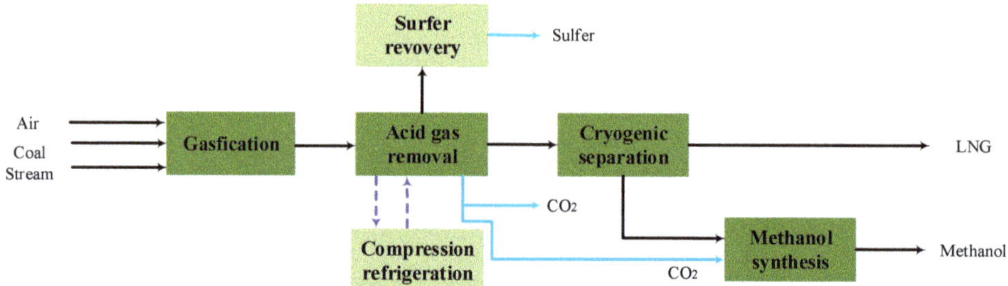

Figure 4. Flowchart of coal to liquefied natural gas (LNG) and methanol (CTLNG-M) coproduction process.

There are different gases that are present as impurities in crude syngas. Amongst them, sulfides can cause deactivation of methanol synthesis catalysts, while carbon dioxide can reduce the conversion of methanol synthesis. These impurities, as well as tar, phenol, and ammonia, can be removed by acid gas removal unit [10]. After that and heat recovery, the molecular sieve process is used to further reduce the content of carbon dioxide and methanol to less than 1 ppm and then meet feed requirements of the cryogenic separation [25].

The new process employs nitrogen cycle refrigeration technology to separate the methane [26]. Nitrogen provides most of the cooling capacity through the adiabatic expansion cycle in the turbine expander. A double column cryogenic distillation process is used for separation of syngas and LNG [27], as shown in Figure 5. In the double column process, the washing column and the CH_4-CO distillation column are packed columns. The top outlet of the washing tower is syngas with methane content less than 1%. The cold energy is recovered through a heat exchanging system before sending to methanol synthesis and be used for exchange heat from input gas.

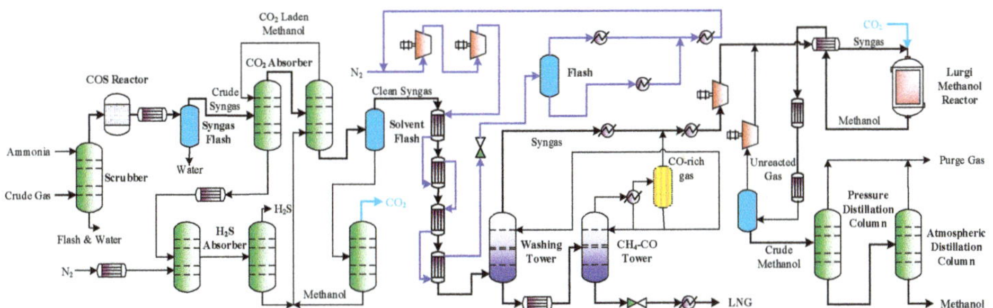

Figure 5. Process flow diagram of CTLNG-M.

The stream extracted from the bottom of the washing tower mainly consists of methane and carbon monoxide. It is sent to the CH_4-CO distillation column for methane separation.

In the CH_4-CO distillation column, the condensed liquid stream at the top of the column is partially used as the reflux, and the other part enters the washing column as recycling stream at the top of the column. The main component of the non-condensable gas at the top of this column is CO with the concentration of 70%. In the bottom of tower, a part of LNG returns to the circulation inside the tower for improving product quality with higher CH_4 content. The other part is cooled to −163 °C through the heat exchanger.

The syngas from the Lurgi gasifier reaches the standard for methanol synthesis through the use of Rectisol and the cryogenic separation unit. After compression, components in syngas react to the product methanol with copper-based catalyst. The main equations are shown as below.

$$CO + 3H_2 = CH_3OH + H_2, \ \Delta H = -90.64 \ kJ/mol. \tag{2}$$

$$CO_2 + 3H_2 = CH_3OH + H_2O, \ \Delta H = -49.47 \ kJ/mol. \tag{3}$$

3. Modeling and Simulation

As has been mentioned above, there are four main units involved in the CTLNG-M process. Namely coal gasification, acid gas removal, cryogenic separation, and methanol synthesis unit. The detailed simulation of coal gasification and the acid gas removal unit can be found in our group's previous work [28–30]. Consequently, this paper gives modelling and simulation results for two added units, as the cryogenic separation and methanol synthesis unit.

The coal quality parameters are shown in Table 2.

Table 2. Proximate and ultimate analyses of Shenmu coal. HHV: high heating value.

	Proximate Analysis (wt.%, ad)				Ultimate Analysis (wt.%, ad)					HHV, MJ kg^{-1}
	M	**FC**	**V**	**A**	**C**	**H**	**O**	**N**	**S**	
Coal	18.6	50.7	34.28	7.02	66.48	4.29	13.16	1.00	0.50	26.73

3.1. Cryogenic Separation Unit

The role of the cryogenic separation unit used in the new process is to obtain purified LNG products [31]. As illustrated in Figure 6, the clean syngas is firstly introduced to an absorption tower for H_2O and methane removal. After that, the resulted syngas (S1) is cooled into liquid phase. Most of the remaining carbon monoxide component in S2 is sent to the methanol synthesis unit. Stream S4 from the bottom of T2-W mainly consists of CO and CH_4, which needs further distillation in T3-D [32]. Stream S5 is then separated in to two parts, one is recycled to the top of T2-W, and another for methanol synthesis. Stream S8 obtained from the bottom of the T3-D is the LNG product.

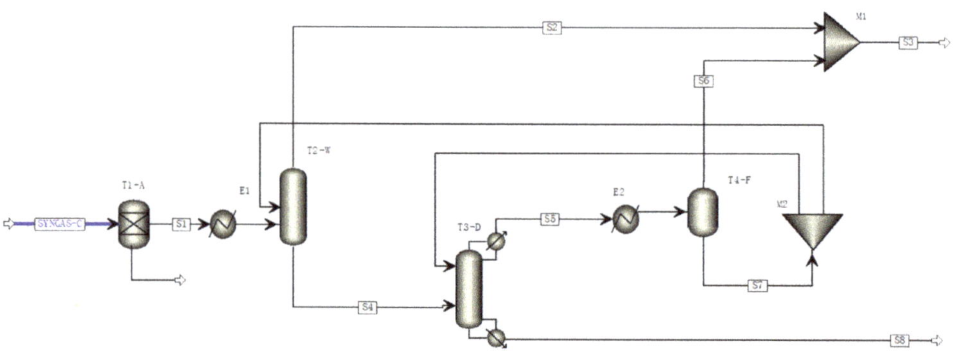

Figure 6. Flow diagram of the cryogenic separation process.

In this paper, the SRK method is used to predict the physical properties of streams. T1-A is modeled by the SEP module, T2-W and T3-F by the RadFrac module and T4-F by the Flash module. The pressure of the top of T2-W is 36 bar, and the theoretical number of tower plates is set to 30, which is twice the minimum theoretical number calculated by Aspen's succinct calculation. The feeding position is stage 14. In the T3-D, its pressure is 34 bar, and drop pressure is 7 bar. In order to maintain the quality of the LNG product, the $CH_4/(CH_4 + C_2)$ ratio of steam gas in the bottom is less than 97.5% (GB/T 19204-2003), which is controlled by adjusting the reflux ratio.

After compression, the nitrogen is liquefied. Through the throttle valve, the high-pressure liquid nitrogen is expanded to a low-pressure state. In this process, the gas absorbs heat from the environment. Therefore, this expansion process provides cooling capacity for the T2-W and the T3-D. The cryogenic system uses two-stage nitrogen circulation expansion refrigeration, which is shown in Figure 7 [33].

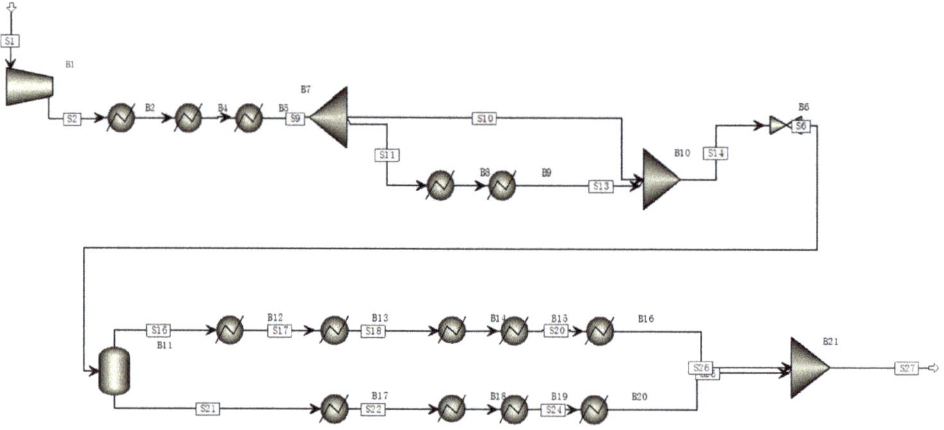

Figure 7. Flow diagram of the nitrogen circulation refrigeration process.

Simulation results are given in Table 3. Steam S4 consists of 38.8% CH_4 and 55.1% CO. Stream S4 is liquefied as the LNG product and outputted from the bottom of T3-D. In the LNG product, the purity of CH_4 can approach up to 97.6%, with the impurity content of CO and C_2 less than 1%. The separation is modeled and simulated compared with the data in the reference [34], with the error less than 5%.

Table 3. Simulation results of the cryogenic separation unit.

Stream	S1	S2	S3	S4	S5	S6	S7	S8
Temperature (°C)	40	−169	−168	−162.3	−140.7	−158	−158	−93.5
Pressure (bar)	36	36	34	35	34	34	34	34
Mole Flow (kmol/h)	28,310	23,215	23,511	8950	8000	315	7685	4790
Mole Fraction (%)								
N_2	0.3	0.3	0.3	0.3	0.7	0.6	0.7	543 ppb
AR	0.1	0.2	0.2	0.3	0.6	0.3	0.6	18 ppm
CO	23.7	28.3	28.5	38.9	85	50.3	86.4	147 ppm
CO_2	24 ppm	0	0	0	0	0	0	0
H_2	58.5	70.7	70.4	4.1	7.3	47.6	5.6	0
CH_4	17	0.6	0.6	55.1	6.4	1.2	6.6	97.6
C_2H_6	0.4	0	0	1.3	0	0	0	2.4
C_2H_4	5 ppm	0	0	15 ppm	0	0	0	27 ppm

3.2. Methanol Synthesis Unit

The flow chart of the methanol synthesis is shown in Figure 8. Methanol synthesis gas is, at first, mixed with the stream of CO_2 from the acid gas removal unit to adjust the H:C ratio. This is because the syngas is rich in hydrogen and additional carbon source for methanol synthesis is needed until that H:C ratio decreased to around 2.1 [35,36].

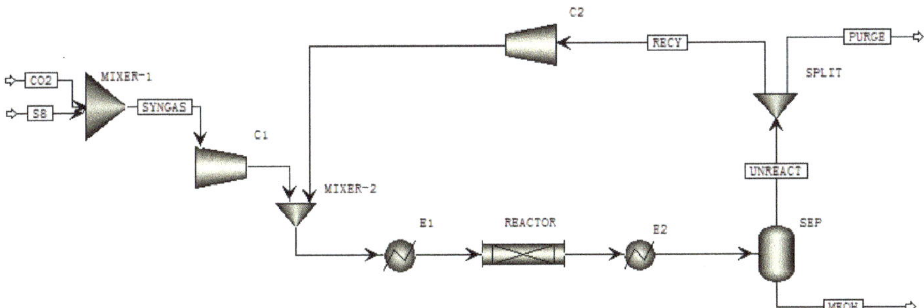

Figure 8. Flow diagram of the methanol synthesis unit.

According to the practical process in Datang Keqi project, the Lurgi low-pressure methanol synthesis method is employed. The methanol syngas enters the reactor with the recycling gas. After heat recovery of the outlet stream, the gas–liquid steam is separated by a separator to recycle the unreacted syngas. The liquid part is the input stream of the methanol rectification unit. A small part of unreacted gas is discharged as purge gas.

In this unit, the PR property method was selected in Aspen for modelling the methanol synthesis unit. The purified syngas is pressurized to 8.2 MPa by the compressor modeled by the COMP module [37]. The synthesis reactor was modeled by a plug flow model of RPlug. Table 4 shows the simulation result of the process.

Table 4. Simulation results of the methanol synthesis unit.

Stream	SYNGAS	CO$_2$	UNREACT	RECY	PURGE	MEOH
Temperature (°C)	45	30	40	40	40	63.5
Pressure (bar)	82	34	71	71	71	6
Mole Flow (kmol/h)	23,408	710	54,875	47,193	1121	6573
Mole Frac (%)						
N$_2$	0.3	2 PPM	0.9	0.9	0.9	134 ppm
AR	0.2	61 PPM	0.6	0.6	0.6	209 ppm
CO	28.6	0	22.3	22.3	22.3	0.5
CO$_2$	2.8	99	4.9	4.9	4.9	1.4
H$_2$	68.7	0	69.9	69.9	69.9	1.4
CH$_4$	0.2	0	0.6	0.6	0.6	284 ppm
CH$_3$OH	0	0	0	0	0	99.7

The methanol synthesis gas with a H:C ratio of 2.1 is compressed to 82 bar and sent to the reactor [28]. In this syngas, the CO$_2$ volume fraction is less than 3%, which is in line with industrial practice. During the process, 86% unreacted gas is recycled back to the reactor to improve the overall conversion. As a result, the CTLNG-M process produces a total of 6573 kmol/h methanol product.

The key parameters in the CTLNG-M process are listed in Table 5. The carbon element conversion ratio in the coal gasification unit is 99.9%. In the acid gas removal unit, the H$_2$S removal ratio is 99.5%, and the pressure is under 50 bar. In the cryogenic separation unit, the CH$_4$ mole fraction is 97.5%, the recycle ratio and purge ratio in the methanol synthesis unit are set as 0.86 and 0.05 separately.

Table 5. Main design parameters in the CTLNG-M simulation.

Unit	Condition	Design Parameters
	Pressure (bar)	48
Coal Gasification	Temperature (°C)	1400
	Carbon Element Conversion Ratio (%)	99.9
Acid Gas Removal	H$_2$S Removal Rate (%)	99.5
	Pressure (bar)	50
Cryogenic Separation	CH$_4$ Mole Fraction (%)	>97.5
	Absorption Series	2
	Temperature (°C)	240
Methanol Synthesis	Recycling Ratio	0.86
	Purge Ratio	0.05

3.3. Parameters Analysis

There are two important operation parameters in the new process. One is the temperature and pressure at the bottom of the gas–liquid separation tower in the cryogenic separation unit, and another one is the unreacted gas recycle ratio in the methanol synthesis unit. These two parameters are highly corelated with the composition of syngas and methanol production. These two parameters have significant impact on the composition of syngas and methanol production, which are analyzed in detail as follows.

Figure 9a,b shows the effect of operating temperature and pressure of T4-S on the composition of syngas out from the cryogenic separation unit. For analysis, we fix the recycling ratio to 0.86.

It can be seen from Figure 9a that when the operating pressure of the tower is 31.1 bar, the H:C ratio of syngas decreases from 2.25 to 2.1, and the content of inert gas increases from 0.68% to 0.88% as the temperature rises from −175 to −155 °C. When the tower operating pressure is 21.1 bar, as shown in Figure 9b, the H:C ratio is from 2.25 to 1.95, and the content of inert gas increased from 0.68% to 0.95%. The trend of syngas composition changing is similar under different operating pressures.

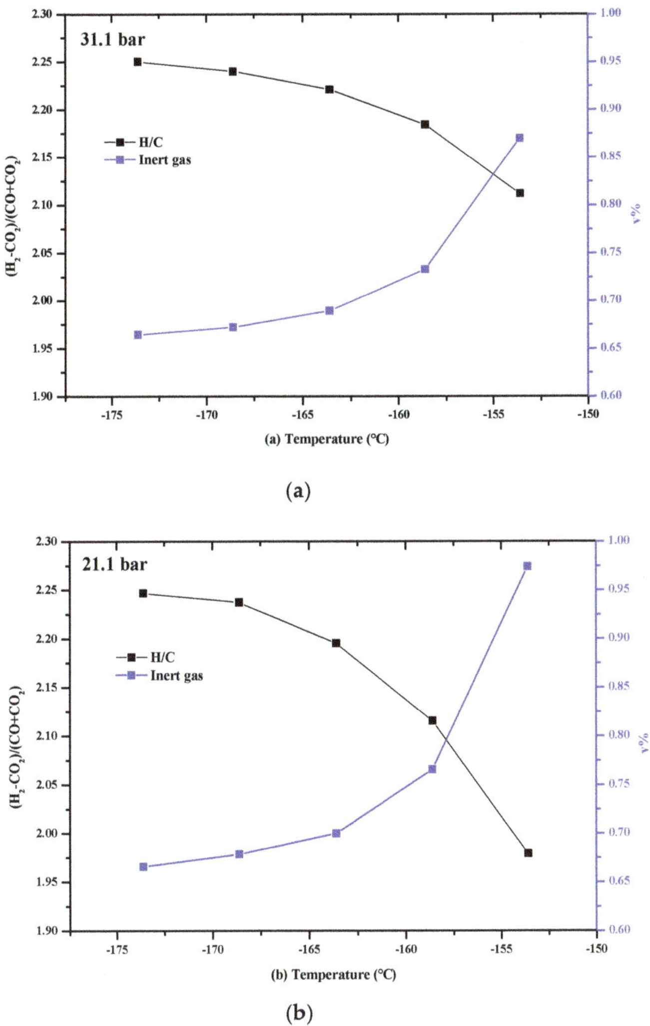

Figure 9. (a,b) Effect of temperature and pressure on the H:C ratio and inert gas concentration.

However, the H:C ratio of syngas for methanol synthesis should not exceed 2.1 [38]. When the H:C ratio decreases to 2.1 and the pressure is 31.1 bar, the increase of inert gas content is less than that at 21.1 bar. Considering that accumulation of inert gas in syngas will reduce methanol production [17], in this paper the operating pressure is set to 31.1 bar, and the operating temperature is set to −158 °C.

We then study the effect of the unreacted gas recycling ratio to the conversion (a) and the compression duty for gas recycling (b), as shown in Figure 10a,b.

The productivity of methanol shows an upward tendency with the recycling ratio increasing, which indicates that a higher carbon utilization efficiency can be achieved by adjusting that ratio. The results confirm a good match with the previous study of Man et al. (2016) [39]. However, with more gas recycled, the units will consume more compression duty, as shown in Figure 10b.

Figure 10a shows that, when the recycle ratio increases from 0.50 to 0.86, the methanol production increases slowly, and when the cycle ratio is more than 0.86, the methanol production increases rapidly and the energy consumption increases rapidly. In order to balance the capacity and energy consumption of the system, in this paper, 0.86 is chosen as the recycle ratio of unreacted methanol gas in this unit.

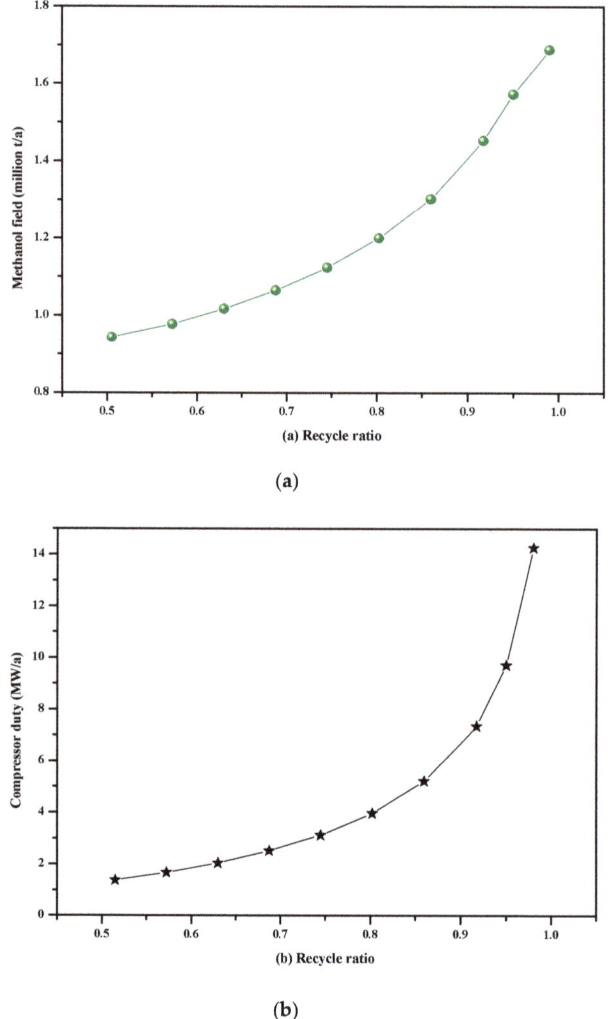

(a)

(b)

Figure 10. (a,b) Effect of recycling ratio on methanol productivity and compressor duty.

All results at key points of the CTLNG-M process are shown in Table 6. After dehydration and cooling, the crude syngas of 42,338 kmol/h is sent to the acid gas removal unit. In this unit, the CO_2 is removed and the content of CO_2 is reduced to 20 ppm and that of the H_2S less than 1 ppm. The clean syngas flow (SNYGAS-C) is 28,310 kmol/h with the H:C ratio of 2.47. This stream enters the cryogenic separation unit and is cooled to −168 °C. CH_4 is separated out of this unit as the LNG product of 4790 kmol/h, denoted by stream S8 in this figure. The yearly output of LNG is 642,000 tons. After separation, CH_4 content in the syngas is reduced to 0.6% before entering the methanol reactor. This input stream is mixed in this new process with the CO_2 stream from the acid gas removal unit. Finally, the syngas has its H:C ratio at 2.1 and the total yearly methanol output of 1.368 million tons.

Table 6. Simulation results at key points of CTLNG-M process.

Stream	Crude Syngas	SYNGAS-C	S3	S8	SYNGAS	CO$_2$	RECY	PURGE	MEOH
Temperature (°C)	40	40	−158	−93.5	45	30	40	40	63.5
Pressure (bar)	39	36	31.1	34	82	34	71	71	6
Mole Flow (kmol/h)	42,338	28,310	23,511	4790	23,408	710	47,193	1121	6573
Mole Frac (%)									
N$_2$	0.2	0.3	0.3	543 ppb	0.3	2 ppm	0.9	0.9	134 ppm
AR	898 ppm	0.1	0.2	18 ppm	0.2	61 ppm	0.6	0.6	209 ppm
CO	15.9	23.7	28.5	147 ppm	28.6	0	22.3	22.3	0.5
CO$_2$	31.4	24 ppm	0	0	2.8	0.99	4.9	4.9	1.4
H$_2$S	0.1	0	0.01	0	0	0	0	0	0
H$_2$	39.1	58.5	70.4	0	68.7	0	69.9	69.9	1.4
CH$_4$	11.8	17	0.6	97.6	0.2	0	0.6	0.6	284 ppm
CH$_3$OH	0	0	0	0	0	0	0	0	99.7
C$_2$H$_6$	0.4	0.4	0	2.4	0	2.3	0	0	0
C$_2$H$_4$	259 ppm	5 ppm	0	27 ppm	0	27 ppm	0	0	0

4. Discussion

The CTLNG-M process input consists of 4.656 million tons/a raw coal, which remains the same amount as a benchmark CTSNG process, meanwhile, the outputs consist of 1,367,800 tons/a of methanol and 642,000 tons/a of LNG. The benchmark has the same input amount of coal and outputs consisting of 2 billion Nm^3 nature gas only. Based on the simulations, we compare these two processes with respect to energy efficiency, carbon element utilization rate, energy consumption, and economic benefit, as given in Table 7. In the following section, we explain the definition of the indexes and analyze the performances of these two processes.

Table 7. System performances parameters of CTLNG-M and CTSNG.

Items	CTLNG-M	CTSNG	Units
Input			
Coal	4656.0	4656.0	10^3 tons/a
Steam	1343.9	1856.8	10^3 tons/a
Electricity	109.5	77.3	MW
Output			
Methanol	1367.8	0	10^3 tons/a
LNG	642.1	0	10^3 tons/a
SNG	0	20.0	10^8 NM^3/a
CO_2 Emission	580.0	710.8	10^3 tons/a
Element Utilization (C)	39.6%	34.7%	-
Product Energy	9157.7	7850.0	GJ/h
Energy Efficiency	53.1%	50.4%	-

4.1. Energy Efficiency

According to the first law of thermodynamics, energy efficiency is defined as the ratio of the energy of effective products (E_0) to the energy of input raw materials (E_i), as given by Equation (4). [40].

$$\eta = \frac{\sum E_O}{\sum E_i} \times 100\%, \tag{4}$$

where E_0 is the product energy (MW) of chemical process and E_i is the raw material energy (MW) of chemical process. In this paper, the energy of raw materials and products is calculated by the high heating value (HHV). In the CTLNG-M process, the source of input energy includes raw coal, electricity, and steam. Thereby, the outputs of energy are LNG and methanol. Methanol is a widely-used platform chemical product with a high calorific value of 22.7 GJ/ton. LNG is an energy product and mainly used for urban gas or power generation. It has a high calorific value of 54.6 GJ/ton. Methane is also used as fuel, and the high heating value of the gas conforms to the natural gas GB17820-2012 standard which is 31.4 MJ/m^3.

According to Equations (3) and (4), the product energy of the CTLNG-M process is 9158 GJ/h, and the energy efficiency is 53.1%. The product energy of the CTSNG process is 7850 GJ/h, and its energy efficiency is 50.4%. It shows that the new CTLNG-M process has a higher efficiency of 3% than that of the conventional CTSNG process.

4.2. Element Utilization Ratio

In a coal-based chemical process, the C element in coal is transformed into a chemical product. Thus, it is important to analyze the C resource utilization efficiency to represent the resource utilization. The element C converted into methanol in coal is defined as the effective C, and the element C discharged in the form of CO_2 or waste residue is defined as the ineffective C. The ratio of the carbon mole flow in the product to the mole flow in the raw material is defined as carbon efficiency λ, which can be represented by Equation (5) [41].

$$\lambda = \frac{\sum F_O}{\sum F_i} \times 100\%, \tag{5}$$

where F_0 is the mole flow of carbon in methanol and LNG products and F_i is the mole flow of carbon in coal.

Figure 11 shows the carbon elemental balance in the new process. It shows that the input molar flow in raw material coal of CTLNG-M system is 27,416 kmol/h. The molar flow of carbon in the crude syngas is 27,141 kmol/h after gasification, and gets 11,749 kmol/h carbon elements when washed with methanol at a low temperature. In the cryogenic separation unit, the molar flow of carbon in the LNG product is 4284 kmol/h. Remaining clean syngas is mixed with the pure CO_2 from the acid gas removal unit and the molar flow of carbon in the methanol syngas is 6967 kmol/h. In this case, CO_2 through the acid gas removal process, is separated into two parts. Partial CO_2 is then removed from gas emission, and recycled in the synthesis process to convert the final methanol product. The methanol syngas re-enters the methanol synthesis unit which contains 6573 kmol/h carbon in the methanol products.

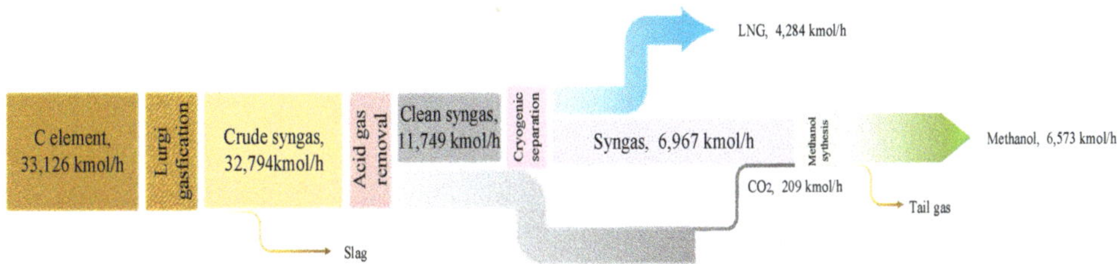

Figure 11. Carbon element flow in the process of GTLNG-M.

According to Equation (5), the carbon element efficiency of the new process is 39.6%. This ratio is 5.2% higher than that of the CTSNG process. This is mainly because CO_2 emission has been partially converted into product. In the conventional process, all syngas has to be converted to only synthesized natural gas (SNG), which requires the H:C ratio of 3.1 using the element balance equation. This is higher than the ratio in the syngas output from the Lurgi gasification as 2.7. It is necessary for the CTSNG process to use the water–gas shift unit to increase the ratio to 3.1 for methanation reaction. In this course, CO_2 emission is increased. However, in the new coproduction process, methanol is present as a suitable product from chemical synthesis through which product methane is separated and cryogenically cooled to directly produce the LNG product. The remaining syngas is only used for methanol synthesis, which requires a lower H:C ratio of 2.1. In this case, the syngas has excessive hydrogen. We then introduce CO_2 into the syngas to adjust the ratio. In this study, 209 kmol/h CO_2 is converted to methanol.

4.3. Energy Consumption Analysis

As has been stated in the above discussion, the CTLNG-M process has a higher energy and carbon utilization ratio than CTSNG process. Moreover, considering the new process is under a coproduction design with an added cryogenic separation unit, which is specially needed at low temperature environment and; therefore, consumes more electricity, quantitative analysis for energy use is a necessity.

The energy consumption is defined as utilities consisting of steam cost and electricity cost. According to our calculation, the steam cost in CTLNG-M is 1.34 million tons/a, and the electricity consumption is 110 MW, while the same in the CTSNG process are, respectively, 1.86 million tons/a and

77 MW. For a more convenient comparison, both steam cost and electricity consumption are converted to the same units as MJ/a, as shown in Figure 12.

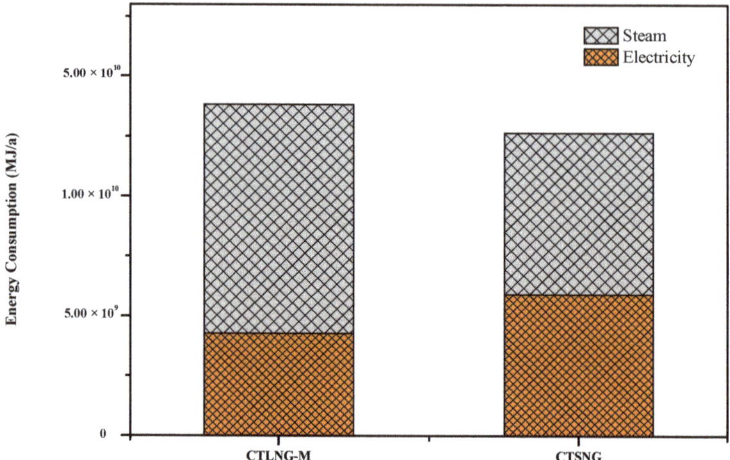

Figure 12. Energy consumption of the CTLNG-M and CTSNG processes.

In Figure 12, the CTLNG-M process consumes 4.3×10^9 MJ of steam and 9.5×10^9 MJ of electricity for a year, and the CTSNG process consumes 5.9×10^9 MJ of steam and 6.7×10^9 MJ of electricity. It shows that the coproduction system has a lower steam cost of about 1.6×10^9 MJ for per year, which is mainly because of a flexible way to integrate heat exchange when there is not only one route for product processing [42]. However, more electricity is consumed in the new process. It is because the nitrogen circulation refrigeration process needs more power assistance, as modelling data indicates. Since there is no power that can be generated within the system, it takes more capital investment, which needs to be further analyzed.

To summarize, the total energy consumption in general increased by 8.7%. The coproduction process has an advantage on utility usage due to integration of a heat exchanger and flexible distribution flow between different product processing. However, in the specific case of CTLNG-M, a higher electricity consumption is due to compression work in the added cryogenic separation unit. In total, the increased electricity consumption cannot be outweighed by the decrease in the steam cost, and the energy demand gap is 1.2×10^9 MJ/a, which indicates more investment on various costs in the new coproduction process and a further economic analysis is needed for profitability measurement.

4.4. Economic Analysis

4.4.1. Total Capital Investment

The total capital investment (TCI) for a given construction project mainly includes fixed capital investment and variable cost. The investment for manufacturing and plant facilities are defined as the fixed capital investment, while those for the plant operation are the working capital [43]. The equipment investment of the system can be calculated by Equations (6) and (7) based on the benchmark investment of the major equipment listed. The total investment can then be derived from the scale factor ((See Tables A1 and A2 in Appendix.).

$$\mathrm{EI} = \sum_j \theta \cdot \mathrm{EI}_j^r \cdot \left(\frac{S_j}{S_j^r}\right)^{sf}, \tag{6}$$

$$\mathrm{TCI} = \mathrm{EI} \cdot \left(1 + \sum_{i=1}^{n} RF_i\right), \tag{7}$$

where EI is the equipment investment (CNY), θ is the localization factor, EI_j^r is the benchmark equipment investment of the j unit, S_j is the scale of the j unit, S_j^r is the base scale of the j unit, TCI is the total investment, and RF_i is the proportionality factor of the investment composition i.

As shown in Figure 13, based on the same input of raw coal, the total investment of the CTSNG project is 16.62 billion CNY, and the CTLNG-M project is 13.66 billion CNY, which is 17.8% lower than CTSNG. This is because the new process eliminates the water–gas shift unit compared to the single-production coal gasification process, so that the carbon emission is less and the amount of gas processed by the acid gas removal unit is decreasing compared to the original CTSNG process. The corresponding investment is also reduced. At the same time, the CTLNG-M process uses a nitrogen expansion refrigeration process, with mature technology and low investment. Therefore, the cryogenic unit equipment and related investment are relatively low, and the total amount of process investment is correspondingly reduced, which is more suitable for CTSNG projects and has economic advantages.

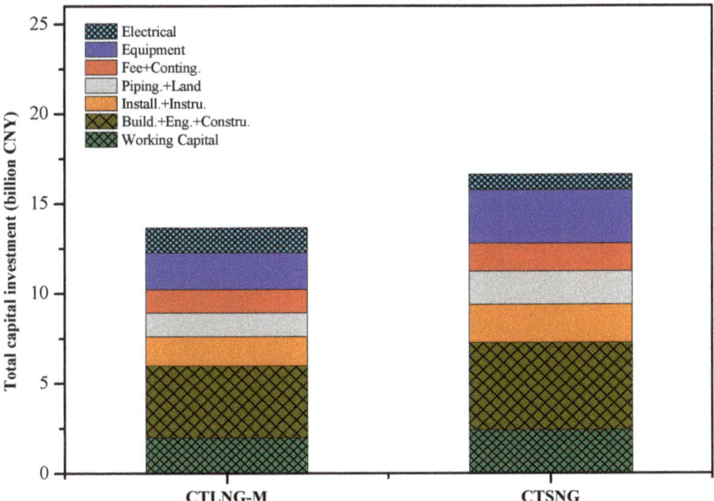

Figure 13. Capital investment of CTLNG-M and CTSNG.

4.4.2. Internal Rate of Return

Internal rate of return (IRR) is another important index for evaluation of economic performance, which takes into account the net present value and the service life of processing route into account [44]. A dynamic evaluation method is taken in this paper, the calculation is as follows.

$$\mathrm{NPV} = \sum_{t=0}^{m} \frac{(CI - CO)_t}{(1 + i)^t}, \tag{8}$$

$$\mathrm{NPV} = \sum_{t=0}^{m} \frac{(CI - CO)_t}{(1 + IRR)^t} = 0, \tag{9}$$

where CI is the net cash inflow in the t year, CO is the net cash outflow; m is the project's life time; i is the benchmark rate of return. NPV stands for net present value (NPV), which refers to the net cash flow generated annually by a technical solution throughout its life cycle. The net cash flow generated each year is converted to the present value at the base time by a specified base discount rate i_0.

Inner rate of return (IRR) can usually be calculated by interpolation method. It represents the discount rate when the cumulative present value of the net cash flow of the project is equal to zero in the whole calculation period. IRR is a dynamic index to evaluate the economic feasibility of new projects. It is usually compared with the base rate of return to determine whether the new chemical process is feasible. In this paper, i is set to 12% [45]. If the IRR is larger than the base rate of return i, the process is economically feasible and achieves the lowest level of return on investment. In addition, with the increase of internal rate of return, the obtained benefit of the process will also increase.

The IRRs of the CTLNG-M and CTSNG processes are compared in Figure 14, which are higher than the industrial criterion of 12%. Specifically, the IRR of the CTSNG process is 13%, which is slightly higher than 12%, and accords with the current status of the CTSNG project. However, the IRR of the CTLNG-M process is 19%, which increased by 6%, so this process has higher profit.

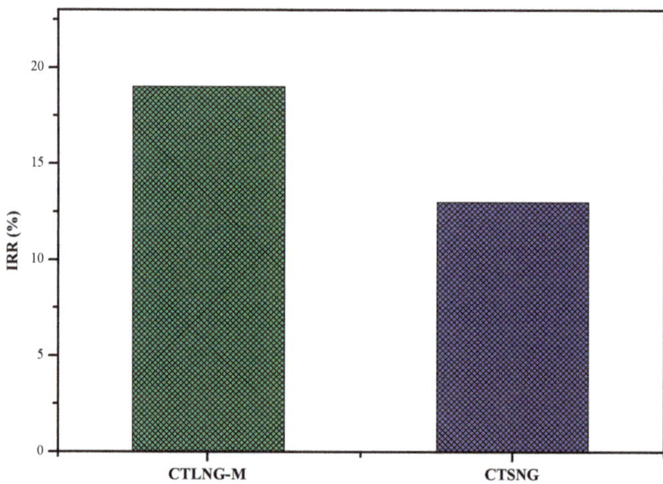

Figure 14. Inner rate of return (IRR) of the CTLNG-M and CTSNG processes.

5. Conclusions

This paper proposes a system of coproduction for LNG and methanol. The aim was to find improvements to the low-earning CTSNG process using the same raw material but producing a low-margin, single SNG product. In the new coproduction process, there are two innovative aspects. On the one hand, the syngas is firstly separated to the LNG product and the lean-methane syngas is then used for methanol synthesis. To realize this improvement, a cryogenic separation unit is added. Besides, the syngas with little CH_4 has a higher hydrogen component than that for methanol synthesis. Thereby, CO_2 is used to supply an additional carbon element to the methanol synthesis. On the other hand, the methanation unit is removed, while the process still outputs a product of the high-valued form of methane as the LNG. In the case study, we modeled and simulated the key units of the CTLNG-M process with 642,000 tons/a LNG and 1.368 million tons/a methanol product, compared to the CTSNG process with the same coal processing coal capacity and 2 billion NM^3/a SNG. In element efficiency analysis, the carbon efficiency of the new process increases from 34.7% to 39.6%, with corresponding decrease of carbon emission by 130,000 tons per year. Because of the additional energy consumption for gas compression, the energy efficiency of the new process is at the same level with the CTSNG process.

In economic analysis, the IRR of the CTSNG process is 13% while the IRR of the CTLNG-M process is 19%. The new process brings much higher economic benefits. This is because the new process produces a higher valued product and saves the carbon resource during methanol synthesis. Moreover, the new process has 17.8% reduction of investment compared to the CTSNG process. Thus, this is a promising solution for coal chemical processes based on Lugri gasification technology, with more economic benefit and less investment.

Author Contributions: J.G. and S.Y. contributed to the conception of the study. J.G. and S.Y. contributed significantly to analysis and manuscript preparation; J.G. performed the data analyses and wrote the manuscript; A.K. helped perform the analysis with constructive discussions.

Funding: We would like to express our appreciation to the National Natural Science Foundation of China (funder ID: http://dx.doi.org/10.13039/501100001809, Grant No. 21736004).

Conflicts of Interest: The authors declare no conflicts of interest.

Appendix A

Table A1. Summary of investment data for main equipment components.

Unit	Scaling Parameter	EI_j^r	S_j	θ	$S_{CTLNG\text{-}M}$	S_{CTSNG}
Gasification	Coal Input	4.84	39.2	0.8	161.7	161.7
Acid Gas Removal	CO_2 Absorption	2.03	2064.4	0.65	1201.1	1174.5
Cryogenic Separation	Output Flow Rate	3.23	21.3	0.5	21.79	0
Methanol Synthesis	Syngas Flow Rate	1.26	10,810	0.65	6628.6	0
Water–Gas Shift	MAF Coal Input (LHV)	3.18	1377	0.67	0	4322.2
Methanation	Syngas Flow Rate	1.26	10,810	0.65	0	8011

Table A2. Ratio factors for capital investment.

Component	Ratio Range (%)	Factor (%)
1. Direct Investment		
Equipment	15~40	21
Installation	6~14	10
Instruments and Controls	2~8	5
Piping	2~20	12
Electrical	2~10	6
Building (including Service)	3~18	15
Land	1~2	1
2. Indirect Investment		
(2.1) Engineering and Supervision	4~21	10
(2.2) Construction Expenses	3~16	9
(2.3) Contractor's Fee	2~6	4
(2.4) Contingency	5~15	7
3. Fixed Capital Investment	Direct + Indirect	100
4. Working Capital	10~20	17
5. Total Capital Investment	Fixed + Flow	117

References

1. BP. BP Statistical Review of World Energy (2017 Full Report). Available online: https://www.bp.com/zh_cn/china/reports-and-publications/_bp_2017-_.html (accessed on 5 July 2017).
2. National Development and Reform Commission. 13th Five-Year Plan for Demonstration of Coal Deep Processing Industry. Available online: http://www.ndrc.gov.cn/fzgggz/fzgh/ghwb/gjjgh/201708/W020170809596996271593.pdf (accessed on 3 March 2017).
3. Ji, H. Study on Peak-Valley Difference of Natural Gas Demand in North China and Its Countermeasures. Master's Thesis, China University of Petroleum, Qingdao, China, 2016.
4. Liu, X.; Kang, S. Discussion on the source of feed gas to synthetic natural gas (SNG). *Coal Chem. Ind.* **2017**, *45*, 58–62.

5. Li, G.; Liu, Z.Y.; Liu, T.; Shan, J.; Fang, Y.T.; Wang, Z.Q. Techno-economic analysis of a coal to hydrogen process based on ash agglomerating fluidized bed gasification. *Energy Convers. Manag.* **2018**, *164*, 552–559. [CrossRef]

6. Guo, Z.; Wang, Q.; Fang, M.; Luo, Z.; Cen, K. Thermodynamic and economic analysis of polygeneration system integrating atmospheric pressure coal pyrolysis technology with circulating fluidized bed power plant. *Appl. Energy* **2014**, *113*, 1301–1314. [CrossRef]

7. Liu, Y.; Qian, Y.; Xiao, H.; Yang, S.Y. Techno-economic and environmental analysis of coal-based synthetic natural gas process in China. *J. Clean. Prod.* **2017**, *166*, 417–424. [CrossRef]

8. Xiao, H.; Liu, Y.; Huang, H.; Yang, S. Modeling and heat integration for water gas shift unit of coal to SNG process. *CIEP J.* **2018**, *37*, 554–560.

9. Li, S.; Jin, H.; Gao, L.; Zhang, X. Exergy analysis and the energy saving mechanism for coal to synthetic/substitute natural gas and power cogeneration system without and with CO_2 capture. *Appl. Energy* **2014**, *130*, 552–561. [CrossRef]

10. Li, S.; Jin, H.; Gao, L. Cogeneration of substitute natural gas and power from coal by moderate recycle of the chemical unconverted gas. *Energy* **2013**, *55*, 658–667. [CrossRef]

11. Yi, Q.; Hao, Y.; Zhang, J. Energy polygeneration systems and CO_2 recycle. In *Advances in Energy System Engineering*; Springer: Cham, Switzerland, 2017; pp. 319–428.

12. Lin, H.; Jin, H.; Gao, L.; Han, W. Techno-economic evaluation of coal-based polygeneration systems of synthetic fuel and power with CO_2 recovery. *Energy Convers. Manag.* **2011**, *52*, 274–283. [CrossRef]

13. Lin, R.; Jin, H.; Gao, L. Review on the evaluation criteria of polygeneration system for power and chemical production. *Gas Turbine Tech.* **2012**, *25*, 1–14.

14. Hao, Y.; Huang, Y.; Gong, M.; Li, W.; Feng, J.; Yi, Q. A polygeneration from a dual-gas partial catalytic oxidation coupling with an oxygen-permeable membrane reactor. *Energy Convers. Manag.* **2015**, *106*, 466–478. [CrossRef]

15. Han, W.; Jin, H.; Lin, R.; Wang, Y. A new methanol-power polygeneration system based on moderate conversion of coal and natural gas. *J. Eng.* **2010**, *31*, 1981–1984.

16. Tu, C.; Li, S.; Gao, L.; Jin, H. Association studies of coal-based methanol/power poly-generation system. *J. Eng. Thermophys.* **2015**, *36*, 32–35.

17. Huang, H.; Yang, S.; Cui, P. Design concept for coal-based polygeneration processes of chemicals and power with the lowest energy consumption for CO_2 capture. *Energy Convers. Manag.* **2018**, *157*, 186–194. [CrossRef]

18. Bai, Z.; Liu, Q.B.; Lei, J.; Li, H.; Jin, H. A polygeneration system for the methanol production and the power generation with the solar–biomass thermal gasification. *Energy Convers. Manag.* **2015**, *102*, 190–201. [CrossRef]

19. Gong, J.; You, F. A new superstructure optimization paradigm for process synthesis with product distribution optimization: Application to an integrated shale gas processing and chemical manufacturing process. *AICHE J.* **2017**, *64*, 123–143. [CrossRef]

20. You, F.; Grossmann, I.E.; Wassick, J.M. Multisite capacity, production, and distribution planning with reactor modifications: MILP model, bilevel decomposition algorithm versus lagrangean decomposition scheme. *Ind. Eng. Chem. Res.* **2011**, *50*, 4831–4849. [CrossRef]

21. He, C.; Feng, X.; Chu, K.H. Process modeling and thermodynamic analysis of Lurgi fixed-bed coal gasifier in a SNG plant. *Appl. Energy* **2013**, *111*, 742–757. [CrossRef]

22. Wu, Z.Q.; Wang, S.Z.; Zhao, J.; Lin, C.; Meng, H. Product distribution during co-pyrolysis of bituminous coal and lignocellulosic biomass major components in a drop-tube furnace. *Energy Fuels* **2015**, *29*, 4168–4180. [CrossRef]

23. An, W.; Tian, P.; Qi, K. Current situation and opportunities & challenges of coal to SNG industry in China. *Coal Chem. Ind.* **2018**, *44*, 7–11.

24. Ma, L. Analysis of the correlation between China's LNG imported price and oil price. *China Petrochem. J.* **2017**, *1*, 138–139.

25. Zhou, Y.; Li, Y. Features of Xianfeng Chemical methane cryogenic separation unit and summaries of its commissioning. *Cryog. Technol.* **2016**, *2*, 15–20.

26. Ghorbani, B.; Shirmohammadi, R.; Mehrpooya, M. A novel energy efficient LNG/NGL recovery process using absorption and mixed refrigerant refrigeration cycles, economic and exergy analyses. *Appl. Therm. Eng.* **2018**, *132*, 283–295. [CrossRef]

27. Zhang, Q.; Li, Z.; Wang, G.; Li, H. Study on the impacts of natural gas supply cost on gas flow and infrastructure deployment in China. *Appl. Energy* **2015**, *162*, 1385–1398. [CrossRef]
28. Man, Y.; Yang, S.; Xiao, H.; Qian, Y. Modeling, simulation and analysis for coal and coke-oven gas to synthetic natural gas. *CIESC J.* **2015**, *66*, 4941–4947.
29. Yang, S.; Qian, Y.; Liu, Y.; Wang, Y.; Yang, S. Modeling, simulation, and techno-economic analysis of Lurgi gasification and BGL gasification for coal-to-SNG. *Chem. Eng. Res. Des.* **2017**, *111*, 355–368. [CrossRef]
30. Yang, S.; Qian, Y.; Ma, D.H.; Wang, Y.F.; Yang, S.Y. BGL gasifier for coal-to-SNG: A comparative techno-economic analysis. *Energy* **2017**, *133*, 158–170. [CrossRef]
31. Yoon, S.; Binns, M.; Park, S.; Kim, J. Development of energy-efficient processes for natural gas liquids recovery. *Energy* **2017**, *128*, 768–775. [CrossRef]
32. Zhao, W.; Zhou, Y. Development of the technology to recover LNG from coal-made synthetic methanol gas. *Cryog. Technol.* **2014**, *7*, 52–55.
33. Liu, J.; Zhang, C.; Cao, L. LNG recovery from fixed bed pressurized syngas. *Mod. Chem. Ind.* **2014**, *34*, 154–156.
34. Lai, X.; Zhang, S.; Hu, M. Development of the process for LNG production from syngas cryogenic separation in coal-to-methanol process. *Cryog. Technol.* **2015**, *2*, 34–38.
35. Xiang, D.; Yang, S.; Mai, Z.; Qian, Y. Comparative study of coal, natural gas, and coke-even gas based methanol to olefins processes in China. *Comput. Chem. Eng.* **2015**, *83*, 176–185. [CrossRef]
36. Xiang, D.; Yang, S.; Liu, X.; Mai, Z.; Qian, Y. Techno-economic performance of the coal-to-olefins process with CCS. *Chem. Eng. J.* **2014**, *240*, 45–54. [CrossRef]
37. Wang, J.; Mao, T.; Sui, J.; Jin, H. Modeling and performance analysis of CCHP (combined cooling, heating and power) system based on co-firing of natural gas and biomass gasification gas. *Energy* **2015**, *93*, 801–815. [CrossRef]
38. Liu, Y.; Qian, Y.; Zhou, H.; Xiao, H. Conceptual design of coal to synthetic natural gas (SNG) process based on BGL gasifier: Modeling and techno-economic analysis. *Energy Fuels* **2017**, *31*, 1023–1034. [CrossRef]
39. Man, Y.; Yang, S.; Qian, Y. Integrated process for synthetic natural gas production from coal and coke-oven gas with high energy efficiency and low emission. *Energy Convers. Manag.* **2016**, *117*, 162–170. [CrossRef]
40. Zhang, M.; Li, G.; Mu, H.; Ning, Y. Energy and exergy efficiencies in the Chinese transportation sector, 1980–2009. *Energy* **2011**, *36*, 770–776. [CrossRef]
41. Ruizmercado, G.J.; Smith, R.L.; Gonzalez, M.A. Sustainability indicators for chemical processes: II. data needs. *Ind. Eng. Chem. Res.* **2012**, *51*, 2329–2353. [CrossRef]
42. Gatti, M.; Martelli, E.; Marechal, F.; Consonni, S. Review, modeling, heat integration, and improved schemes of Rectisol-based processes for CO_2 capture. *Appl. Therm. Eng.* **2014**, *70*, 1123–1140. [CrossRef]
43. Salkuyeh, Y.K.; Elkamel, A.; Thé, J.; Fowler, M. Development and techno-economic analysis of an integrated petroleum coke, biomass, and natural gas polygeneration process. *Energy* **2016**, *113*, 861–874. [CrossRef]
44. Krawczyk, P.; Howaniec, N.; Smoliński, A. Economic efficiency analysis of substitute natural gas (SNG) production in steam gasification of coal with the utilization of HTR excess heat. *Energy* **2016**, *114*, 1207–1213. [CrossRef]
45. NBS (National Bureau of Statistics). *China Energy Statistics Yearbook*; China Statistics Press: Beijing, China, 2013; pp. 22–25. (In Chinese)

Article

Statistical Process Monitoring of the Tennessee Eastman Process Using Parallel Autoassociative Neural Networks and a Large Dataset

Seongmin Heo and Jay H. Lee *

Department of Chemical and Biomolecular Engineering, Korea Advanced Institute of Science and Technology (KAIST), 291 Daehak-ro, Yuseong-gu, Daejeon 34141, Korea
* Correspondence: jayhlee@kaist.ac.kr

Received: 3 May 2019; Accepted: 21 June 2019; Published: 1 July 2019

Abstract: In this article, the statistical process monitoring problem of the Tennessee Eastman process is considered using deep learning techniques. This work is motivated by three limitations of the existing works for such problem. First, although deep learning has been used for process monitoring extensively, in the majority of the existing works, the neural networks were trained in a supervised manner assuming that the normal/fault labels were available. However, this is not always the case in real applications. Thus, in this work, autoassociative neural networks are used, which are trained in an unsupervised fashion. Another limitation is that the typical dataset used for the monitoring of the Tennessee Eastman process is comprised of just a small number of data samples, which can be highly limiting for deep learning. The dataset used in this work is 500-times larger than the typically-used dataset and is large enough for deep learning. Lastly, an alternative neural network architecture, which is called parallel autoassociative neural networks, is proposed to decouple the training of different principal components. The proposed architecture is expected to address the co-adaptation issue of the fully-connected autoassociative neural networks. An extensive case study is designed and performed to evaluate the effects of the following neural network settings: neural network size, type of regularization, training objective function, and training epoch. The results are compared with those obtained using linear principal component analysis, and the advantages and limitations of the parallel autoassociative neural networks are illustrated.

Keywords: process monitoring; nonlinear principal component analysis; parallel neural networks; autoassociative neural network; big data

---

1. Introduction

Statistical process monitoring is one of the most intensely-studied problems for the modern process industry. With the rising need for sustainable operation, it has been attracting extensive research effort in the last few decades [1,2]. The key step of statistical process monitoring is to define normal operating regions by applying statistical techniques to data samples obtained from the process system. Typical examples of such techniques include principal component analysis (PCA) [3–6], partial least squares [7–9], independent component analysis [10,11], and support vector machine [12,13]. Any data sample that does not lie in the normal operating region is then classified as a fault, and its root cause needs to be identified through fault diagnosis.

Recently, deep learning and neural networks have been widely used for the purpose of statistical process monitoring, where both supervised and unsupervised learning algorithms have been implemented. In designing process monitoring systems in a supervised manner, various types of neural networks have been used, such as feedforward neural networks [14], deep belief networks [15], convolutional neural networks [16], and recurrent neural networks [17]. In the case of unsupervised learning, the autoassociative neural network (also known as an autoencoder), which has been proposed as a nonlinear generalization of PCA [18], is typically used [19–21]. While the traditional statistical approaches typically rely only on the normal operating data to develop the process monitoring systems, most of the deep learning-based process monitoring studies have adopted supervised learning approaches. However, in the real industrial processes, it is difficult to obtain a large number of data samples for different fault types, which can be used for the training of deep neural networks. Thus, it is important to examine rigorously the potential of autoassociative neural networks as a basis for the design of process monitoring systems.

In the process systems area, the Tennessee Eastman (TE) process, a benchmark chemical process introduced by Downs and Vogel [22], has been a popular test bed for process monitoring techniques. There already exist a few studies where the process monitoring systems for this process are designed on the basis of autoassociative neural networks [23–25]. However, considering the complexity of the neural network training, these studies have two limitations. First, a rigorous case study has not been performed to evaluate the effects of different neural network settings, such as neural network hyperparameters and training objective functions, which can have great impact on the performance of the process monitoring systems. Furthermore, a few thousand normal training samples were used to train neural networks with much more parameters, ranging from a hundred thousand to a million parameters. A larger dataset is required to investigate the effectiveness of unsupervised deep learning for the statistical process monitoring. In addition to the above limitations, there is another issue that is directly related to the structure of autoassociative neural networks. It has been reported that there is a high chance for the principal components, which are extracted using autoassociative neural networks, to be redundant due to the co-adaptation in the early phase of neural network training [18]. The objective of this work is to address these limitations.

The rest of the article is organized as follows. First, the concept of linear PCA is briefly explained, and how it can be used for the statistical process monitoring is discussed. Then, the information on autoassociative neural networks is provided, and the parallel autoassociative neural network architecture, which was proposed in our previous work [26] to alleviate the co-adaptation issue mentioned above, is described. This is followed by the description of the statistical process monitoring procedure using autoassociative neural networks. Finally, a comprehensive case study is designed and performed to evaluate the effects of different neural network settings on the process monitoring performance. The dataset used in this study has 250,000 normal training samples, which is much larger than the ones considered in the previous studies.

2. Principal Component Analysis and Statistical Process Monitoring

2.1. Linear Principal Component Analysis

Let us first briefly review the concept of linear PCA. PCA is a statistical technique that decorrelates the original variables, resulting in a set of uncorrelated variables called principal components. Let x be a sample vector of m variables and X be a data matrix whose rows represent n sample vectors. Assuming that each column of X has zero mean and unit variance, the singular value decomposition can be applied to the sample covariance matrix:

$$\frac{1}{n-1}X^T X = P\Lambda P^T \tag{1}$$

where Λ is a diagonal matrix that contains the eigenvalues of the sample covariance matrix on its main diagonal and P denotes an orthogonal matrix whose columns are the eigenvectors of the sample covariance matrix. In the context of PCA, P is called the loading matrix since its column vectors can be used to extract principal components from the original data as follows:

$$T = XP \tag{2}$$

where T represents the score matrix whose elements are the principal component values.

Let λ_i be the diagonal element in the ith row of Λ, and let us assume that Λ is arranged in a descending order (i.e., $\lambda_1 \geq \lambda_2 \geq \cdots \geq \lambda_m$) so that the jth column of P corresponds to the direction with the jth largest variance in the principal component space. Then, we can partition the loading matrix into two blocks as below:

$$P = \begin{bmatrix} P_{PC} & P_R \end{bmatrix} \tag{3}$$

where P_{PC} and P_R contain the first l columns and the remaining columns of P, respectively. These two submatrices can be respectively used to map the original data onto the lower dimensional principal component space and residual space (of dimension l and dimension $n\text{-}l$, respectively):

$$\begin{aligned} T_{PC} &= XP_{PC} \\ T_R &= XP_R \end{aligned} \tag{4}$$

where T_{PC} and T_R are the first l columns and the remaining columns of T, respectively. In what follows, we explain how linear PCA can be used for statistical process monitoring.

2.2. Statistical Process Monitoring Using Linear PCA

In the PCA-based process monitoring, a new data sample is first projected onto the lower dimensional principal component space and the residual space [27]. Then, it is evaluated whether the new sample lies in the normal operating range in both spaces. Hotelling's T^2 and Q (or squared prediction error) statistics are typically used to define the normal operating range in the principal component space and the residual space, respectively, for such evaluation. These statistics can be computed by the following equations:

$$\begin{aligned} T^2 &= xP_{PC}\Lambda_{PC}^{-1}P_{PC}^T x^T \\ Q &= xP_R P_R^T x^T \end{aligned} \tag{5}$$

where Λ_{PC} represents a diagonal matrix formed by the first l rows and columns of Λ.

The upper control limit for the T^2 statistic is given as [28]:

$$T_\alpha^2 = \frac{l(n^2-1)}{n(n-1)}F_\alpha(l, n-l) \tag{6}$$

where $F_\alpha(l, n-l)$ represents the α percentile of the F-distribution with l and $n\text{-}l$ degrees of freedom. The Q statistic has the upper limit of the following form [29]:

$$Q_\alpha = g\chi_\alpha^2(h) \tag{7}$$

where:

$$\begin{aligned} g &= \theta_2/\theta_1 \\ h &= \theta_1^2/\theta_2 \\ \theta_i &= \sum_{k=l+1}^{m} \lambda_k^i, \quad i = 1,2 \end{aligned} \tag{8}$$

and $\chi_\alpha^2(h)$ is the α percentile of the χ^2-distribution with h degrees of freedom.

If a very large number of data samples is available, the mean and covariance estimated from the data will be very close to the true values of the underlying probability distribution. In this case, the upper control limit for T^2 statistic takes the following form [30]:

$$T_\alpha^2 = \chi_\alpha^2(l) \tag{9}$$

while the upper control limit for the Q statistic can be approximated by the following equation [29]:

$$Q_\alpha = g\chi_\alpha^2(h)$$
$$g = \frac{\sigma_R^2}{2\mu_R}$$
$$h = \frac{2\mu_R^2}{\sigma_R^2} \tag{10}$$

where μ_R and σ_R are the mean and standard deviation of the squared prediction errors (i.e., Q values) obtained from the training dataset.

If any of the above control limits is violated, the new sample is classified as a fault. Once a fault is detected, its root cause needs to be identified. The contribution plot is typically used to this end, where the contribution of each variable to the T^2 or Q statistic is calculated [31–33].

It is also important to select a proper number of principal components to be retained in the PCA model. For such selection, various criteria are available in the literature including cumulative percent variance [34], residual percent variance [35], parallel analysis [36], and cross-validation [37]. For a detailed discussion on this subject, the readers are referred to the work by Valle et al. [38].

3. Statistical Process Monitoring Using Autoassociative Neural Networks

3.1. Nonlinear Principal Component Analysis Using Autoassociative Neural Networks

We now describe neural network-based nonlinear principal component analysis (NLPCA). Kramer [18] proposed to use a special type of neural network, called the autoassociative neural network, for NLPCA. As shown in Figure 1, an autoassociative neural network consists of five layers: input, mapping, bottleneck, demapping, and output layers. Its goal is to learn the identity mapping function to reconstruct its input data at the output layer. The problem of learning identity mapping becomes non-trivial if the dimension of the bottleneck layer, f, is smaller than that of the original data, m.

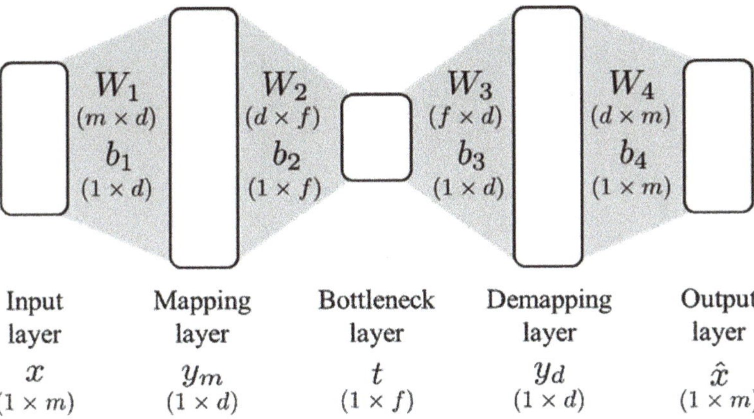

Figure 1. Network architecture of the autoassociative neural network.

The first three layers of autoassociative neural network approximate the mapping functions, which project the original data onto the lower dimensional principal component space, while the last two layers approximate the demapping functions, which bring back the projected data to the original data space. The mathematical model of the autoassociative neural network has the following form:

$$
\begin{aligned}
y_m &= a(xW_1 + b_1) \\
t &= y_m W_2 + b_2 \\
y_d &= a(tW_3 + b_3) \\
\hat{x} &= y_d W_4 + b_4
\end{aligned}
\tag{11}
$$

where x, y_m, t, y_d, and $\hat{x}$ represent the vectors of input, mapping, bottleneck, demapping, and output layers, respectively. W and b are weight matrices and bias vectors, respectively. The dimensions of all the matrices and vectors are summarized in Figure 1. a denotes the nonlinear activation function. The objective of autoassociative neural network training is to find optimal parameter values (i.e., optimal values of W and b) that minimize the difference between the input and the output, i.e.:

$$
E = \frac{\sum\limits_{i=1}^{n} \sum\limits_{j=1}^{m} \left(x_{ij} - \hat{x}_{ij}\right)^2}{nm}
\tag{12}
$$

which is also called the reconstruction error.

3.2. Alternative Neural Network Architecture: Parallel Autoassociative Neural Networks

It was pointed out by Kramer [18] that principal components extracted from an autoassociative neural network can be redundant, as multiple principal components are aligned together in the early stage of network training. To this end, in our previous work [26], we proposed an alternative neural network architecture to address this limitation, which decouples the training of different principal components. Such decoupling can be achieved by alternating the network architecture of the autoassociative neural network as shown in Figure 2.

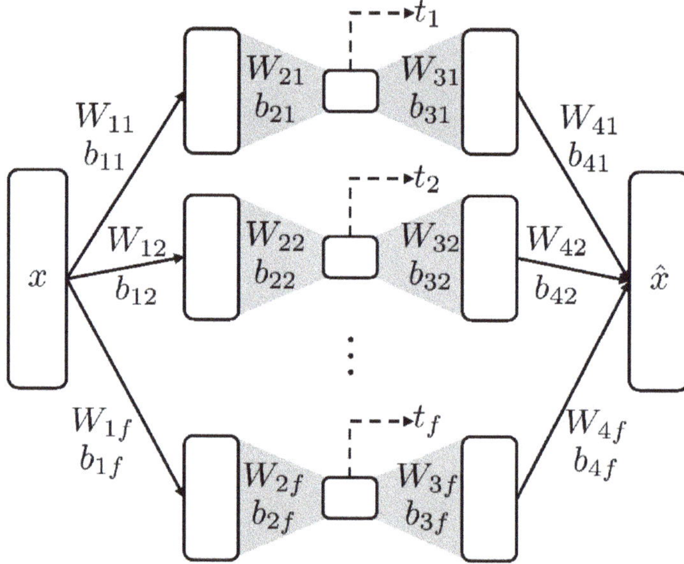

Figure 2. Alternative neural network architecture for parallel extraction of principal components.

In this network architecture, all the hidden layers are decomposed into f sub-layers to form f decoupled parallel subnetworks. Each subnetwork extracts one principal component directly from the input and reconstructs the pattern that it captures. Then, the outputs from all the subnetworks are added up to reconstruct the input. The mathematical model of this network architecture has the same form as the one in Equation (11) with the structural changes to the weight matrices and bias vectors as below:

$$
\begin{aligned}
W_1 &= \begin{bmatrix} W_{11} & W_{12} & \cdots & W_{1f} \end{bmatrix} \\
W_2 &= \begin{bmatrix} W_{21} & 0 & \cdots & 0 \\ 0 & W_{22} & \cdots & 0 \\ \vdots & \vdots & \ddots & \vdots \\ 0 & 0 & \cdots & W_{2f} \end{bmatrix} \\
W_3 &= \begin{bmatrix} W_{31} & 0 & \cdots & 0 \\ 0 & W_{32} & \cdots & 0 \\ \vdots & \vdots & \ddots & \vdots \\ 0 & 0 & \cdots & W_{3f} \end{bmatrix} \\
W_4 &= \begin{bmatrix} W_{41} \\ W_{42} \\ \vdots \\ W_{4f} \end{bmatrix} \\
b_1 &= \begin{bmatrix} b_{11} & b_{12} & \cdots & b_{1f} \end{bmatrix} \\
b_2 &= \begin{bmatrix} b_{21} & b_{22} & \cdots & b_{2f} \end{bmatrix} \\
b_3 &= \begin{bmatrix} b_{31} & b_{32} & \cdots & b_{3f} \end{bmatrix} \\
b_4 &= b_{41} + b_{42} + \cdots + b_{4f}
\end{aligned}
\tag{13}
$$

The readers are referred to [26] for more detailed information on the parallel autoassociative neural networks (e.g., the systematic approach for the network decoupling and the advantages of the decoupled parallel neural networks). The proposed network architecture has two potential advantages over the existing one, which are relevant to the statistical process monitoring. First, due to the decoupling, the proposed network architecture is expected to extract more independent (i.e., less correlated) principal components and result in smaller reconstruction errors compared to the existing architecture. If we can achieve smaller reconstruction error using the same number of principal components, it can imply that the more essential information of the original data is captured. Thus, there is a potential for small reconstruction error to translate into high process monitoring performance. The other advantage is that the proposed network architecture requires much fewer parameters compared to the existing architecture, given that the networks have the same size (i.e., the same number of hidden layers and nodes). As a result, the proposed architecture is expected to be more robust to network overfitting, which can lead to more consistent process monitoring performance. Furthermore, the proposed architecture is more suitable for online implementation since it can compute the values of the T^2 and Q statistics more quickly than the existing architecture.

3.3. Objective Functions for Autoassociative Neural Network Training

Besides the reconstruction error in Equation (12), there exist several objective functions available for the autoassociative neural network training. Here, we provide a brief description of two alternative objective functions: hierarchical error and denoising criterion. Hierarchical error was proposed by Scholz and Vigário [39] to develop a hierarchy (i.e., relative importance) among the nonlinear principal components, which does not generally exist for the principal components obtained by using the reconstruction error as the objective function. In linear PCA, it can be shown that the maximization of the principal component variance is equivalent to the minimization of the residual variance. Motivated by this, the following hierarchical reconstruction error can be defined:

$$E_H = \sum_{k=1}^{f} \alpha_k E_k \tag{14}$$

where E_k represents the reconstruction error calculated by using only the first k nodes in the bottleneck layer and α_k is a hyperparameter that balances the trade-off among the different error terms. The problem of selecting the optimal values of α_k can be computationally expensive, especially in the case of a large bottleneck layer (i.e., large number of principal components). It was illustrated that setting the values of α_k to one can robustly balance the trade-off among the different error terms [39].

The denoising criterion was proposed by Vincent et al. [40] to extract more robust principal components. To apply the denoising criterion, the corrupted input $\tilde{x}$ is generated by adding a noise, such as Gaussian noise and masking noise, to the original input x. Then, the autoassociative neural network is trained such that it can recover the original input from the corrupted input. It was shown that, using the denoising criterion, autoassociative neural networks were able to learn a lower dimensional manifold that captures more essential patterns in the original data. The three objective functions are schematically summarized in Figure 3.

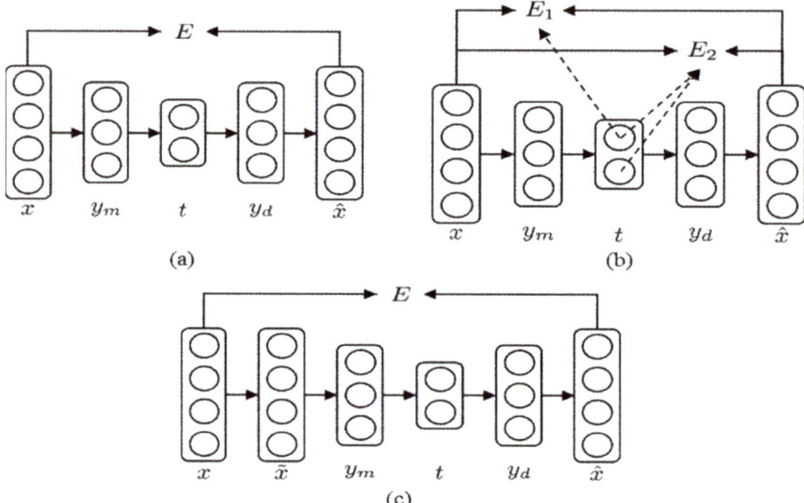

Figure 3. Schematic representation of different objective functions: (**a**) reconstruction error; (**b**) hierarchical error; (**c**) denoising criterion.

3.4. Statistical Process Monitoring Using NLPCA

We can design a similar procedure for statistical process monitoring using autoassociative neural networks. First, the original data matrix, which contains only the normal operating data, is partitioned into two disjoint sets, one for the training and the other for the testing of neural networks. Then, an autoassociative neural network is trained using the training dataset. Once the network training is complete, a new data sample is provided to the trained autoassociative neural network to compute principal components and residuals. The T^2 and Q statistics can then be calculated as follows:

$$T^2 = \sum_{k=1}^{f} \frac{t_k^2}{\sigma_k^2}$$
$$Q = (x - \hat{x})(x - \hat{x})^T \tag{15}$$

where t_k is the value of the kth principal component for the new data sample and σ_k represents the standard deviation of the kth principal component calculated from the training dataset.

Note that the upper control limits presented in the previous section are obtained by assuming that the data follow a multivariate normal distribution. In the case of linear PCA, if the original data are normal random vectors, the principal components and residuals also have multivariate normal distributions. Thus, the limits in Equations (6)–(10) can be directly applied to the statistical process monitoring using linear PCA. However, in the case of NLPCA, it is not guaranteed that the principal components follow a multivariate normal distribution since they are obtained by nonlinear transformations. Therefore, in this work, we take an alternative approach, where the upper control limits for two statistics are directly calculated from the data without assuming a specific type of probability distribution, given that a large dataset is available. For example, with 100 normal training data samples, the second largest T^2 (or Q) value is selected to be the upper control limit to achieve the false alarm rate of 0.01.

4. Process Monitoring of the Tennessee Eastman Process

Let us now evaluate the performance of the NLPCA-based statistical process monitoring with the Tennessee Eastman (TE) process as an illustrative example. The TE process is a benchmark chemical process [22], which involves five major process units (reactor, condenser, compressor, separator, and stripper) and eight chemical compounds (from A–H). A data sample from this process is a vector of 52 variables, and there are 21 programmed fault types. In this work, Faults 3, 9, and 15 are not considered since they are known to be difficult to detect due to no observable change in the data statistics [41]. The large dataset provided by Rieth et al. [42] is used in this study, and the data structure is summarized in Table 1. Note that Fault 21 is not included in this dataset, and thus not considered in this study. This dataset includes data samples from 500 simulation runs as the training data of each operating mode (normal and 20 fault types) and from another 500 simulation runs as the test data of each operating mode. From each simulation run, 500 data samples were obtained for training, while 960 data samples were recorded for testing. Different types of faults were introduced to the process after Sample Numbers 20 and 160 for the fault training and fault testing, respectively.

Table 1. Number of samples in each data subset.

	Normal Training	**Normal Test**	**Fault Training**	**Fault Test**
Simulation runs	500	500	500/fault type	500/fault type
Samples/run	500	960	20 normal	160 normal
			480 faulty	800 faulty

All the neural networks were trained for 1000 training epochs with the learning rate of 0.001. The rectified linear unit (ReLU) was used as the nonlinear activation function, which is defined as $\max(0,x)$. The ADAM optimizer [43] and the Xavier initialization [44] were used for the network training. The results reported here are the average values of 10 simulation runs.

In what follows, we first check the validity of the upper control limits estimated using the data only. Then, the performance of the process monitoring using NLPCA is evaluated by analyzing the effects of various neural network settings. Finally, the performance of the NLPCA-based process monitoring is compared with the linear-PCA-based process monitoring.

4.1. Upper Control Limit Estimation

Let us first compare the upper control limits calculated from the F- and χ^2-distributions and from the data distribution. The objective of this analysis is to show that the dataset used in this study is large enough so that the upper control limits can be well approximated from the data. Another dataset, which contains only 500 normal training samples, is also used for illustration purposes and was obtained from http://web.mit.edu/braatzgroup/links.html. This dataset and the large dataset from Rieth et al. [42] will be denoted as the S (small) dataset and L (large) dataset, respectively.

Equations (9) and (10) were used to calculate the upper control limits for the L dataset on the basis of the F- and χ^2-distributions, while those for the S dataset were computed using Equations (6)–(8). Linear PCA was used to calculate the upper control limits in the principal component space and the residual space with $\alpha = 0.99$, and the results are tabulated in Table 2. Note that, for the L dataset, the control limits obtained directly from the data had almost the same values as the ones from the F- and χ^2-distributions, while large deviations were observed for the S dataset. Thus, in the subsequent analyses, the control limits obtained directly from the data will be used for both linear PCA and nonlinear PCA.

Table 2. Upper control limits calculated from the probability distributions and the data.

	L Dataset					
	T^2 Statistic			**Q Statistic**		
l	**F-Distribution**	**Data**	**Difference**	**χ^2-Distribution**	**Data**	**Difference**
5	15.09	15.17	0.53%	54.89	54.99	0.20%
10	23.21	23.34	0.56%	42.06	41.85	0.51%
15	30.58	30.63	0.16%	34.46	34.41	0.15%
20	37.57	37.55	0.07%	27.23	27.23	0.02%
25	44.32	44.25	0.17%	19.82	19.90	0.44%
30	50.90	50.82	0.17%	12.99	13.03	0.31%
35	57.35	57.28	0.12%	7.13	7.21	1.14%
	S Dataset					
	T^2 Statistic			**Q Statistic**		
l	**F-Distribution**	**Data**	**Difference**	**χ^2-Distribution**	**Data**	**Difference**
5	15.43	14.47	6.20%	57.86	56.32	2.67%
10	24.05	21.41	11.00%	43.52	41.12	5.51%
15	32.10	28.14	12.34%	33.67	31.65	6.00%
20	39.93	36.62	8.31%	25.55	24.11	5.65%
25	47.70	43.38	9.06%	18.60	17.05	8.32%
30	55.46	49.55	10.65%	12.54	11.52	8.16%
35	63.27	55.58	12.15%	7.19	6.96	3.08%

4.2. Neural Network Hyperparameters

In this work, two neural network architectures were used to evaluate the NLPCA-based process monitoring. The NLPCA methods utilizing the networks shown in Figures 1 and 2 will be referred to as sm-NLPCA (simultaneous NLPCA) and p-NLPCA (parallel NLPCA), respectively. The performance of the process monitoring was evaluated by two indices, fault detection rate (FDR) and false alarm rate (FAR). The effects of the following hyperparameters were analyzed:

- Number of hidden layers
- Number of mapping/demapping nodes
- Number of nonlinear principal components

We designed four different types of neural networks to evaluate the effects of the hyperparameters listed above. Types 1 and 2 had five layers (three hidden layers), while seven layers (five hidden layers) were used for Types 3 and 4. In Types 1 and 3, the numbers of mapping/demapping nodes were fixed at specific values, while they were proportional to the number of principal components to be extracted for Types 2 and 4. The number of parameters for different network types are summarized in Table 3. The numbers for the network structures represent the number of nodes in each layer starting from the input layer. Note that, for the same network type, p-NLPCA always had fewer parameters compared to sm-NLPCA due to the network decoupling.

Table 3. Number of parameters for different neural network types.

Neural Network Structure		Number of Parameters	
		sm-NLPCA	p-NLPCA
Network type 1	52-100-f-100-52	$210f + 10{,}652$	$f + 10{,}852$
Network type 2	52-100f-f-100f-52	$200f^2 + 10{,}601f + 52$	$10{,}801f + 52$
Network type 3	52-100-50-f-50-100-52	$101f + 20{,}752$	$f + 10{,}852 + 10{,}000/f$
Network type 4	52-100f-50f-f-50f-100f-52	$10{,}100f^2 + 10{,}701f + 52$	$20{,}801f + 52$

Tables 4 and 5 show the process monitoring results for Types 1 and 2 and Types 3 and 4, respectively. Note that, in this analysis, only the average value of FDR (over all the fault types) is reported for brevity.

Table 4. Process monitoring results with varying the neural network size (with five layers).

Network Type 1: 52-100-f-100-52								
	sm-NLPCA				p-NLPCA			
	T^2		Q		T^2		Q	
f	FDR	FAR	FDR	FAR	FDR	FAR	FDR	FAR
5	0.4019	0.0085	0.7340	0.0144	0.4360	0.0091	0.7332	0.0136
10	0.5060	0.0089	0.7357	0.0146	0.5191	0.0094	0.7366	0.0141
15	0.4614	0.0097	0.7459	0.0156	0.5807	0.0094	0.7383	0.0146
20	0.4975	0.0093	0.7487	0.0174	0.5773	0.0094	0.7443	0.0160
Network Type 2: 52-100f-f-100f-52								
	sm-NLPCA				p-NLPCA			
	T^2		Q		T^2		Q	
f	FDR	FAR	FDR	FAR	FDR	FAR	FDR	FAR
5	0.4754	0.0098	0.7440	0.0221	0.4501	0.0094	0.7445	0.0190
10	0.5861	0.0081	0.7889	0.0661	0.5433	0.0087	0.7507	0.0229
15	0.6232	0.0081	0.8427	0.1698	0.5850	0.0083	0.7603	0.0295
20	0.6254	0.0081	0.9038	0.3610	0.5953	0.0091	0.7737	0.0364

Table 5. Process monitoring results with varying the neural network size (with seven layers).

Network Type 3: 52-100-50-f-50-100-52								
	sm-NLPCA				p-NLPCA			
	T^2		Q		T^2		Q	
f	FDR	FAR	FDR	FAR	FDR	FAR	FDR	FAR
5	0.4393	0.0086	0.7382	0.0141	0.3748	0.0088	0.7413	0.0143
10	0.4942	0.0085	0.7376	0.0142	0.4701	0.0090	0.7600	0.0140
15	0.5078	0.0088	0.7479	0.0157	0.4911	0.0099	0.7572	0.0134
20	0.5264	0.0088	0.7423	0.0146	0.5060	0.0093	0.7429	0.0122
Network Type 4: 52-100f-50f-f-50f-100f-52								
	sm-NLPCA				p-NLPCA			
	T^2		Q		T^2		Q	
f	FDR	FAR	FDR	FAR	FDR	FAR	FDR	FAR
5	0.4439	0.0079	0.8018	0.0895	0.3769	0.0090	0.7500	0.0254
10	0.5821	0.0069	0.9796	0.8005	0.4683	0.0095	0.7744	0.0470
15	0.6057	0.0075	0.9997	0.9928	0.4916	0.0089	0.7841	0.0537
20	0.6032	0.0078	1.0000	0.9993	0.5283	0.0093	0.8138	0.0961

The main trends to note are:

- The FDR in the residual space always showed a higher value than that in the principal component space, which matches the results reported in the literature where different techniques were utilized [45,46].
- In all network types, the FDR in the principal component space was improved with diminishing rates as the number of principal components increased.
- On the other hand, the number of principal components, which resulted in the best FDR value in the residual space, was different for different types of neural networks. As the size of the network became larger, the FAR in the residual space increased significantly, and sm-NLPCA completely failed when Network Type 4 was used, classifying the majority of the normal test data as faults. The main reason for this observation was the overfitting of the neural networks. Despite the network overfitting, the FDR in the principal component space was increased by adding more nodes in the mapping/demapping layers, while the FAR in the principal component space was not affected by such addition.

Regarding the last point, in the case of the demapping functions, the input had a lower dimension than the output, which made the problem of approximating demapping functions ill-posed. Thus, it can be speculated that the network overfitting mainly occurred during the reconstruction of the data (i.e., demapping functions were overfitted), leaving the results in the principal component space unaffected by the network overfitting. In addition to this, it was observed that, by including more nodes in the mapping/demapping layers, the average standard deviation of the principal components was increased by a factor of 2~8. This implies that, in Network Types 2 and 4, the normal operating region in the principal component space was more "loosely" defined (i.e., the normal data cluster had a larger volume) compared to Network Types 1 and 3, which can make the problem of approximating the demapping functions more ill-posed. It can also be a reason why the FAR in the principal component space was consistently low regardless of the network size and the degree of network overfitting.

Table 6 shows the FDR values obtained by adjusting the upper control limits such that the FAR became 0.01 for the normal test data. The following can be clearly seen:

- sm-NLPCA performed better than p-NLPCA in the principal component space, while p-NLPCA was better in the residual space.

Table 6. Fault detection rates with adjusted upper control limits (FAR = 0.01 for normal test data).

| | Network Type 1 | | | | Network Type 2 | | | |
| | sm-NLPCA | | p-NLPCA | | sm-NLPCA | | p-NLPCA | |
f	T^2	Q	T^2	Q	T^2	Q	T^2	Q
5	0.4082	0.7246	0.4406	0.7256	0.4767	0.7214	0.4526	0.7268
10	0.5098	0.7265	0.5208	0.7285	0.5917	0.7262	0.5472	0.7293
15	0.4629	0.7341	0.5823	0.7289	0.6284	0.7257	0.5891	0.7303
20	0.4992	0.7316	0.5791	0.7313	0.6305	0.7253	0.5975	0.7331

| | Network Type 3 | | | | Network Type 4 | | | |
| | sm-NLPCA | | p-NLPCA | | sm-NLPCA | | p-NLPCA | |
f	T^2	Q	T^2	Q	T^2	Q	T^2	Q
5	0.4441	0.7297	0.3787	0.7314	0.4523	0.7225	0.3804	0.7234
10	0.4997	0.7294	0.4737	0.7507	0.5920	0.6951	0.4706	0.7245
15	0.5116	0.7350	0.4916	0.7490	0.6128	0.6887	0.4967	0.7261
20	0.5297	0.7327	0.5079	0.7377	0.6096	0.6866	0.5307	0.7251

By comparing the results from Network Types 1 and 3, adding additional hidden layers was shown to improve the FDR in the principal component space for sm-NLPCA and the FDR in the residual space for p-NLPCA. However, the effects of such addition cannot be evaluated clearly for Network Types 2 and 4. Thus, in what follows, we apply the neural network regularization techniques to Network Types 2 and 4 and evaluate the effects of such techniques on the performance of NLPCA-based process monitoring.

4.3. Neural Network Regularization

In this analysis, we consider three different types of neural network regularization: dropout and L_1 and L_2 regularizations. Dropout is a neural network regularization technique, where randomly-selected nodes and their connections are dropped during the network training to prevent co-adaptation [47]. L_1 and L_2 regularizations prevent network overfitting by putting constraints on the L_1 norm and L_2 norm of the weight matrices, respectively.

The process monitoring results for Network Types 2 and 4 are tabulated in Tables 7 and 8, respectively. In the case of Network Type 2, dropout did not address the problem of overfitting for p-NLPCA, and therefore, the results obtained using dropout are not presented here.

Table 7. Process monitoring results with neural network regularization (Network Type 2).

sm-NLPCA											
No Regularization				L_1 Regularization				L_2 Regularization			
T^2		Q		T^2		Q		T^2		Q	
FDR	FAR	FDR	FAR	FDR	FAR	FDR	FAR	FDR	FAR	FDR	FAR
f											
10	0.5861	0.0081	0.7889	0.0661	0.4870	0.0096	0.7358	0.0155	0.5151	0.0093	0.7358	0.0249
15	0.6232	0.0081	0.8427	0.1698	0.4867	0.0099	0.7411	0.0274	0.5288	0.0095	0.7403	0.0175
20	0.6254	0.0081	0.9038	0.3610	0.5352	0.0100	0.7448	0.0208	0.5867	0.0088	0.7443	0.0204

p-NLPCA											
No Regularization				L_1 Regularization				L_2 Regularization			
T^2		Q		T^2		Q		T^2		Q	
FDR	FAR	FDR	FAR	FDR	FAR	FDR	FAR	FDR	FAR	FDR	FAR
f											
10	0.5433	0.0087	0.7507	0.0229	0.5201	0.0087	0.7532	0.0234	0.5065	0.0087	0.7503	0.0237
15	0.5850	0.0083	0.7603	0.0295	0.5711	0.0087	0.7649	0.0321	0.5532	0.0086	0.7647	0.0333
20	0.5995	0.0091	0.7737	0.0364	0.5905	0.0090	0.7769	0.0401	0.5868	0.0090	0.7796	0.0420

Table 8. Process monitoring results with neural network regularization (Network Type 4).

sm-NLPCA															
No Regularization				L_1 Regularization				L_2 Regularization				Dropout			
T^2		Q		T^2		Q		T^2		Q		T^2		Q	
FDR	FAR	FDR	FAR	FDR	FAR	FDR	FAR	FDR	FAR	FDR	FAR	FDR	FAR	FDR	FAR
f															
10	0.5821	0.0069	0.9796	0.8005	0.4086	0.0088	0.7360	0.0142	0.4881	0.0089	0.7535	0.0292	0.5091	0.0082	0.7435	0.0141
15	0.6057	0.0075	0.9997	0.9928	0.5507	0.0098	0.7375	0.0143	0.5697	0.0087	0.7685	0.0412	0.4866	0.0081	0.7600	0.0197
20	0.6032	0.0078	1.0000	0.9993	0.5870	0.0087	0.7331	0.0128	0.6159	0.0087	0.7975	0.0817	0.5346	0.0083	0.7852	0.0405

p-NLPCA															
No Regularization				L_1 Regularization				L_2 regularization				Dropout			
T^2		Q		T^2		Q		T^2		Q		T^2		Q	
FDR	FAR	FDR	FAR	FDR	FAR	FDR	FAR	FDR	FAR	FDR	FAR	FDR	FAR	FDR	FAR
f															
10	0.4683	0.0095	0.7744	0.0470	0.4469	0.0090	0.7695	0.0363	0.4721	0.0085	0.7715	0.0452	0.4988	0.0091	0.7577	0.0208
15	0.4916	0.0089	0.7841	0.0537	0.4671	0.0092	0.7774	0.0412	0.4940	0.0094	0.7903	0.0638	0.5120	0.0095	0.7602	0.0191
20	0.5283	0.0093	0.8138	0.0961	0.5114	0.0096	0.7774	0.0347	0.5433	0.0086	0.8034	0.0766	0.5481	0.0094	0.7714	0.0216

- In most cases, neural network regularization degraded the process monitoring performance in the principal component space with p-NLPCA of Network Type 4 regularized by dropout being the only exception.

- On the other hand, it dramatically reduced the FAR values in the residual space for sm-NLPCA, while such reduction was not significant for p-NLPCA (the FAR in the residual space even increased for Network Type 2). This indicates that overfitting was a problem for the residual space detection using sm-NLPCA.

Table 9 shows the process monitoring results obtained by adjusting the upper control limits as mentioned in the previous analysis. Overall, p-NLPCA performed better than sm-NLPCA, while sm-NLPCA was better than p-NLPCA in the principal component space when Network Type 4 was used. By comparing the results provided in Tables 6 and 9, it can be seen that having more nodes in the mapping/demapping layers is only beneficial to the process monitoring in the principal component space. Putting more nodes implies the increased complexity of the functions approximated by neural networks. Thus, it makes the problem of approximating demapping functions more ill-posed and has the potential to be detrimental to the performance of process monitoring in the residual space. Although neural network regularization techniques can reduce the stiffness and complexity of the functions approximated by neural networks [48], they seem to be unable to define better boundaries for one-class classifiers.

Table 9. Process monitoring results with network regularization and adjusted upper control limits (FAR = 0.01 for the normal test data).

	Network Type 2															
	sm-NLPCA								p-NLPCA							
	No Regularization		L_1		L_2		Dropout		No Regularization		L_1		L_2		Dropout	
f	T^2	Q	T^2	Q	T^2	Q	T^2	Q	T^2	Q	T^2	Q	T^2	Q	T^2	Q
10	0.5917	0.7262	0.4884	0.7254	0.5173	0.7248	-	-	0.5472	0.7293	0.5239	0.7309	0.5113	0.7280	-	-
15	0.6284	0.7257	0.4872	0.7266	0.5304	0.7268	-	-	0.5891	0.7303	0.5746	0.7316	0.5577	0.7300	-	-
20	0.6305	0.7253	0.5357	0.7236	0.5904	0.7248	-	-	0.5975	0.7331	0.5932	0.7319	0.5897	0.7321	-	-
	Network Type 4															
	sm-NLPCA								p-NLPCA							
	No Regularization		L_1		L_2		Dropout		No Regularization		L_1		L_2		Dropout	
f	T^2	Q	T^2	Q	T^2	Q	T^2	Q	T^2	Q	T^2	Q	T^2	Q	T^2	Q
10	0.5920	0.6951	0.4145	0.7279	0.4922	0.7249	0.5148	0.7357	0.4706	0.7245	0.4517	0.7285	0.4780	0.7235	0.5015	0.7372
15	0.6128	0.6887	0.5519	0.7294	0.5737	0.7244	0.4928	0.7419	0.4967	0.7261	0.4705	0.7309	0.4936	0.7245	0.5134	0.7422
20	0.6096	0.6866	0.5908	0.7277	0.6194	0.7225	0.5404	0.7492	0.5307	0.7251	0.5133	0.7338	0.5476	0.7256	0.5499	0.7483

4.4. Network Training Objective Function

Let us now evaluate the effects of different objective functions on the process monitoring performance. For illustration purposes, only Network Type 4 was considered in this analysis. All the values of α_k were set to one for the hierarchical error, and the corrupted input was generated by using a Gaussian noise of zero mean and 0.1 standard deviation for the denoising criterion. L_2 regularization and L_1 regularization were used to prevent overfitting for sm-NLPCA and p-NLPCA, respectively. Tables 10 and 11 summarize the process monitoring results obtained by using the autoassociative neural networks trained with different objective functions.

The following are the major trends to note:

- The monitoring performance of sm-NLPCA in the principal component space became more robust by using the hierarchical error, showing similar FDR values regardless of the number of principal components.
- However, the FDR value in the principal component space scaled better with the reconstruction error objective function. On the other hand, the monitoring performance of p-NLPCA in the principal component space became more sensitive to the number of principal components with the hierarchical error as the objective function. As a result, the FDR value in the principal component space was improved when the number of principal components was 20.

- While the hierarchical error provided a slight improvement to the monitoring performance in the residual space for sm-NLPCA, it degraded the performance of p-NLPCA in the residual space.
- The denoising criterion was beneficial to both NLPCA methods in the residual space, improving the FDR values when the upper control limits were adjusted. The monitoring performance of p-NLPCA in the principal component space was not affected by using the denoising criterion, while that of sm-NLPCA deteriorated.

Table 10. Process monitoring results with different neural network training objective functions.

						sm-NLPCA						
	Reconstruction Error				**Hierarchical Error**				**Denoising**			
	T^2		Q		T^2		Q		T^2		Q	
f	FDR	FAR	FDR	FAR	FDR	FAR	FDR	FAR	FDR	FAR	FDR	FAR
10	0.4881	0.0089	0.7535	0.0292	0.5869	0.0099	0.7312	0.0137	0.4203	0.0096	0.7347	0.0147
15	0.5697	0.0087	0.7685	0.0412	0.5940	0.0084	0.7367	0.0161	0.5253	0.0090	0.7393	0.0150
20	0.6159	0.0087	0.7975	0.0817	0.5918	0.0082	0.7628	0.0398	0.5821	0.0084	0.7483	0.0181
						p-NLPCA						
	Reconstruction Error				**Hierarchical Error**				**Denoising**			
	T^2		Q		T^2		Q		T^2		Q	
f	FDR	FAR	FDR	FAR	FDR	FAR	FDR	FAR	FDR	FAR	FDR	FAR
10	0.4469	0.0090	0.7695	0.0363	0.3854	0.0092	0.7705	0.0428	0.4260	0.0093	0.7374	0.0121
15	0.4671	0.0092	0.7774	0.0412	0.4736	0.0085	0.7877	0.0572	0.4749	0.0095	0.7445	0.0123
20	0.5114	0.0096	0.7774	0.0347	0.5397	0.0089	0.7975	0.0658	0.5122	0.0093	0.7469	0.0121

Table 11. Process monitoring results with different neural network training objective functions and adjusted upper control limits (FAR = 0.01 for normal test data).

		sm-NLPCA				
	Reconstruction Error		**Hierarchical Error**		**Denoising**	
f	T^2	Q	T^2	Q	T^2	Q
10	0.4922	0.7249	0.5874	0.7241	0.4223	0.7258
15	0.5737	0.7244	0.5985	0.7259	0.5292	0.7295
20	0.6194	0.7225	0.5975	0.7281	0.5879	0.7315
		p-NLPCA				
	Reconstruction Error		**Hierarchical Error**		**Denoising**	
f	T^2	Q	T^2	Q	T^2	Q
10	0.4517	0.7285	0.3890	0.7235	0.4291	0.7323
15	0.4705	0.7309	0.4797	0.7258	0.4766	0.7393
20	0.5133	0.7338	0.5429	0.7245	0.5146	0.7412

4.5. Neural Network Training Epochs

In this analysis, the effects of neural network training epochs are analyzed. To this end, Network Type 4 with 15 principal components was trained, and the neural network parameters were saved at every 10 epochs. Figure 4 shows how different values evolve as the neural networks are trained. For the reference case, the neural networks were trained without any regularization and with the reconstruction error as the objective function. It can be clearly seen that the network overfitting (which is captured by the difference between the solid and dashed black lines) resulted in high FAR values in the residual space (which is captured by the difference between the solid and dashed red lines), while it did not affect the FAR values in the principal component space (which is captured by the difference between the solid and dashed blue lines).

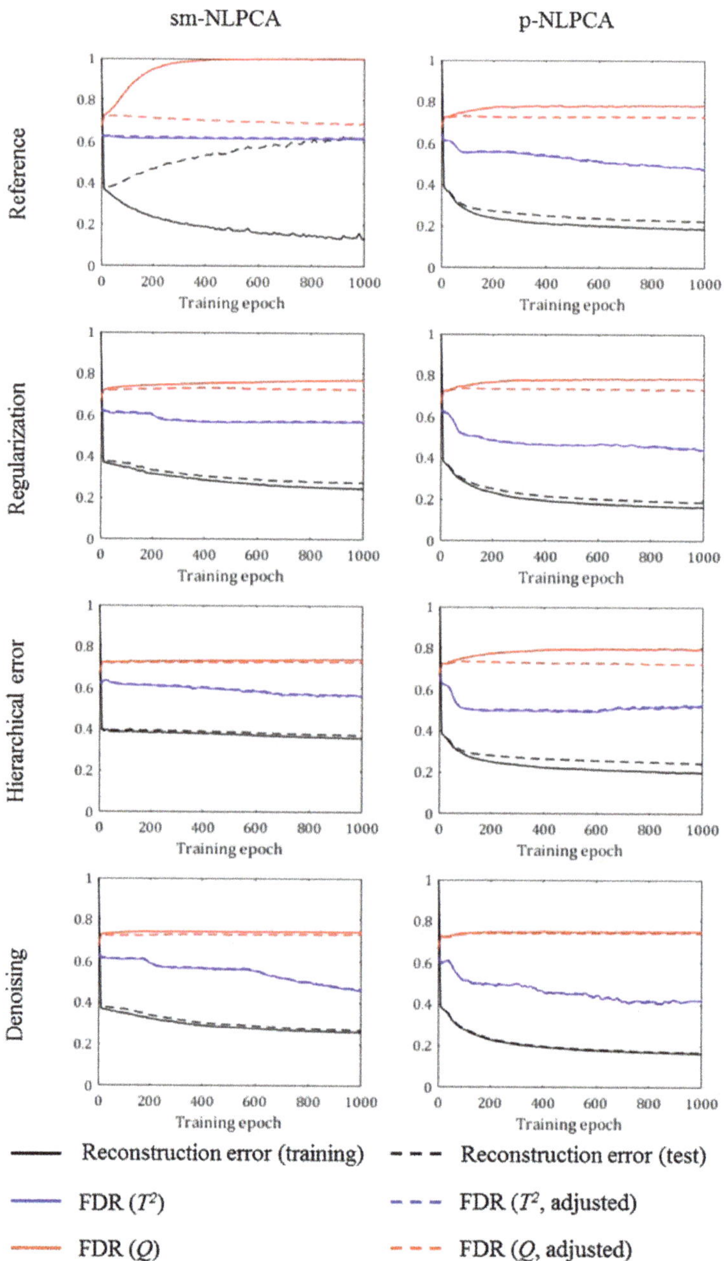

Figure 4. Change in the process monitoring results with respect to the training epoch.

Note the following:

- Despite the decrease in the reconstruction error over a wide range of training epochs, the adjusted FDR value in the residual space increased during only the first few training epochs and was kept almost constant in the rest of the training.
- In some cases, there was even a tendency that the adjusted FDR value in the principal component space decreased as the network was trained more.

Thus, during the network training, it was required to monitor both reconstruction error and process monitoring performance indices, and early stopping needed to be applied as necessary to ensure high monitoring performances. Nonetheless, from the above observations, it can be concluded that the objective functions available in the literature, which focus on the reconstruction ability of the autoassociative neural networks, may not be most suitable for the design of one-class classifiers. This necessitates the development of alternative training algorithms of autoassociative neural networks to improve the performance of neural-network-based one-class classifiers.

4.6. Comparison with Linear-PCA-Based Process Monitoring

Let us finally compare the NLPCA-based process monitoring with the linear-PCA-based one. For the linear-PCA-based process monitoring, based on the parallel analysis [35], the number of principal components to be retained was selected as 15. The same number of principal components was used in the NLPCA-based process monitoring. For sm-NLPCA, the following setting was used: Network Type 2, no regularization, reconstruction error as the objective function. For p-NLPCA, the following setting was used: Network Type 3, no regularization, reconstruction error as the objective function.

Table 12 tabulates the process monitoring results obtained using three different PCA methods. Note that the upper control limits were adjusted to have FAR values of 0.01. The main trends observed were:

- Compared to the linear PCA, the process monitoring results in both spaces were improved slightly by using sm-NLPCA. The most significant improvements were obtained for Faults 4 and 10 in the principal component space and Faults 5 and 10 in the residual space.
- On the other hand, p-NLPCA showed a lower performance than linear PCA in the principal component space, while the performance in the residual space was significantly improved. The adjusted FDR value in the residual space from p-NLPCA was higher than that from the linear PCA for all the fault types, with Faults 5, 10, and 16 being the most significant ones.

Let us consider two cases that illustrate the advantages of p-NLPCA over the linear PCA as the basis for the process monitoring system design. Figure 5 shows the Q statistic values (black solid line) for one complete simulation run with Fault 5, along with the upper control limit (dashed red line). Although both linear PCA and p-NLPCA detected the fault very quickly (fault introduced after Sample Number 160 and detected at Sample Number 162), in the case of the linear PCA, the Q statistic value dropped below the upper control limit after around Sample Number 400. The Q statistic value calculated from p-NLPCA did not decrease much, indicating that the fault was not yet removed from the system.

Table 12. Detailed process monitoring results obtained using linear PCA and two NLPCA methods (FAR = 0.01 for normal test data).

Fault ID	Linear PCA		sm-NLPCA		p-NLPCA	
	T^2	Q	T^2	Q	T^2	Q
1	0.9926	0.9965	0.9931	0.9963	0.9891	0.9967
2	0.9848	0.9864	0.9849	0.9863	0.9841	0.9870
4	0.0916	0.9939	0.1375	0.9830	0.0939	0.9983
5	0.2404	0.1192	0.2461	0.2028	0.3312	0.7922
6	0.9896	1.0000	0.9939	1.0000	0.9968	1.0000
7	1.0000	0.9998	0.9994	1.0000	0.9990	1.0000
8	0.9637	0.9541	0.9634	0.9651	0.8827	0.9656
10	0.1960	0.1586	0.2436	0.1896	0.0860	0.2195
11	0.2953	0.6901	0.2995	0.6670	0.1415	0.7141
12	0.9785	0.9563	0.9773	0.9769	0.9008	0.9810
13	0.9353	0.9425	0.9378	0.9436	0.8974	0.9452
14	0.9549	0.9995	0.9457	0.9995	0.7661	0.9996
16	0.0706	0.0974	0.0944	0.1131	0.0361	0.1547
17	0.7177	0.8816	0.7173	0.8779	0.6200	0.8922
18	0.9270	0.9337	0.9270	0.9345	0.9131	0.9358
19	0.0819	0.0845	0.0431	0.1114	0.0222	0.1003
20	0.2535	0.4395	0.2775	0.4420	0.1271	0.4737
Average	0.6278	0.7196	0.6342	0.7288	0.5757	0.7739

The contribution plots of Q statistic for Fault 1, which involves a step change in the flowrate of the A feed stream, are provided in Figure 6. In the case of the linear PCA, the variables with the highest contribution to the Q statistic were Variables 4 and 6, which are the flowrates of Stream 4 (which contains both A and C) and the reactor feed rate, respectively. Note that although these variables were also highly affected by the fault, they were not the root cause of the fault. On the other hand, in the case of p-NLPCA, the variables with the highest contribution to the Q statistic were Variables 1 and 44, which both represent the flowrate of the A feed stream, the root cause of the fault. Thus, it can be concluded that the process monitoring using p-NLPCA showed some potential to perform better at identifying the root cause of the fault introduced to the system.

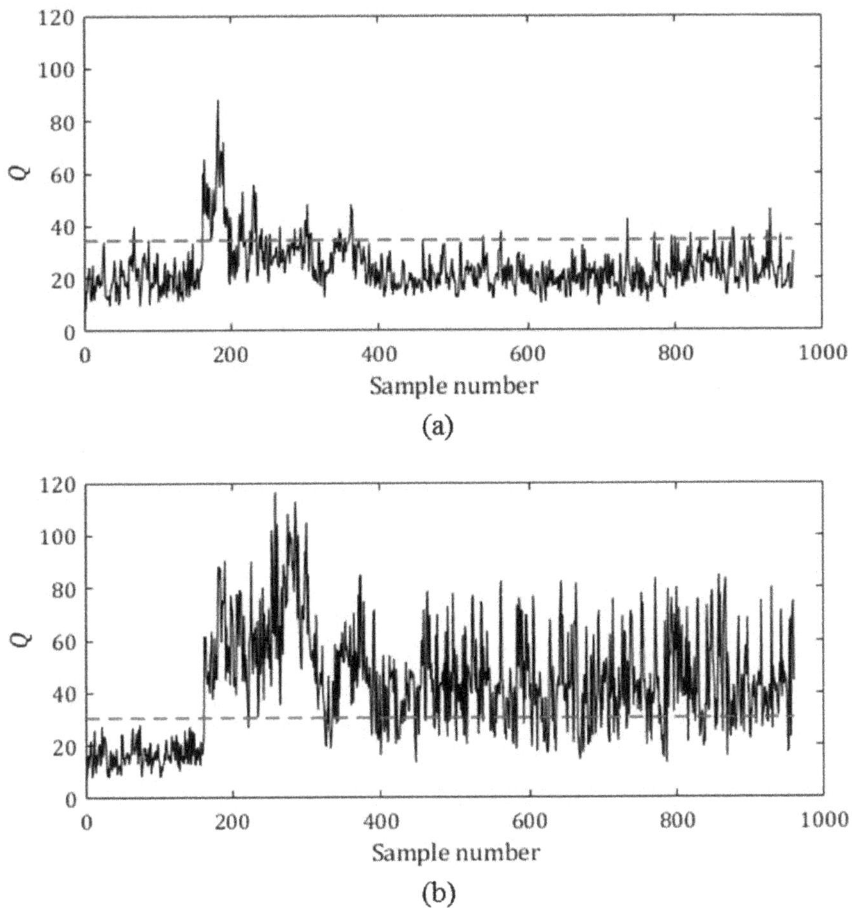

Figure 5. Process monitoring results in the residual space for Fault 5: (a) linear PCA; (b) p-NLPCA.

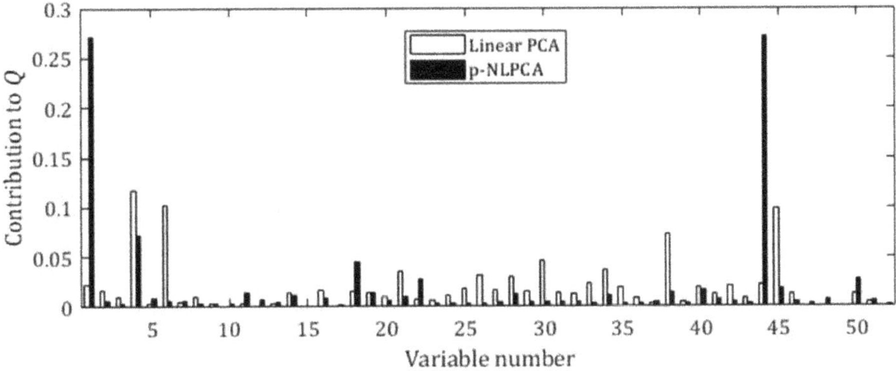

Figure 6. Contribution plots of Q statistics for Fault 1.

5. Conclusions

The statistical process monitoring problem of the Tennessee Eastman process was considered in this study using autoassociative neural networks to define normal operating regions. Using the large dataset allowed us to estimate the upper control limits for the process monitoring directly from the data distribution and to train relatively large neural networks without overfitting. It was shown that the process monitoring performance was very sensitive to the neural network settings such as neural network size and neural network regularization. p-NLPCA was shown to be more effective for the process monitoring than the linear PCA and sm-NLPCA in the residual space, while its performance was worse than the others in the principal component space. p-NLPCA also showed the potential of better fault diagnosis capability than the linear PCA, locating the root cause more correctly for some fault types.

There still exist several issues that need to be addressed to make autoassociative neural networks more attractive as a tool for statistical process monitoring. First, a systematic procedure needs to be developed to provide a guideline for the design of optimal autoassociative neural networks to be used for the statistical process monitoring. Furthermore, a new neural network training algorithm may be required to extract principal components that are more relevant to the process monitoring tasks. Finally, the compatibility of different techniques to define the upper control limits, other than the T^2 and Q statistics, needs to be extensively evaluated.

Author Contributions: Conceptualization, S.H. and J.H.L.; formal analysis, S.H.; funding acquisition, J.H.L.; investigation, S.H.; methodology, S.H.; project administration, J.H.L.; software, S.H.; supervision, J.H.L.; visualization, S.H.; writing, original draft, S.H.; writing, review and editing, J.H.L.

Funding: This work was supported by the Advanced Biomass R&D Center (ABC) of the Global Frontier Project funded by the Ministry of Science and ICT (ABC-2011-0031354).

Conflicts of Interest: The authors declare no conflict of interest.

References

1. Qin, S.J. Survey on data-driven industrial process monitoring and diagnosis. *Annu. Rev. Control* **2012**, *36*, 220–234. [CrossRef]
2. He, P.H.; Wang, J. Statistical process monitoring as a big data analytics tool for smart manufacturing. *J. Process Control* **2018**, *67*, 35–43. [CrossRef]
3. Yu, J. Local and global principal component analysis for process monitoring. *J. Process Control* **2012**, *22*, 1358–1373. [CrossRef]
4. Rato, T.; Reis, M.; Schmitt, E.; Hubert, M.; De Ketelaere, B. A systematic comparison of PCA-based statistical process monitoring methods for high-dimensional, time-dependent processes. *AIChE J.* **2016**, *62*, 1478–1493. [CrossRef]
5. Hamadache, M.; Lee, D. Principal component analysis based signal-to-noise ratio improvement for inchoate faulty signals: Application to ball bearing fault detection. *Int. J. Control Autom.* **2017**, *15*, 506–517. [CrossRef]
6. Khatib, S.; Daoutidis, P.; Almansoori, A. System decomposition for distributed multivariate statistical process monitoring by performance driven agglomerative clustering. *Ind. Eng. Chem. Res.* **2018**, *57*, 8283–8398. [CrossRef]
7. Kruger, W.; Dimitriadis, G. Diagnosis of process faults in chemical systems using a local partial least squares approach. *AIChE J.* **2008**, *54*, 2581–2596. [CrossRef]
8. Li, G.; Qin, S.J.; Zhou, D. Geometric properties of partial least squares for process monitoring. *Automatica* **2010**, *46*, 204–210. [CrossRef]
9. Jia, Q.; Zhang, Y. Quality-related fault detection approach based on dynamic kernel partial least squares. *Chem. Eng. Res. Des.* **2016**, *106*, 242–252. [CrossRef]
10. Lee, J.; Kang, B.; Kang, S. Integrating independent component analysis and local outlier factor for plant-wide process monitoring. *J. Process Control* **2011**, *21*, 1011–1021. [CrossRef]
11. Zhu, J.; Ge, Z.; Song, Z. Non-Gaussian industrial process monitoring with probabilistic independent component analysis. *IEEE Trans. Autom. Sci. Eng.* **2017**, *14*, 1309–1319. [CrossRef]

12. Erfani, S.M.; Rajasegarar, S.; Karunasekera, S.; Leckie, C. High-dimensional and large-scale anomaly detection using a linear one-class SVM with deep learning. *Pattern Recogn.* **2016**, *58*, 121–134. [CrossRef]

13. Onel, M.; Kieslich, C.A.; Pistikopoulos, E.N. A nonlinear support vector machine-based feature selection approach for fault detection and diagnosis: Application to the Tennessee Eastman process. *AIChE J.* **2019**, *65*, 992–1005. [CrossRef]

14. Heo, S.; Lee, J.H. Fault detection and classification using artificial neural networks. *IFAC–PapersOnLine* **2018**, *51*, 470–475. [CrossRef]

15. Zhang, Z.; Zhao, J. A deep belief network based fault diagnosis model for complex chemical processes. *Comput. Chem. Eng.* **2017**, *107*, 395–407. [CrossRef]

16. Wu, H.; Zhao, J. Deep convolutional neural network model based chemical process fault diagnosis. *Comput. Chem. Eng.* **2018**, *115*, 185–197. [CrossRef]

17. Choi, E.; Schuetz, A.; Stewart, W.F.; Sun, J. Using recurrent neural network models for early detection of heart failure onset. *J. Am. Med. Assoc.* **2017**, *24*, 361–370. [CrossRef]

18. Kramer, M.A. Nonlinear principal component analysis using autoassociative neural networks. *AIChE J.* **1991**, *37*, 233–243. [CrossRef]

19. Chong, Y.S.; Tay, Y.H. Abnormal Event Detection in Videos Using Spatiotemporal Autoencoder. In Proceedings of the International Symposium on Neural Networks, Hokkaido, Japan, 21–26 June 2017; pp. 189–196.

20. Ma, J.; Ni, S.; Xie, W.; Dong, W. Deep auto-encoder observer multiple-model fast aircraft actuator fault diagnosis algorithm. *Int. J. Control Autom.* **2017**, *15*, 1641–1650. [CrossRef]

21. Ribeiro, M.; Lazzaretti, A.E.; Lopes, H.S. A study of deep convolutional auto-encoders for anomaly detection in videos. *Pattern Recogn. Lett.* **2018**, *105*, 13–22. [CrossRef]

22. Downs, J.; Vogel, E. A plant-wide industrial process control problem. *Comput. Chem. Eng.* **1993**, *17*, 245–255. [CrossRef]

23. Yan, W.; Guo, P.; Gong, L.; Li, Z. Nonlinear and robust statistical process monitoring based on variant autoencoders. *Chemom. Intell. Lab.* **2016**, *158*, 31–40. [CrossRef]

24. Jiang, L.; Ge, Z.; Song, Z. Semi-supervised fault classification based on dynamic sparse stacked auto-encoders model. *Chemom. Intell. Lab.* **2017**, *168*, 72–83. [CrossRef]

25. Zhang, Z.; Jiang, T.; Li, S.; Yang, Y. Automated feature learning for nonlinear process monitoring—An approach using stacked denoising autoencoder and k-nearest neighbor rule. *J. Process Control* **2018**, *64*, 49–61. [CrossRef]

26. Heo, S.; Lee, J.H. Parallel neural networks for improved nonlinear principal component analysis. *Comput. Chem. Eng.* **2019**, *127*, 1–10. [CrossRef]

27. Qin, S.J. Statistical process monitoring: Basics and beyond. *J. Chemom.* **2003**, *17*, 480–502. [CrossRef]

28. Tracy, N.D.; Young, J.C.; Mason, R.L. Multivariate control charts for individual observations. *J. Qual. Technol.* **1992**, *24*, 88–95. [CrossRef]

29. Nomikos, P.; MacGregor, J.F. Multivariate SPC charts for monitoring batch processes. *Technometrics* **1995**, *37*, 41–59. [CrossRef]

30. Box, G.E.P. Some theorems on quadratic forms applied in the study of analysis of variance problems, I. Effect of inequality of variance in the one-way classification. *Ann. Math. Stat.* **1954**, *25*, 290–302. [CrossRef]

31. Miller, P.; Swanson, R.E.; Heckler, C.E. Contribution plots: A missing link in multivariate quality control. *Appl. Math. Comp. Sci.* **1998**, *8*, 775–792.

32. Conlin, A.K.; Martin, E.B.; Morris, A.J. Confidence limits for contribution plots. *J. Chemom.* **2000**, *14*, 725–736. [CrossRef]

33. Westerhuis, J.A.; Gurden, S.P.; Silde, A.K. Generalized contribution plots in multivariate statistical process monitoring. *Chemom. Intell. Lab.* **2000**, *51*, 95–114. [CrossRef]

34. Malinowski, E.R. *Factor Analysis in Chemistry*; Wiley: New York, NY, USA, 2002.

35. Cattell, R.B. The scree test for the number of factors. *Multivar. Behav. Res.* **1966**, *1*, 245–276. [CrossRef] [PubMed]

36. Horn, J.L. A rationale and test for the number of factors in factor analysis. *Psychometrika* **1965**, *30*, 179–185. [CrossRef] [PubMed]

37. Eastment, H.T.; Krzanowski, W.J. Cross-validatory choice of the number of components from a principal component analysis. *Technometrics* **1982**, *24*, 73–77. [CrossRef]

38. Valle, S.; Li, W.; Qin, S.J. Selection of the number of principal components: The variance of the reconstruction error criterion with a comparison to other methods. *Ind. Eng. Chem. Res.* **1999**, *38*, 4389–4401. [CrossRef]

39. Scholz, M.; Vigário, R. Nonlinear PCA: A new hierarchical approach. In Proceedings of the 10th European Symposium on Artificial Neural Networks, Computational Intelligence and Machine Learning (ESANN 2002), Bruges, Belgium, 24–26 April 2002; pp. 439–444.

40. Vincent, P.; Larochelle, H.; Lajoie, I.; Bengio, Y.; Manzagol, P.A. Stacked denoising autoencoders: Learning useful representations in a deep network with a local denoising criterion. *J. Mach. Learn. Res.* **2010**, *11*, 3371–3408.

41. Zhang, Y. Enhanced statistical analysis of nonlinear processes using KPCA, KICA and SVM. *Comput. Chem. Eng.* **2009**, *64*, 801–811. [CrossRef]

42. Rieth, C.A.; Amsel, B.D.; Tran, R.; Cook, M.B. Additional Tennessee Eastman Process Simulation Data for Anomaly Detection Evaluation; Harvard Dataverse, Cambridge, MA, United States. Available online: https://doi.org/10.7910/DVN/6C3JR1 (accessed on 5 January 2019).

43. Kingma, D.; Ba, J. Adam: A Method for Stochastic Optimization. *arXiv* **2014**, arXiv:1412.6980.

44. Glorot, X.; Bengio, Y. Understanding the difficulty of training deep feedforward neural networks. In Proceedings of the 13th International Conference on Artificial Intelligence and Statistics, Montreal, QC, Canada, 13–15 March 2010; pp. 249–256.

45. Russel, E.L.; Chiang, L.H.; Braats, R.D. Fault detection in industrial processes using canonical variate analysis and dynamic principal component analysis. *Chemom. Intell. Lab.* **2000**, *51*, 81–93. [CrossRef]

46. Gajjar, S.; Kulahci, M.; Palazoglu, A. Real-time fault detection and diagnosis using sparse principal component analysis. *J. Process Control* **2018**, *67*, 112–128. [CrossRef]

47. Srivastava, N.; Hinton, G.; Krizhevsky, A.; Sutskever, I.; Salakhutdinov, R. Dropout: A simple way to prevent neural networks from overfitting. *J. Mach. Learn. Res.* **2014**, *15*, 1929–1958.

48. Girosi, F.; Jones, M.; Poggio, T. Regularization theory and neural networks architectures. *Neural Comput.* **1995**, *7*, 219–269. [CrossRef]

Article

An Intensified Reactive Separation Process for Bio-Jet Diesel Production

Miriam García-Sánchez [1], Mauricio Sales-Cruz [2], Teresa Lopez-Arenas [2], Tomás Viveros-García [1] and Eduardo S. Pérez-Cisneros [1,*]

[1] Departamento de Ingeniería de Procesos e Hidráulica, Universidad Autónoma Metropolitana-Iztapalapa, Av. San Rafael Atlixco No. 186, Col. Vicentina, Ciudad de México C.P. 09340, Mexico; ing.mgarcia11@gmail.com (M.G.-S.); tvig@xanum.uam.mx (T.V.-G.)

[2] Departamento de Procesos y Tecnología, Universidad Autónoma Metropolitana-Cuajimalpa, Avenida Vasco de Quiroga No. 4871, Colonia Santa Fe, Delegación Cuajimalpa de Morelos, Ciudad de México C.P. 05300, Mexico; asales@correo.cua.uam.mx (M.S.-C.); mtlopez@correo.cua.uam.mx (T.L.-A.)

[*] Correspondence: espc@xanum.uam.mx; Tel.: +52-55-5804-4900

Received: 17 June 2019; Accepted: 23 September 2019; Published: 25 September 2019

Abstract: An intensified three-step reaction-separation process for the production of bio-jet diesel from tryglycerides and petro-diesel mixtures is proposed. The intensified reaction-separation process considers three sequentially connected sections: (1) a triglyceride hydrolysis section with a catalytic heterogeneous reactor, which is used to convert the triglycerides of the vegetable oils into the resultant fatty acids. The separation of the pure fatty acid from glycerol and water is performed by a three-phase flash drum and two conventional distillation columns; (2) a co-hydrotreating section with a reactive distillation column used to perform simultaneously the deep hydrodesulphurisation (HDS) of petro-diesel and the hydrodeoxigenation (HDO), decarbonylation and decarboxylation of the fatty acids; and (3) an isomerization-cracking section with a hydrogenation catalytic reactor coupled with a two-phase flash drum is used to produce bio-jet diesel with the suitable fuel features required by the international standards. Intensive simulations were carried out and the effect of several operating variables of the three sections (triglyceride-water feed ratio, oleic acid-petro-diesel feed ratio, hydrogen consumption) on the global intensified process was studied and the optimal operating conditions of the intensified process for the production of bio-jet diesel were achieved.

Keywords: bio-jet diesel; co-hydrotreating; hydrodesulphurisation; hydrodeoxigenation; reactive distillation

---

1. Introduction

The area of process system engineering (PSE) has been rapidly developing since the 1950s, reflecting the remarkable growth of the oil, gas, petrochemical and biotechnological industries and their increasing economical and societal impact. It can be said that the roots of this field go back to the Sargent's pioneering school in United Kingdom [1,2]. Modelling, simulation and optimization (MSO) of large-scale (product or process) systems is a core technology to deal with the complexity and connectivity of chemical processes and their products at multiple scales [3,4]. These technologies have been implemented in easy-to-use software systems that allow the systematic solution for problem solving practitioners. The systematic (explicit or implicit) generation, evaluation and integration of a comprehensive set of design alternatives is considered a key concern for the optimal design. This systematic integration associates the PSE strongly with its traditional focus in complete plants for both, *process intensification* [5,6] and to *chemical product design* [7]. Specifically, process intensification involves the development of novel apparatuses and techniques that, in comparison with those commonly used, are expected to bring enhancements in manufacturing and processing, substantially reducing the equipment-size/production-capacity ratio, energy consumption, or waste production, and ultimately resulting in cheaper, sustainable technologies [5]. Also, it is known that the whole scope of process intensification generally can be divided into two areas: (i) *process-intensifying equipment*, such as novel reactors, and exhaustive mixing, heat and mass transfer devices and, (ii) *process-intensifying methods*, such as new or hybrid separations, integration of reaction and separation, heat exchange or phase transition (i.e., reactive distillation), techniques using alternative energy sources (light, ultrasound, etc.), and new process-control methods (such as intentional unsteady-state operation).

The concerns about energy demand are obliging the oil-based fuels consumer countries to reconsider their energy policies to promote the investigation on trustworthy alternatives to conventional fuels. Thus, the bio-jet diesel has arisen as an alternative for petro-diesel jet fuels used in the aviation enterprises. Specifically, for the jet fuel, the International Air Transport Association (IATA) estimated that the consumption of the jet diesel would increase every year by 5% till 2030 [8]. Also, due to the growing of the flight demand and the strong regulations to diminish the CO_2, IATA established a carbon neutral reduction up to 50% by 2050. In this way, the bio-jet fuel obtained from vegetable oils of from mixtures of vegetable oils and petro-diesel, can be contemplated as one of the most favourable solutions to satisfy the global demand. To now, there has been identified four main routes to obtain the bio-jet diesel: (i) oil-to-jet (deoxygenation of triglyceride and consequent hydrocracking), [9] (ii) gas-to-jet (gasification/Fischer-Tropsch reaction followed by hydrocracking), [10] (iii) alcohol-to-jet (dehydration of alcohols and successive oligomerization), [11] and (iv) sugar-to-jet (several catalytic conversions of sugars) [12]. Table 1 shows the key conversion steps, the catalyst used, the companies producing the jet-fuel and the feedstocks considered for each route to obtain bio-jet fuel.

For the conversion of oil to bio-jet fuel (OTJ), different type of oil feedstock has various converting pathways. The common pathways include the hydrogenated esters and fatty acids (HEFA) and catalytic hydrothermolysis (CH). The feedstocks for HEFA are non-edible vegetable oils, used cooking oil, and animal fats. While the feedstocks for CH are algal oils or oil plant. HEFA is a process to hydrotreat the triglycerides, saturated or unsaturated fatty acids in the non-edible vegetable oils, used cooking oils and animal fats to produce bio-jet fuels. The process is generally divided into two steps. The first step is converting unsaturated fatty acids and triglycerides into saturated fatty acid by catalytic hydrogenation, the triglycerides occur a β-hydrogen elimination reaction to yield a fatty acid during the process. The saturated fatty acid is converted to C_{15}–C_{18} straight chain alkanes by hydrodeoxygenation and decarboxylation. The co-products are propane, H_2O, CO and CO_2. The early-developed catalysts for this step are noble metals supported with zeolites or oxides, and later shifted to other transition metals, such as Ni, Mo, Co, Mo or their supported bimetallic composites due to catalyst deactivation by poisoning, production of cracking species and process costs. The second step of HEFA is the cracking and isomerization reactions: the deoxygenated straight chain alkanes are further selectively hydrocracked and deep isomerized to generate highly branched alkanes mixed liquid fuels. The common catalysts for this step are Pt, Ni or other precious metals supported by activated carbon, Al_2O_3, zeolite molecular sieves. The bio-jet fuels produced by HEFA, as high energy biofuels, can be directly used in flight engine even without blending. The fuel has high thermal stability, good cold flow behaviours, high cetane number, and low tailpipe emissions, while has low aromatic content, which would cause fuel low lubricity and fuel leakage problems. Another pathway to convert algal or oil plant to jet fuel is CH, which is also named hydrothermal liquefaction (HTL). The integrated process has three steps, including pretreatment of triglycerides, CH conversion and post-refining steps. The pretreatments including conjugation, cyclization, and cross-linking, which are aiming to improve the molecular structures. The products undergo a cracking and hydrolysis reaction with the help of water and catalysts. Then it occurs catalytic decarboxylation and dehydration during the CH process. Last, post-refining hydrotreating and fractionation are designed to convert straight-chain, branched and cyclo-olefins into alkanes. The conversion process of lignocellulosic biomass to jet-fuel have advantages in lowering cost, feasible availability and no competition with food supplies. Hydroprocessed depolymerized cellulosic jet (HDCJ) is an oil upgrading technology to convert bio-oils produced from the pyrolysis or hydrothermal of the lignocellulose into a jet fuel by hydrotreating. The main technology for bio-oil upgrading is a two-step hydroprocessing. First, the bio-oil is hydrotreated with the help of catalyst under mild conditions. Organic could be used to promote hydrodeoxygenation of bio-oil and overcome coke formation. Second, conventional hydrogenation setup and catalyst were used under high temperature for obtaining hydrocarbon fuel. The HDCJ process could produce high aromatic content, low oxygen content and few impurities jet fuel. However, there is high hydrogen consumption and deoxygenation requirements in this process, which can make a considerable expense. Moreover, the short catalyst lifetime and modest hydrocarbon yields can be challenges for used in aviation.

For the conversion of gas to bio-jet fuel (GTJ) the Fischer-Tropsch (FT) process has been commercially implemented. Fischer-Tropsch (FT) is a process to produce liquid hydrocarbon fuels from syngas. The common process for FT could be divided into six procedures: feedstock pretreatments, biomass gasification, gas conditioning, acid gas removal, FT synthesis and syncrude refining. The FT synthesis can also be divided into high temperature FT and low temperature FT. The temperature for high temperature FT is around 310–340 °C, and the products are main gasoline, solvent oil and olefins; the temperature for low temperature FT is around 210–260 °C, the products are main kerosene, diesel oil, lubricating base oil and naphtha fractions. Too low temperature of FT will format high quantities of methane as a by-product. Typical pressures of FT process are in the range of one to several tens of atmospheres. The high pressures will result the formation of long-chain alkanes. The FT fuel is free of sulphur, nitrogen, has high specific energy, high thermal stability, and cause low emissions when used for aviation. However, the disadvantages for the fuel are low aromatic content and less energy density, which would also cause a low efficiency and high production cost for the process.

For the conversion of sugar to bio-jet fuel (STJ) the biotechnological process of convert sugars to alkane-type fuels directly instead of firstly converting to ethanol intermediate, which is called Direct Sugar to Hydrocarbons (DSHC) has been implemented commercially (see Table 1). The feedstock for the DSHC are similar to the feedstock of ethanol production, including the sugar cane, beets and maize. DSHC is a process to produces alkane-type fuels directly from sugars via fermentation. It is different from the alcohol to jet pathway, which needs an alcohol intermediate. The technology is developed based on the development of genetic engineering and screening technologies that enable to modify the way microbes metabolize sugar. A complete conversion process of DSHC, involving six major steps: pretreatment and conditioning, enzymatic hydrolysis, hydrolysate clarification, biological conversion, product purification and hydroprocessing has been proposed [13]. The DSHC has a low energy input due to the low temperature of the fermentation, while the fuel blend is limited (10%) and not meet some performances standards and it is also identified as more suitable for production high-value chemicals.

Finally, the conversion route to convert alcohols to bio-jet fuel (ATJ) use several processes depending of the feedstock. The process of production hydrocarbons in the jet fuel range from the alcohols generally undergoes a four-step upgrading process. First is the alcohol dehydration to generate olefins, then the olefins are oligomerised in the presence of catalysts to produce a middle distillate. Next, the middle distillates are hydrogenated to produce the jet-fuel-ranged hydrocarbons and a final step is the distillation to purify the bio-jet fuel product. Commercial production always use ethanol, butanol and isobutanol to be the intermediate to converse biomass to jet fuel. The economics of alcohol to jet process is mainly affected by the way to produce alcohol, while the biochemistry way has relatively smaller minimum jet fuel selling price (MJSP), the sugar cane and starch are suitable feedstock from an economics prospective. Complete reviews of the technologies described above considering the techno-economic analysis are given in [14–17].

Table 1. Summary of bio-jet fuel production pathways. Data obtained from [14].

Category	Pathway	Key Conversion Steps	Catalyst	FeedStock	Companies
Alcohol to Jet (ATJ)	Ethanol to Jet	Ethanol dehydration	Al_2O_3, Transition metal oxides	Sugar cane Corn grain	Terrabon; Swedish biofuels; Coskata
	Butanol to Jet	Butanol dehydration	Zirconia, Solid acid catalyst	Lignocellulose	Gevo; Byogy; Solazyme
Oil to Jet (OTJ)	Hydroprocessed renewable Jet (HEFA)	Catalytic hydrogenation / Cracking and isomerization	Noble metals, Transition metals Pt, Ni, Precious metals	Camelina oil Spybean oil Jatropha oil Waste oils Animal fat Microalgae	UOP; SG biofuels; Neste oil; PEMEX; Syntroleum-Tyson Food PetroChina
	Catalytic hydrothermolysis (CH)	Catalytic hydrothermolysis Decarboxylation-Hydrotreating	Zinc Acetate Nickel	Camelina oils Lignocellulose	Aemetis/Chevron Lummus global
	Hydrotreated Depolymerized Cellulosic Jet (HDCJ)	Hydrodeoxygenation	MoCx/C, Pd-Mo	Lignocellulose	Kior/Hunt; Refining/Petrotech; Envergent; Dinamotive
Gas to Jet (ATJ)	FT Synthesis	FT Process	Fe, Co, Ni and Ru	Lignocellulose	Syntrolleum; SynFuels; Shell
Sugar to Jet (ATJ)	Direct sugar biological to hydrocarbons	Acid Condensation Hydrodeoxygenation	Acid catalyst Ru/C	Sugar cane Lignocellulose	Amyris/Total; Solazyme, LS9

Considering the above technologies, the (i) oil-to-jet and (ii) gas-to-jet are contemplated as the most convincing alternatives in the short term. In fact, the bio-jet fuels produced with these technologies are now allowed by ASTM specification D7566-14 for blending into commercial jet fuel at concentrations up to 50% [18]. Recently [19], a hydrotreating reactive distillation column (RDC) for the production of green-diesel from sulphured petro-diesel and non-edible vegetable oils mixtures was proposed. From this work, it was concluded that the deep hydrodesulphurisation (HDS) of petro-diesel is strongly affected by feeding high amounts of triglycerides to the RDC, while it is not affected when only fatty acids are fed. Also, an integrated reactive separation process for the production of jet-diesel was developed but the assumptions about the conversion of tryglycerides to fatty acids and the yield of isomerization-cracking of the linear hydrocarbon chains lead to an oversimplified reactive separation process [20]. Therefore, in this work, from a process intensification context, an innovative intensified three-step reactive separation process for bio-jet diesel production is proposed.

Thus, the intensified three-step reactive separation process consist of: (1) a triglyceride hydrolysis section where a catalytic heterogeneous reactor is used to convert the triglycerides to the resultant fatty acids, followed by a three-phase separation device and a sequence of two distillation columns to obtain pure fatty acid and glycerol; (2) a hydrotreating section with a reactive distillation column used to simultaneously carried out the deep hydrodesulphurisation (HDS) of petro-diesel and the hydrodeoxigenation, decarbonylation and decarboxylation of the fatty acids; and (3) an isomerization-cracking section with a hydrogenation fixed bed reactor connected to a two phase flash separator to produce bio-jet diesel with the required fuel properties. It should be pointed out that the conversion and yield assumptions used in [20] are not considered here and the appropriate reaction kinetics found in the open literature [21,22] is used to perform the rigorous simulation of the intensified process.

2. The Intensified Reactive Separation Process

The production of bio-jet fuel can be described by the subsequent reactions (see Figure 1): (i) hydrogenation of the C=C bonds present in the tryglycerides, (ii) hydrogenolysis of the tryglycerides to produce the respective fatty acids, (iii) deoxygenation of the resultant fatty acids to obtain n-paraffins and (iv) hydroisomerization and hydrocracking of the n-paraffins to produce a mixture containing mainly isomerized shorter chains (i-C_8–i-C_{16}) that are appropriate as jet fuel. The first three reactions can be carried out using a metal catalyst or a MoS_2-based catalyst. For the hydroisomerization and hydrocracking reaction (iv), a metal/acid bifunctional catalyst is recommended. Figure 1 shows the reaction pathways for the hydroconversion of tryglycerides into bio-jet fuel. It has been shown [23] that the direct hydrogenation of tryglycerides may be very expensive due to the price of the hydrogen at high pressures. Thus, the conventional hydrolysis of tryglycerides with high pressure steam is preferred. An alternative to the hydrogenation and hydrolysis at high pressure is to use a heterogeneous hydrolysis catalytic reactor [24] at moderated pressures. Thus, in the present work this technological alternative is considered.

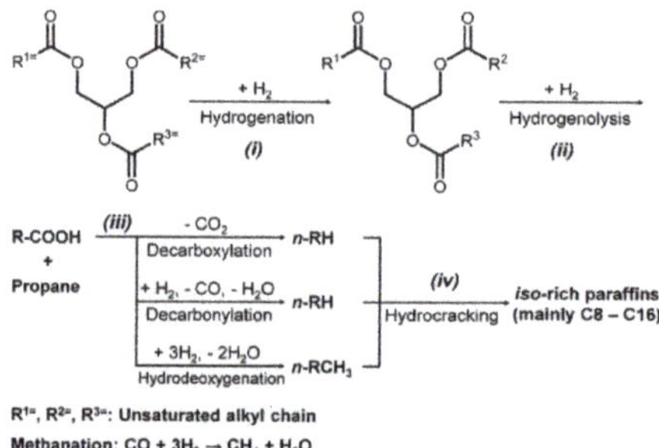

Figure 1. Reaction pathways of the hydroconversion of tryglycerides into biojet fuel.

Figure 2 shows a simplified flow sheet of the intensified three-step reactive separation process for bio-jet diesel production. The first section (hydrolysis) consists of two pumps operating from 1 to 30 atm and two heat exchangers where the exit temperature is set to 280 °C. The heterogeneous hydrolysis reactor operates at 30 atm and it consists of two 25 m length tubes with 0.5 m of diameter. Further, a sequence of a three-phase flash and a two-phase drum are used to separate the fatty acid-glycerol-water mixture and the unconverted glycerides that are recycled to the hydrolysis reactor. Finally, two distillation columns are used to separate the pure fatty acid from a low concentration glycerol-water mixture and to produce pure glycerol. The second section, hydrodesulphurisation-hydrodeoxigenation (HDS-HDO) consists of a reactive distillation column (RDC) where a mixture of petro-diesel and fatty acid is fed to the RD column at stage 9 and excess hydrogen is fed at the bottom. The gases produced by the HDS-HDO reactions are released from the two-phase condenser. The exit liquid mixture of C_{11}–C_{12} linear hydrocarbon chains is mixed with the exit stream containing C_{17}–C_{18} hydrocarbons produced by the HDO reactions and the ultra-clean petro-diesel products. The third section consists of a heterogeneous hydro-conversion reactor where the long linear hydrocarbon chains (C_{11}–C_{18}) are isomerized and cracked to attain the bio-jet diesel.

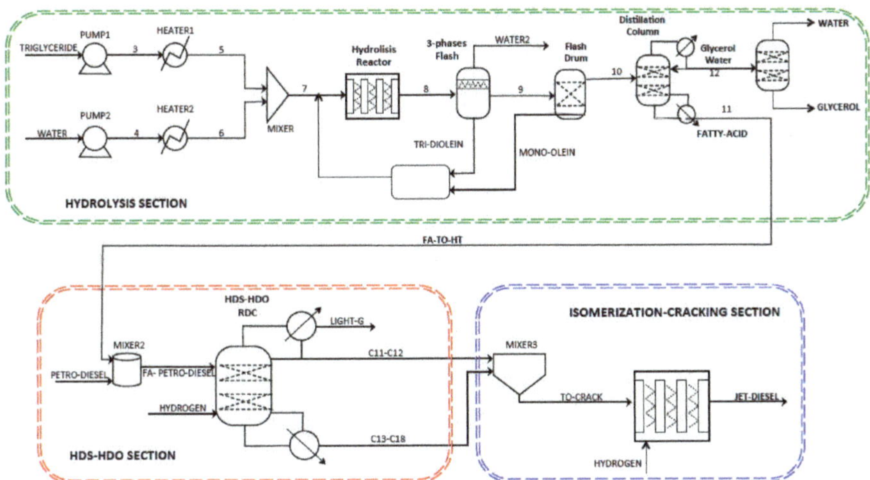

Figure 2. Simplified three-step reaction-separation process for bio-jet diesel production.

2.1. The Triglycerides Hydrolysis Section

Vegetable oils and fats have been considered as one of the most used renewable raw materials in the chemical industry. These can be hydrolyzed to produce free fatty acids (FFA) with a high degree of purity to be used in the synthesis of chemically pure compounds [25]. However, in the present work, the proposed technological alternative to produce bio-jet fuel considers non-edible oils, waste oils and animal fats as feedstock, which are not in conflict with food resources. Fatty acids are used in a wide variety of industries, for example in the pharmaceutical and cosmetics industry. Also, fatty acids can be utilised to produce n-alkane chains through a decarboxylation process [26]. These hydrocarbons work properly in internal combustion engines as substitutes for petro-diesel. Actually, the hydrolysis of fats and vegetable oils, composed mainly of triglycerides, has been practiced in the industry for many years. In general, the hydrolysis of the esters occurs through the acyl-oxygen break [27] with an excess of water at high temperature or using an appropriate acid catalyst to hydrolyze the glycerol pillar in the ester group of any triglyceride (TG), diglyceride (DG) or monoglyceride (MG) [28]. The result of the hydrolisis reactions is the production of three moles of FFA and one mole of glycerol (Gly). In this work, triolein is considered as the main triglyceride compound and oleic acid as the correspondent fatty acid. The three consecutive reversible reactions can be written as:

$$\text{triolein} + H_2O \Leftrightarrow \text{diolein} + \text{oleic acid} \tag{1}$$

$$\text{diolein} + H_2O \Leftrightarrow \text{mono-olein} + \text{oleic acid} \tag{2}$$

$$\text{mono-olein} + H_2O \Leftrightarrow \text{glycerol} + \text{oleic acid} \tag{3}$$

Commonly, the pure fatty acids are obtained from the reaction of vegetable oils and/or animal fats with superheated steam. The commercial hydrolisis conditions are around 100–260 °C and 10–7000 kPa using a 0.4–1.5 wt % water-to-oil ratio. Several variants of this technology have been used by industry [29,30]. In this work, the hydrolisis kinetics reported in the open literature [21] is used for the numerical simulation of the catalytic reactor using tungstated zirconia (WZ) and the solid composite SAC-13 as catalyst.

2.2. The HDS-HDO Reactive Distillation Section

The hydrotreatment of non-edible vegetable oils, waste oils or animal fats, this is, oils and fats that are not used for food and others medical applications, to produce renewable fuels has several advantages: (i) flexibility in the disposal of raw materials due to the great variety of oils available from vegetables on the earth; (ii) the process can be carried out using the existing infrastructure in petro-refineries and (iii) the bio-fuels produced can be used in conventional internal combustion machines since these bio-fuels have properties similar to those obtained from mineral oil [31]. The hydrotreatment of vegetable oils can be accomplished using traditional catalysts, for example with $NiMo/Al_2O_3$ catalysts. The hydrotreatment mainly produces n-paraffins through the hydrodecarboxylation, hydrodecarbonylation and hydrodeoxigenation reactions. From the refining point of view, the hydrodecarboxylation reaction is better than the hydrodeoxygenation reaction since it consumes less hydrogen. However, the large amount of CO (or CO_2) generated represents a problem for refining, since CO_2 can form carbonic acid with liquid water [32]. This means that the risk of carbonic corrosion of the reactive separation equipment is a key design problem. In general, vegetable oils can be hydrotreated as pure compounds or can be mixed with petro-diesel to be co-hydrotreated, in such a way that the hydrodesulphurisation (HDS) and hydrodeoxygenation (HDO) reactions are carried out simultaneously in a single unit. Therefore, the reactions considered for the HDS of the sulphured petro-diesel can be written in a simplified form as:

$$\text{Thiophene (Th)} + 2\,H_2 \rightarrow \text{Butadiene} + H_2S \tag{4}$$

$$\text{Benzothiophene (BT)} + 3\,H_2 \rightarrow \text{Ethylbenzene} + H_2S \tag{5}$$

$$\text{DBT} + 2\,H_2 \rightarrow \text{Biphenyl} + H_2S \tag{6}$$

For the HDS reactions only the hydrogenolysis reaction pathway is considered (Equation (6)). The simplified HDO reactions of the oleic acid can be written as:

$$C_{18}H_{34}O_2 + 4\,H_2 \rightarrow n\text{-}C_{18}H_{38} + 2\,H_2O \quad \text{(hydrodeoxygenation)} \tag{7}$$

$$C_{18}H_{34}O_2 + 2\,H_2 \rightarrow n\text{-}C_{17}H_{36} + H_2O + CO \quad \text{(decarbonylation)} \tag{8}$$

$$C_{18}H_{34}O_2 + H_2 \rightarrow n\text{-}C_{17}H_{36} + CO_2 \quad \text{(decarboxylation)} \tag{9}$$

The reaction rate equations for the HDS reactions can be obtained from [19] and, due to the absence of reliable kinetic expressions for the complex HDO reactions, a 99% conversion of oleic acid is assumed.

2.3. The Isomerization-Hydrocracking Section

In general the processes of isomerization and cracking through hydrogenation occur simultaneously. When hydrogenation favours isomerization it is called hydro-isomerization and when it promotes cracking it is called hydro-cracking. Depending on the characteristics of the hydrocarbon chains to be hydrotreated, the selection of the appropriate catalyst for this purpose can be made in a wide range of possibilities [33]. Specifically, for the production of jet diesel, the catalysts are bifunctional and are characterized by having acid sites that allow the selective function of isomerization and cracking simultaneously. Experimental evidence [34] indicate that the conversion of the n-paraffins can be described by the following reactions network:

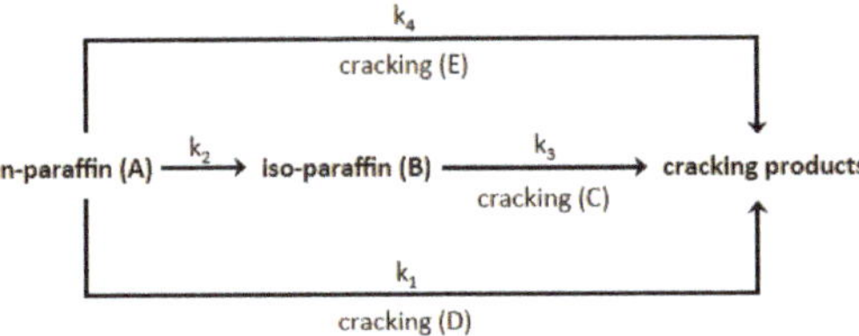

where each single reaction is irreversible and follows a pseudo first-order kinetic. (A) is the concentration of the n-paraffin; (B) is the concentration of the iso-paraffin; (C), (D) and (E) are the concentrations of the cracking products. Following the linear reaction sequence path: A–>B–>C the hydro-conversion reactions could be written as:

$$n\text{-}C_{11}H_{24} \rightarrow i\text{-}C_{11}H_{24} \tag{10}$$

$$n\text{-}C_{12}H_{26} \rightarrow i\text{-}C_{12}H_{26} \tag{11}$$

$$n\text{-}C_{13}H_{28} \rightarrow i\text{-}C_{13}H_{28} \tag{12}$$

$$n\text{-}C_{14}H_{30} \rightarrow i\text{-}C_{14}H_{30} \tag{13}$$

$$n\text{-}C_{16}H_{34} \rightarrow i\text{-}C_{16}H_{34} \tag{14}$$

$$n\text{-}C_{17}H_{36} \rightarrow i\text{-}C_{17}H_{36} + H_2 \rightarrow i\text{-}C_8H_{18} + i\text{-}C_9H_{20} \tag{15}$$

$$n\text{-}C_{18}H_{38} \rightarrow i\text{-}C_{18}H_{38} + H_2 \rightarrow i\text{-}C_8H_{18} + i\text{-}C_{10}H_{22} \tag{16}$$

The kinetic equations for the isomerization and cracking of the linear hydrocarbon chains were taken from [16] and a catalyst of Pt supported on a nano-crystalline large-pore BEA zeolite is used [28].

3. Results and Discussion

In order to determine the best operating conditions of the intensified three-step hydrotreating reactive-separation process, the effect of different design and operating variables on the performance of the global process was investigated and analysed through intensive simulations. The intensive simulations of the intensified hydrotreating reactive separation process were carried out in the commercial Aspen-Plus simulator environment. Thus, heat and mass transfer phenomena and mixing issues were not taken into account.

3.1. Hydrolysis Section

Different amounts of triolein-water feed ratio were used for the simulation of the intensified process. The numerical results shown in Table 2 are for the reference case (production of 70 kmol/h of pure oleic acid). From such numerical results it is found that total conversion of triolein is attained at 553 K and excess water (265 kmol/h). This water flow corresponds to 1/9 triolein/water feed ratio. The three-phase flash that separates water, oleic acid-water-glycerol and triolein-diolein mixture operates at 5 atm and 410 K, and the flash drum for the separation of the water-oleic acid-glycerol-unconverted glycerides operates at 1 atm. The two distillation columns to produce pure oleic acid and pure glycerol operate at 1 atm. It should be pointed out that the non-ideality of the reactive polar mixture is modelled using the RK-ASPEN equation of state.

Table 2. Simulation results for the reference case (hydrolysis section).

Stream	Triolein	Water	7	8	9	Glycerol	11
Mole Flow kmol/h							
Triolein	30	0.0	30	5.21	5.21	0.0	0.0
Oleic acid	0.0	0.0	0.0	70.83	70.83	1.11×10^{-26}	70.00
Water	0.0	265	265	194.16	47.00	9.27×10^{-3}	1.44×10^{-33}
Glycerol	0.0	0.0	0.0	21.89	21.55	21.54	1.67×10^{-16}
Diolein	0.0	0.0	0.0	0.62	1.06×10^{-14}	0.0	0.0
Monolein	0.0	0.0	0.0	2.25	2.25	0.0	0.0
Mass fraction							
Triolein	1.0	0.0	0.8476	0.1474	0.1634	0.0	0.0
Oleic acid	0.0	0.0	0.0	0.6385	0.7077	1.58×10^{-27}	1.00
Water	0.0	1.0	0.1524	0.1166	0.0300	8.41×10^{-5}	1.29×10^{-36}
Glycerol	0.0	0.0	0.0	0.0643	0.0702	0.9999	7.60×10^{-19}
Diolein	0.0	0.0	0.0	0.0123	2.29×10^{-16}	0.0.	0.0
Monolein	0.0	0.0	0.0	0.0257	0.0284	0.0.	0.0
Total Flow (kg/h)	26,563.47	4774.049	31,337.52	31,337.52	28,255.39	1985.16	19,772.67
Temperature (K)	298.15	298.15	494.26	553.15	410.28	557.18	631.04
Pressure (atm)	1.0	1.0	30.0	30.0	5.0	1.0	1.0

Effect of the Triolein-Water Feed Ratio

The effect of the triolein-water feed ratio was studied in order to verify the maximum conversion of triolein to oleic acid. Figure 3a shows the effect of the feed ratio on the hydrolysis reactor exit composition. It can be observed that as the feed ratio increases, the triolein composition decreases to zero at 1/7 triolein-water feed ratio, approximately. However, it should be noted that, as the feed ratio increases, the hydrolysis reactor exit mixture contains less fatty acid (oleic acid). This is expected due to the excess of water fed to the reactor. Also, it can be observed that after a feed ratio of 1/7, the final reactor exit mixture contains only water, oleic acid and glycerol. Figure 3b shows the effect of the triolein-water feed ratio on the distillation column bottom flow (stream 11 in Figure 2). It can be noted in Figure 3b that as the feed ratio increases, the amount of oleic acid produced increases to a constant value of 88 (kmol/h). The production of 70 (kmol/h) at the bottom of the distillation column at a feed ratio of 1/9 has been taken as a reference study case.

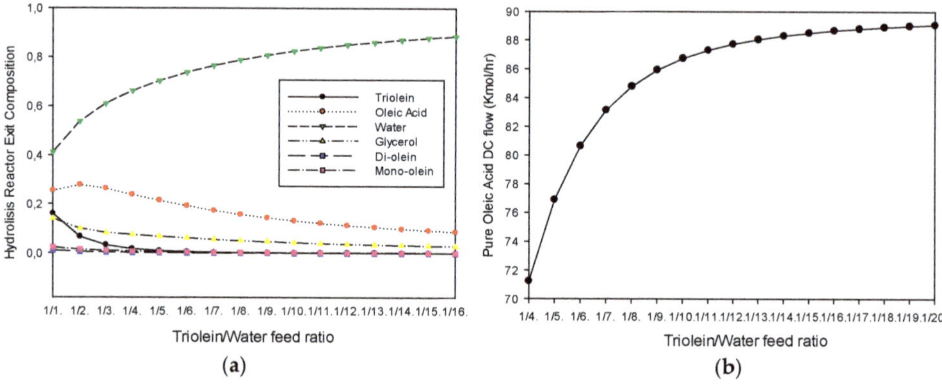

Figure 3. (a) Effect of the triolein-water feed ratio on the hydrolysis reactor exit composition. (b) Effect of the triolein-water feed ratio on the distillation column bottom flow (stream 11).

3.2. HDS-HDO Section

In order to perform the deep hydrodesulphurisation of the sulphured petro-diesel and the reactions involved in the hydrodeoxigenation of the fatty acid, a two-zone reactive distillation column (RDC) is recommended [13]. Figure 4 shows the sizing details of the RDC, as well as the basic information for the simulation of the reference case. The RDC consists of 14 stages with two sections of reactive stages (5–7, 9–11) operating at 30 atm. It can be observed in Figure 4 that the liquid streams C11–C12 and C13–C18 contains mainly the larger linear hydrocarbon chains. Table 3 displays the numerical simulation results for the HDS-HDO section with 70 kmol/h of oleic acid and 100 (kmol/h) of petro-diesel fed at stage 9. Hydrogen feed was set to 400 (kmol/h) in order to accomplish full conversion of oleic acid and deep HDS of petro-diesel.

Table 3. Simulation results of the HDS-HDO section in ASPEN-PLUS (reference case).

Stream	Petro-Diesel	Hydrogen	Oleic Acid (11)	Light-G	C11–C12	C13–C18
Mole Flow kmol/h						
H_2	0.0	400	0.0	202.2064	0.0415	14.4201
H_2S	0.0	0.0	0.0	9.8951	0.0157	0.0884
Th	0.8699	0.0	0.0	0.8363	0.0108	0.0228
BT	0.8699	0.0	0.0	0.0798	0.0122	0.7779
DBT	9.9999	0.0	0.0	6.73×10^{-7}	9.72×10^{-7}	7.61×10^{-4}
Biphenil	0.0	0.0	0.0	0.0390	0.0126	9.9476
n-C_{16}	5.8899	0.0	0.0	7.79×10^{-5}	9.84×10^{-5}	5.8898
n-C_{14}	0.1500	0.0	0.0	5.84×10^{-5}	3.57×10^{-5}	0.14990
n-C_{13}	0.8899	0.0	0.0	3.46×10^{-3}	1.24×10^{-3}	0.8852
n-C_{12}	31.6599	0.0	0.0	0.9127	0.2000	30.5477
n-C_{11}	49.6699	0.0	0.0	8.9238	1.2100	39.5333
Oleic-Acid	0.0	0.0	70	0.0	0.0	0.0
n-C_{18}	0.0	0.0	0.0	4.06×10^{-6}	1.17×10^{-5}	23.3332
n-C_{17}	0.0	0.0	0.0	5.05×10^{-5}	1.02×10^{-4}	46.6665
Water	0.0	0.0	0.0	68.7544	0.2980	0.9470
CO_2	0.0	0.0	0.0	23.2247	0.0224	0.08615
CO	0.0	0.0	0.0	23.2884	7.72×10^{-3}	0.03714
Mole fraction						
H_2	0.0	1.0	0.0	0.5979	0.0226	0.0831
H_2S	0.0	0.0	0.0	0.0292	8.55×10^{-3}	5.10×10^{-4}
Th	0.0087	0.0	0	2.47×10^{-3}	5.90×10^{-3}	1.31×10^{-4}
BT	0.0087	0.0	0.0	2.36×10^{-4}	6.6.E-03	4.44×10^{-3}
DBT	0.1	0.0	0.0	1.99×10^{-9}	5.30×10^{-7}	4.41×10^{-6}
Biphenyl	0.0	0.0	0.0	1.15×10^{-4}	6.84×10^{-3}	0.0573
n-C_{16}	0.0589	0.0	0.0	2.31×10^{-7}	5.36×10^{-5}	0.0339
n-C_{14}	0.0015	0.0	0.0	1.73×10^{-7}	1.94×10^{-5}	8.64×10^{-4}
n-C_{13}	0.0089	0.0	0.0	1.02×10^{-5}	6.78×10^{-4}	5.10×10^{-3}
n-C_{12}	0.3166	0.0	0.0	2.69×10^{-3}	0.1087	0.1762
n-C_{11}	0.4967	0.0	0.0	0.0263	0.6608	0.2280
Oleic-Acid	0.0	0.0	1.0	0.0	0.0	0.0
n-C_{18}	0.0	0.0	0.0	1.20×10^{-8}	6.37×10^{-6}	0.1346
n-C_{17}	0.0	0.0	0.0	1.49×10^{-7}	5.54×10^{-5}	0.2692
Water	0.0	0.0	0.0	0.2033	0.1626	5.46×10^{-3}
CO2	0.0	0.0	0.0	0.0686	0.0122	4.97×10^{-4}
CO	0.0	0.0	0.0	0.0688	4.20×10^{-3}	2.15×10^{-4}
Total Flow (Kg/h)	16,717.09	806.35	19,772.67	5296.10	235.52	31,764.49
Temperature (K)	513	533	513	458.15	458.15	746.89
Pressure (atm)	30	30	30	30	30	30

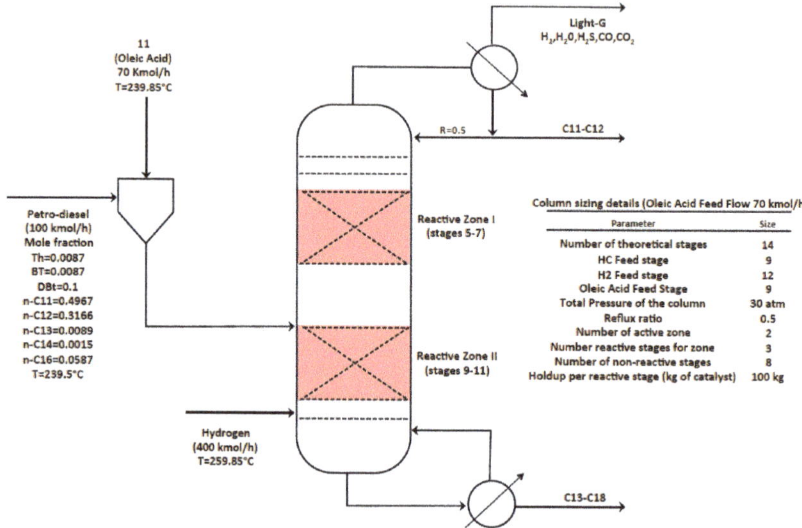

Figure 4. Hydrotreating two-zone reactive distillation column. Oleic acid premixed with sulphured petro-diesel fed at stage 9.

3.2.1. Effect of the Oleic Acid-Petro-Diesel Feed Ratio

The temperature profile along the reactive distillation column (RDC) assuming that oleic acid is premixed with the petro-diesel and fed in stage 9 is shown in Figure 5a. The co-hydrotreating process starts with a low concentration of oleic acid (10 kmol/h) in the mixture fed and it was continuously augmented up to 70 (kmol/h). It can be seen in Figure 5a that as the content of oleic acid increases, the temperature at the reactive zone I is reduced. The temperature reduction can be explained by considering that the boiling point of oleic acid is higher than the boiling point of most hydrocarbons present in the sulphured petro-diesel. The effect of the oleic acid composition in the feed on the deep HDS of the sulphured petro-diesel is shown in Figure 5b. It can be noted in Figure 5b that, even for a high content of oleic acid (70 kmol/h), the deep HDS of petro-diesel is accomplished. The analysis of these results leads to the conclusion that the performance of the reactive distillation column (RDC), initially used for deep hydrodesulphurisation of petro-diesel, is not affected by the introduction of a vegetable oil with a high content of fatty acid (oleic acid).

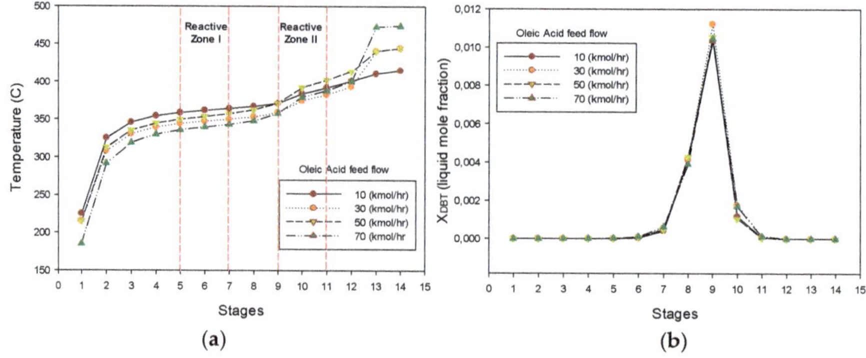

Figure 5. (**a**) Temperature profile along the RDC. (**b**) DBT liquid composition profile.

3.2.2. Hydrogen Consumption and Liquid Water Production

The molar flow of hydrogen required for the HDS and HDO reactions as a function of the oleic acid content in the feed is shown in Figure 6a. It can be noted in Figure 6a that, for an oleic acid feed flow greater than 10 (kmol/h), the temperature of the condenser is reduced to 215 °C, and for a feed flow of 70 (kmol/h), the required hydrogen is increased to 400 (kmol/h) and the temperature of the condenser should be diminished to 185 °C. It should be pointed out that, if the temperature of the condenser is not decreased, the numerical convergence of the RDC simulation is not reached. Thus, a control loop between the source of vegetable oil (oleic acid) and the hydrogen supplied should be linked to the temperature of the condenser. The liquid water composition profiles along the RDC for different oleic acid feed flows are shown in Figure 6b. It can be noted in Figure 6b that the concentration of liquid water increases from reactive zone I (steps 5–7) to the upper part of the RDC. Liquid water is mainly produced in reactive zone II (reactive stage 9–11) by the hydrodeoxygenation and decarbonylation reactions and its molar fraction increases continuously from stage 9 to the upper part of the RDC, as the concentration of oleic acid in the feed is augmented. Therefore, in order to minimize catalyst deactivation and corrosion of the RDC equipment, the amount of oleic acid in the mixture fed should not be higher than 70 (kmol/h).

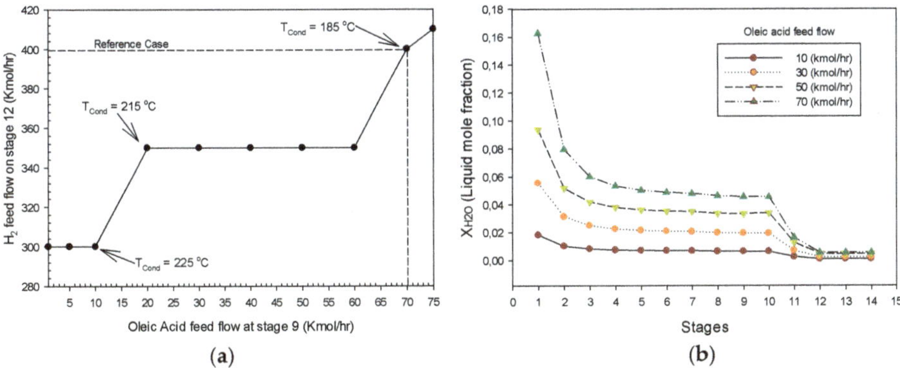

Figure 6. (**a**) Hydrogen feed flow required on stage 12. Oleic acid-petro-diesel mixture fed at stage 9. (**b**) Water liquid composition profile.

3.2.3. Hydrocarbon Distribution and Release of Generated Gases

Figure 7a shows the hydrocarbon liquid composition distribution. It can be noted in Figure 7a that the bottom exit stream contains mainly the linear hydrocarbons C_{17}, C_{18}, C_{16}, C_{14}, and C_{13}. At the top of the RDC a rich liquid mixture of C_{11} and C_{12} linear hydrocarbons is obtained. It should be pointed out that such linear hydrocarbons (C_{11}–C_{18}) are further mixed and fed to the isomerization-hydrocracking reactor (see Figure 2). Figure 7b shows the gas composition profile along the RDC. In the case of the hydrotreating RDC, the gases produced are mainly driven in the vapour phase at each equilibrium stage and released from the partial condenser at the top of the RDC. It can be noted in Figure 7b that the gases produced by HDS and HDO reactions: CO_2, CO, steam and H_2S and the unconverted excess of hydrogen are released at the top of the column. It is interesting to observe that the vapour composition of C_{11} and C_{12} decreases at the top of the column.

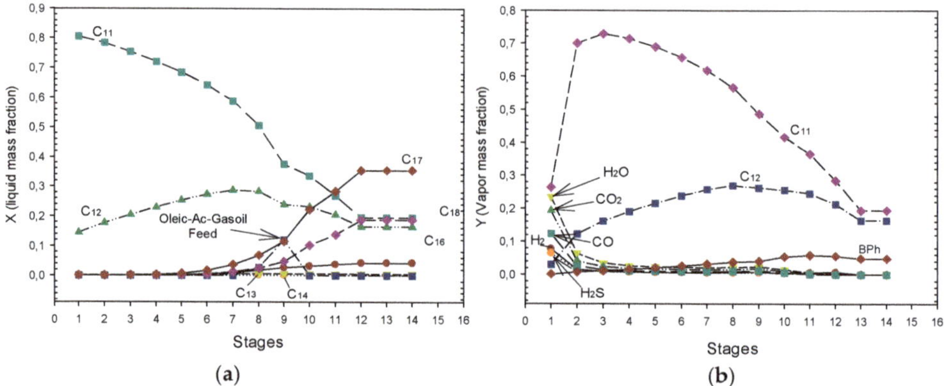

Figure 7. (**a**) RDC profiles for oleic acid-petro-diesel blend fed at stage 9 at 30 atm. (**b**) Gas compositions profile along the RDC. Reproduced with permission from García-Sánchez et al., Proceedings of the 28th European Symposium on Computer Aided Process Engineering; Published by Elsevier B.V., 2018. [20].

3.2.4. Operability and Controllability of the HDO-HDS Reactive Distillation Section

Reactive distillation column mathematical models are highly nonlinear, and multiple steady states (MSS) solutions have been reported by many researchers [35–37]. However, none of these works have addressed the HDO-HDS reactive distillation process, most of them studied the MTBE and TAME cases. In the present work, the MSS solutions are only briefly described. Two multiplicity types can be found: input and output multiplicities, but a combined input–output multiplicity may also be present. Output multiplicity occurs when one set of input variables (manipulated variables) results in two or more unique and independent sets of output variables (measured variables). In chemical reactors and reactive distillation columns there are usually three steady states associated to ignition (high conversion), extinction (low conversion) and medium unstable conversion. On the other hand, input multiplicity occurs when two or more unique sets of input variables produce the same output condition. This type of multiplicity has important implications for close loop control since it is related to the so-called zero dynamics of the system, which is associated with unusual, unexpected, or inverse responses of the outputs after a step change, has been applied to the inputs. As RDC share some common features with chemical reactors and conventional (non-reactive) distillation columns, the RDC behavior may exhibit input–output multiplicity with three steady states or output multiplicity, with a large number of different steady states induced by ignition and extinction of every single reactive column tray [38]. Thus, in the present work, from the intensive simulation results, it was found that the reflux ratio is a key parameter for an appropriate RDC design and operation. Additionally, an increment in the hydrocarbon feed (sulphured diesel) flowrate leads to lower conversions of the organo-sulfur and fatty acid compounds with the reduction of the H_2/HC feed ratio. In addition, it was found that, the DBT conversion was highly affected by the variations of the reboiler heat duty, while the fatty acid conversion was practically constant (~99%). However, DBT conversion design target (99.9%) was achieved at four different reboiler duties, indicating the existence of input-output multiplicities. It was also determined that the amount of catalyst loaded at the reactive zones must be greater than 8000 kg in order to achieve an ultra low sulfphur diesel (ULSD) production and a complete conversion of the fatty acids. This is, if a suitable excess of H_2 is present in the reaction zones, the behavior of the RDC is very likely to a HDS conventional catalytic reactor. Therefore, it may be said that the multiplicities found in the intensive simulations are highly related to the specific phenomena (reaction-separation) involved. It should be pointed out that the accurate determination of the thermodynamic properties of all species participating in the reacting mixture must be taken carefully. The correct determination of the boiling points, critical properties, etc. of all species involved in the reactive separation process and the computation of the complex phase equilibrium in the intensified reactive separation process is a key step. This is, for example, for the HDO-HDS section, the inaccuracy of the property values (boiling point of the fatty acids) can lead to multiple steady states and unstable operation and control of the reactive distillation process. The analysis of the intensive simulation results suggest that, under optimal design and operating conditions, reactive distillation can be considered as a viable technological alternative to produce bio-jet diesel trough a RDC co-hydrotreating process.

3.3. The Isomerization-Cracking Section

The numerical simulation results with the final composition of the bio-jet diesel obtained in the hydro-conversion reactor are shown in Table 4. It should be mentioned that an excess of hydrogen must be fed to the reactor in turn to accomplish the complete isomerization and cracking of the longer hydrocarbon chains. The total pressure of the reactor is set at 80 atm and a 5:1 hydrogen-hydrocarbon feed ratio was considered. The exit stream from the isomerization-cracking reactor is further flashed to eliminate the hydrogen excess and the light hydrocarbons produced during the cracking step. The isomerization-cracking kinetics and reactor arrangement were taken from the open literature [34]. From Table 3 it can be observed that the bio-jet-diesel produced contains mainly i-C_8 to i-C_{16} hydrocarbons indicating that a light bio-jet diesel is produced.

Table 4. Simulation results of the isomerization-cracking section in ASPEN-PLUS (reference case).

Component	To Iso-Crack	Bio-Jet Diesel	Component	To Iso-Crack	Bio-Jet Diesel
	Mole Flow (kmol/h)		Mole Fraction		
n-C_{16}	5.8899	0.0	n-C_{16}	0.03968	0.0
n-C_{14}	0.1500	0.0	n-C_{14}	0.00101	0.0
n-C_{13}	0.8899	0.0	n-C_{13}	0.00600	0.0
n-C_{12}	30.7472	0.0	n-C_{12}	0.20716	0.0
n-C_{11}	40.7460	0.0	n-C_{11}	0.27453	0.0
n-C_{18}	23.3333	0.0	n-C_{18}	0.15721	0.0
n-C_{17}	46.6666	0.0	n-C_{17}	0.31441	0.0
i-C_{16}	0.0	5.8899	i-C_{16}	0.0	0.02697
i-C_{14}	0.0	0.15	i-C_{14}	0.0	0.00069
i-C_{13}	0.0	0.8899	i-C_{13}	0.0	0.00407
i-C_{12}	0.0	30.7472	i-C_{12}	0.0	0.14077
i-C_{11}	0.0	40.7460	i-C_{11}	0.0	0.18655
i-C_{10}	0.0	23.3333	i-C_{10}	0.0	0.10682
i-C_9	0.0	46.6666	i-C_9	0.0	0.21365
i-C_8	0.0	69.9999	i-C_8	0.0	0.32048
Total Flow (kmol/h)	148.423	218.4228			
Temperature (K)	653				

4. Conclusions

An intensified three-step hydrotreating reaction-separation process for the production of bio-jet diesel from triolein and petro-diesel mixtures has been developed. Through intensive simulations the effect of different operating variables (triglyceride-water feed ratio, oleic acid-petro-diesel feed ratio, hydrogen consumption) on the performance of the intensified reactive separation process was studied. By analysing the simulation results, it can be established that, a 1/9 triolein-water feed ratio guarantee the complete conversion of triolein to oleic acid at moderated pressures (30 atm) at the hydrolysis section. For the HDS-HDO section, it can be mentioned that a mixture containing up to 70 (kmol/h) of oleic acid in the hydrocarbon feed of the RDC is the optimal mixture composition in order to achieve the deep hydrodesulphurisation of petro-diesel. Also, with this mixture composition, any change to the basic structure (reactive and non-reactive stages) of the RDC is not required. It should be pointed out that the HDS-HDO RD column should operate at moderated pressures (30 atm) in order to allow the fast vaporization of the light gases produced and avoid the corrosion problem with carbonic acid generated by the reaction of CO or CO_2 with water. For the isomerization and cracking section, a low-pressure flash separator, after the hydro-conversion reactor is required to eliminate the undesirable side products. Therefore, it can be concluded that the key design and operating parameters for the production of the bio-jet diesel are: (i) in the hydrolysis section, the water excess and the total pressure for the heterogeneous catalytic hydrolysis reactor; (ii) for the HDS-HDO section, if high molar flows of fatty acid are considered, it is mandatory to have more reactive stages in the HDS-HDO reactive distillation column in order to achieve ultra-clean (no-sulphur) petro-diesel at the bottom of the column. Finally, the absence of light compounds in the exit stream of the heterogeneous isomerization-cracking reactor is required in order to circumvent undesirable side products and reach the suitable fuel properties required by the international standards. Finally, the economic and sustainability analysis of the intensified reactive separation process to produce bio-jet diesel is being carried out.

Author Contributions: Conceptualization, M.S.-C. and E.S.P.-C.; Formal analysis, T.V.-G.; Investigation, E.S.P.-C.; Methodology, M.G.-S. and E.S.P.-C.; Software, T.L.-A.; Validation, M.S.-C. and T.V.-G.; Writing—original draft, M.G.-S.; Writing—review & editing, T.L.-A. and E.S.P.-C.

Funding: This research was funded by CONACyT PhD scholarship of Miriam García-Sánchez.

Conflicts of Interest: The authors declare no conflict of interest.

References

1. Sargent, R.W.H. Integrated design and optimization of processes. *Chem. Eng. Prog.* **1967**, *63*, 71–78.
2. Sargent, R.W.H. Forecasts and trends in systems engineering. *Chem. Eng.* **1972**, *262*, 226.
3. Grossmann, I.E.; Westerberg, A.W. Research challenges in process systems engineering. *AIChE J.* **2000**, *46*, 1700–1703. [CrossRef]
4. Pantelides, C.C. New challenges and opportunities for process modelling. In *Computer Aided Chemical Engineering*; Gani, R., Jorgensen, S.B., Eds.; Elsevier: Amsterdam, The Netherlands, 2001.
5. Moulijn, J.A.; Stankiewicz, A.; Grievink, J.; Gorak, A. Process intensification and process systems engineering: A friendly symbiosis. *Comput. Chem. Eng.* **2008**, *32*, 3–11. [CrossRef]
6. Zhang, L.; Babi, D.K.; Gani, R. New vistas in chemical product-Process design . *Ann. Rev. Chem. Biomol. Eng.* **2016**, *7*, 557–582. [CrossRef]
7. Gani, R. Chemical product design: Challenges and opportunities. *Comput. Chem. Eng.* **2004**, *28*, 2441–2457. [CrossRef]
8. Rosillo-Calle, F.; Teelucksingh, S.; Thrän, D.; Seiffert, M. IEA Bioenergy. 2012. Available online: http://task40.ieabioenergy.com/iea-publications/task-40-library/ (accessed on 22 July 2019).
9. McCall, M.J.; Kocal, J.A.; Bhattacharyya, A.; Kalnes, T.N.; Brandvold, T.A. Production of Aviation Fuel from Renewable Feedstocks. U.S. Patent 8,039,682B2, 18 October 2011.
10. Lamprecht, D. Fischer–Tropsch Fuel for Use by the U.S. Military as Battlefield-Use Fuel of the Future. *Energy Fuels* **2007**, *21*, 1448–1453. [CrossRef]
11. Luning Prak, D.J.; Jones, M.H.; Trulove, P.; McDaniel, A.M.; Dickerson, T.; Cowart, J.S. Physical and Chemical Analysis of Alcohol-to-Jet (ATJ) Fuel and Development of Surrogate Fuel Mixtures. *Energy Fuels* **2015**, *29*, 3760–3769. [CrossRef]
12. Olcay, H.; Subrahmanyam, A.V.; Xing, R.; Lajoie, J.; Dumesic, J.A.; Huber, G.W. Production of renewable petroleum refinery diesel and jet fuel feedstocks from hemicellulose sugar streams. *Energy Environ. Sci.* **2013**, *6*, 205–216. [CrossRef]
13. Davis, R.; Biddy, M.J.; Tan, E.; Tao, L.; Jones, S.B. *Biological Conversion of Sugars to Hydrocarbons Technology Pathway*; Pacific Northwest National Lab.(PNNL): Richland, WA, USA, 2013.
14. Wang, W.C.; Tao, L.; Markham, J.; Zhang, Y.; Tan, E.; Batan, L.; Warner, E.; Biddy, M. Bio-Jet Fuel Conversion Technologies. Review of Bio-Jet Fuel Conversion Technologies at 2016. NREL/TP-5100-66291. Available online: http://www.nrel.gov/publications (accessed on 13 January 2018).

15. Douvartzides, S.L.; Charisiou, N.D.; Papageridis, K.N.; Goula, M.A. Green Diesel: Biomass Feedstocks, Production Technologies, Catalytic Research, Fuel Properties and Performance in Compression Ignition Internal Combustion Engines. *Energies* **2019**, *12*, 809. [CrossRef]

16. Wang, W.C.; Tao, L. Bio-Jet fuel conversion technologies. *Renew. Sustain. Energy Rev.* **2016**, *53*, 801–822. [CrossRef]

17. Martinez-Hernandez, E.; Ramírez-Verduzco, L.F.; Amezcua-Allieri, M.A.; Aburto, J. Process simulation and techno-Economic analysis of bio-Jet fuel and green diesel production—Minimum selling prices. *Chem. Eng. Res. Des.* **2019**, *146*, 60–70. [CrossRef]

18. *American Society for Testing and Materials; Standard Specification for Aviation Turbine Fuel Containing Synthesized Hydrocarbons*; ASTM International: West Conshohocken, PA, USA, 2014.

19. Perez-Cisneros, E.S.; Sales-Cruz, M.; Lobo-Oehmichen, R.; Viveros-García, T. A reactive distillation process for co-hydrotreating of non-Edible vegetable oils and petro-Diesel blends to produce green diesel fuel. *Comput. Chem. Eng.* **2017**, *105*, 105–122. [CrossRef]

20. García-Sánchez, M.; Sales-Cruz, M.; Lopez-Arenas, T.; Viveros-García, T.; Ochoa-Tapia, A.; Lobo-Oehmichen, R.; Pérez-Cisneros, E.S. An Integrated Reactive Separation Process for Co-Hydrotreating of Vegetable Oils and Gasoil to Produce Jet Diesel. In Proceedings of the 28th European Symposium on Computer Aided Process Engineering, Graz, Austria, 10–13 June 2018; pp. 839–844. [CrossRef]

21. Ngaosuwan, K.; Lotero, E.; Suwannakarn, K.; Goodwin, J.G.; Praserthdam, P. Hydrolysis of Triglycerides Using Solid Acid Catalysts. *Ind. Eng. Chem. Res.* **2009**, *48*, 4757–4767. [CrossRef]

22. Calemma, V.; Peratello, S.; Perego, C. Hydroisomerization and hydrocracking of long chain n-Alkanes on Pt/amorphous SiO$_2$–Al$_2$O$_3$ catalyst. *Appl. Catal. A* **2000**, *190*, 207–218. [CrossRef]

23. Triantafyllidis, K.; Lappas, A.; Stöcker, M. *The Role of Catalysis for the Sustainable Production of Bio-Fuels and Bio-Chemicals*; Elsevier Science: Amsterdam, The Netherlands, 2013.

24. Namdev, P.D.; Patil, T.A.; Raghunathan, T.S.; Shankar, H.S. Thermal Hydrolysis of Vegetable Oils and Fats. 3. An Analysis of Design Alternatives. *Ind. Eng. Chem. Res.* **1988**, *27*, 739–743. [CrossRef]

25. Metzger, J.O.; Bornscheuer, U. Lipids as renewable resources: Current state of chemical and biotechnological conversion and diversification. *Appl. Microbiol. Biotechnol.* **2006**, *71*, 13–22. [CrossRef]

26. Kubičková, I.; Snåre, M.; Eränen, K.; Mäki-Arvela, P.; Murzin, D.Y. Hydrocarbons for diesel fuel via decarboxylation of vegetable oils. *Catal. Today* **2005**, *106*, 197–200. [CrossRef]

27. Moquin, P.H.L.; Temelli, F. Kinetic modeling of hydrolysis of canola oil in supercritical media. *J. Supercrit. Fluids* **2008**, *45*, 94–101. [CrossRef]

28. Noureddini, H.; Harkey, D.W.; Gutsman, M.R. A continuous process for the glycerolysis of soybean oil. *J. Am. Oil Chem. Soc.* **2004**, *81*, 203–207. [CrossRef]

29. Twitchell, E. Benzenestearosulphonic acid and other sulphonic acids containing the stearic radical. *J. Am. Chem. Soc.* **1900**, *22*, 22–26. [CrossRef]

30. Barnebey, H.L. Continuous Fat Splitting Plants Using the Colgate-Emery Process. *J. Am. Chem. Soc.* **1948**, *25*, 95–99. [CrossRef]

31. Stumborg, M.; Wong, A.; Hogan, E. Hydroprocessed vegetable oils for diesel fuel improvement. *Bioresour. Technol.* **1996**, *56*, 13–18. [CrossRef]

32. Egeberg, R.; Knudsen, K.; Nyström, S.; Lind, E.; Efraimsson, K. Industrial-Scale production of renewable diesel. *Pet. Technol. Q.* **2011**, *16*, 59–65.

33. Scherzer, J.; Gruia, A.J. *Hydrocracking Science and Technology*; Marcel Dekker: New York, NY, USA, 1996.

34. Kim, M.Y.; Kim, J.K.; Lee, M.E.; Lee, S.; Choi, M. Maximizing Biojet Fuel Production from Triglyceride: Importance of the Hydrocracking Catalyst and Separate Deoxygenation/Hydrocracking Steps. *ACS Catal.* **2017**, *7*, 6256–6267. [CrossRef]

35. Baur, R.; Taylor, R.; Krishna, R. Bifurcation analysis for TAME synthesis in a reactive distillation column: Comparison of pseudo-Homogeneous and heterogeneous reaction kinetics models. *Chem. Eng. Process.* **2003**, *42*, 211–221. [CrossRef]

36. Wang, S.J.; Wong, D.S.H.; Lee, E.K. Effect of interaction multiplicity on control system design for a MTBE reactive distillation column. *J. Process Control* **2003**, *13*, 503–515. [CrossRef]

37. Yang, B.; Wu, J.; Zhao, G.; Wang, H.; Lu, S. Multiplicity analysis in reactive distillation column using ASPEN PLUS. *Chin. J. Chem. Eng.* **2006**, *14*, 301–308. [CrossRef]
38. Cárdenas-Guerra, J.C.; López-Arenas, T.; Lobo-Oehmichen, R.; Pérez-Cisneros, E.S. A reactive distillation process for deep hydrodesulfurization of diesel: Multiplicity and operation aspects. *Comput. Chem. Eng.* **2010**, *34*, 196–209. [CrossRef]

 *processes*

Article

Journey Making: Applying PSE Principles to Complex Curriculum Designs

Ian Cameron * and Greg Birkett

School of Chemical Engineering, The University of Queensland, Brisbane 4072, Australia; g.birkett@uq.edu.au
* Correspondence: itc@uq.edu.au

Received: 7 November 2019; Accepted: 16 March 2020; Published: 23 March 2020

Abstract: Since the 1950s, Process Systems Engineering (PSE) concepts have traditionally been applied to the process industries, with great effect and with significant benefit. However, the same general approaches and principles in designing complex process designs can be applied to the design of higher education (HE) curricula. Curricula represent intended learning journeys, these being similar to the design of process flowsheets. In this paper, we set out the formal framework and concepts that underlie the challenges in design of curricula. The approaches use generic and fundamental concepts that can be applied by any discipline to curriculum design. We show how integration of discipline-specific concepts, across time and space, can be combined through design choices, to create learning journeys for students. These concepts are captured within a web-based design tool that permits wide choices for designers to build innovative curricula. The importance of visualization of curricula is discussed and illustrated, using a range of tools that permit insight into the nature of the designs. The framework and tool presented in this paper have been widely used across many disciplines, such as science, engineering, nursing, philosophy and pharmacy. As a special issue in memory of Professor Roger W.H. Sargent; we show these new developments in curriculum design are similar to the development of process flowsheets. Professor Sargent was not only an eminent research leader and pioneer, but an influential educator who gave rise to a new area in Chemical Engineering, influencing its many directions for more than 50 years.

Keywords: process systems engineering; design; higher education; curricula; visualization

1. Introduction

Curriculum design stands at the heart of all education. It is a multiscale challenge across different time and space scales—from whole-of-curriculum design considerations, to distinct learning units or modules, down to the day-to-day learning elements at the lowest level of consideration. It spans sequential stages of learning—from early learners, primary, secondary, tertiary and continuing professional development (CPD). By necessity, curriculum designs seek to embody stated intended outcomes for learners that address knowledge domains, application of knowledge, and personal- and professional-attribute development.

It might seem strange to some that curriculum design could be intimately related to Process Systems Engineering (PSE) thinking and application. In what follows, we show the development of complex curricula from the basic underlying concepts and building blocks that mirror many aspects of PSE. In doing so, we emphasize that PSE possesses a much broader interpretation and application than has traditionally been adopted.

In 1967, Roger Sargent wrote in *Chemical Engineering Progress* a review on "Integrated Design and Optimization of Processes". He stated the following: "Although we are in sight of a truly integrated approach to design of complete processes, a great deal of work remains to be done. With the need for more sophisticated analysis of larger complexes, it is more important than ever to join hands with those working in the fields of control engineering, operational research, numerical analysis and computer science" [1].

The discipline area of PSE arose from the application of systems engineering concepts to industrial processes [2,3]. The 1960s was a period of rapid digitalization in industry, affording significant advances in modeling, control, optimization and new computer-based numerical methods. The focus on "engineering" of "systems" that were primarily within the "process" industries drove PSE as a new focus within Chemical Engineering. The concept of "engineering" as ingenuity in design, using thinking and practices around a set of things that work together ('system'), can be applied to the general idea of any "process". That liberates PSE from the narrow confines of industrial and manufacturing sectors.

The 1967 statements of Roger Sargent, adapted to the case of curricular design, ring true. The tasks require an integrated approach that ensures the final curriculum design is "fit-for-purpose". It is a complex set of tasks dealing with many interconnected learning units, their attributes and intended outcomes. The appropriate sequencing of learning, as well as generating deep insights into the nature and behavior of the design, is essential. It is also a task where the skills and insights of numerous people are necessary to arrive at designs that deliver the requisite outcomes. In short, it bears many resemblances to traditional PSE thinking and practices.

Curriculum design practice has a long and important history. The rapid expansion of human knowledge in all professional domains has increased the need for learning designs that must meet the demand of current and future work demands in ever-changing environments. That is a long-standing challenge, and one that continues to challenge educators.

Early work by Dewey [4] and others, such as Tyler [5], set the scene for modern curriculum considerations. The expanding digitalization trends across society, with the creation and growth of the internet, have brought the need to use enabling information technologies, visualizations and user-centered web systems, to improve curriculum designs and their deployment. These information and communication technologies (ICT) can enhance the curriculum design process. The design process relies on structured information and its use to create educational pathways for learners.

In what follows, we outline why curriculum design and deployment is important, the basic concepts and processes that help engineer the learning system and our ability to assess designs in both qualitative and quantitative ways. Section 2 deals with design purpose and practice, before turning in Section 3 to the fundamental concepts and building blocks that permit designers to assemble a range of desired learning pathways. Section 4 shows the principles in practice, using an actual case study in Chemical Engineering at The University of Queensland. The final section considers how many other disciplines have derived benefit from a systems design environment that was pioneered within Chemical Engineering.

2. Curriculum Design: Purpose and Practice

2.1. Purpose

Curriculum designs set out the key learning pathways that an educational organization places before a learner. That is, curricula represent learning journeys, hence the title of this paper: "Journey Making".

"Curriculum" is related to the Latin word "currere", meaning "to run". As such, it speaks of a *pathway* or *a course* traversed by participants to reach a goal. End-goals and intermediate goals are crucial outcomes within curricula. The attainment of such goals is incremental.

In the last few decades, the underlying principles of curricula have been revisited and emphasized by prominent educational researchers across many disciplines [6,7]. These researchers and practitioners have enhanced earlier understandings of the role of curricula, the educational psychology around learning and the science of learning [8,9].

Barnett and Coate [6] recognized "that curricula have distinctive, but integrative components, as well as allowing for different weightings of each domain within any one curriculum" (p. 70).

It is not just academic researchers who are interested in curriculum design and outcomes. Due to the professional nature of engineering registration, practice and graduate mobility, global organizations such as the International Engineering Alliance (IEA), as represented by the Washington Accord [10], "establish and enforce internationally bench-marked standards for engineering education and expected competence for engineering practice".

In the case of the IEA, 29 countries that span the globe are signatories to that agreement. The professional engineering bodies or accreditation agencies in each country administer the accreditation of programs and curricula for higher education institutions that produce graduate engineers. In the USA, the Accreditation Board of Education & Technology (ABET, Baltimore, MD, USA) administers undergraduate degree programs [11]. Other organizations such as the European Network for Accreditation of Engineering Education (ENAEE), administer the EUR-ACE accreditation system for engineering graduates [12].

This means that, for accreditation of engineering programs across most of the globe, there are necessary learning outcomes that must be seen within the curriculum design, and importantly the evidence gathered to show that graduate engineers display those outcomes to an acceptable level. As well as the professional incentives for curriculum design, there are important institutional incentives in terms of developing curricula that sets apart one institution from another. This is often seen in terms of the learning pathways that students travel through during their undergraduate and graduate programs.

This raises the importance of excellence in curricular design and delivery. PSE-type approaches can provide the rigor necessary to achieve innovative, accredited curricula that provide flexibility in learning pathways to reach intended outcomes. PSE principles related to output requirements, integrated system and quality control through measurement and assessment strategies naturally fit into curriculum considerations.

2.2. Practice

Curriculum design practice across the higher education (HE) sector is extremely varied in the processes adopted to imagine what needs to be done, as well as the tools that might aid in developing, understanding and displaying designs. Many institutions struggle to properly design and document outcomes. Many designs are done by a small group of discipline experts with little 'buy-in' from colleagues. Many academics are not aware of the design principles and how their specific learning units integrate into the whole curriculum.

In recent years, a number of tools have been developed to help disciplines address the design activities in a structured manner. These primarily are mapping tools for learning outcomes (LOs), and they are limited in scope for doing comprehensive curriculum design across all higher education disciplines [13–15]. Some systems such as SOFIA have more pathway features [16]. Early comprehensive work in this space was done by the authors within Chemical Engineering [17]. PSE approaches can help in organizing curriculum around key elements that are combined to address intended learning outcomes and tracking those outcomes through the curriculum. This can help with the design process.

3. Curriculum Design: Underlying Concepts, Building Blocks and Pathways

In this section, we investigate the fundamental concepts upon which curriculum design rests. We consider the following questions:

1. What are the overarching outcomes that underlie curricular designs?
2. What are the fundamental building blocks that help generate learning designs?
3. What are the vital pathways that are to be considered and embedded as part of the design?

3.1. Basic Concepts

What are we trying to achieve in curriculum design? This is a crucial question to consider before undertaking any design activities. It mirrors the focus of Requirements Engineering in seeking to define a functional specification of the needs and desires of stakeholders [18]. The general equivalent in curriculum is the Intended Learning Outcomes (ILOs). A survey of the key literature, including national and international accrediting agencies and research monographs, shows that educational program requirements are in three major areas [11,19–21]. Figure 1 displays the schema of Barnett and Coate [6] that covers these requirements.

1. *Knowing*: this is the engagement of learning with knowledge;
2. *Acting*: this is the performative character in learning, namely putting knowledge to work in various circumstances and changing contexts;
3. *Being*: this is the personal and professional development of the learner. It addresses self-relationships, the educational setting and the outside world.

There is an intentional overlap in this schema, emphasising that all three elements combine synergistically toward the development of graduate capabilities. This schema is not confined to engineering disciplines, but has universal application.

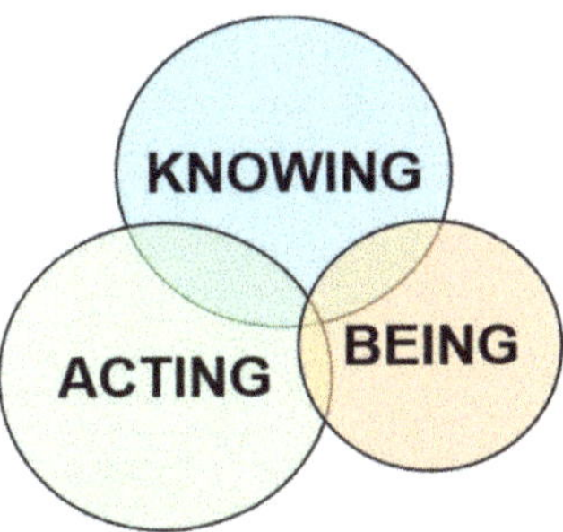

Figure 1. Educational schema as the major focus of educational designs.

Figure 2 shows the engineering program outcomes schema for Engineers Australia, the peak professional body in Australia. The accreditation frameworks in the USA and Europe, including The Washington Accord signatories, have a similar schema.

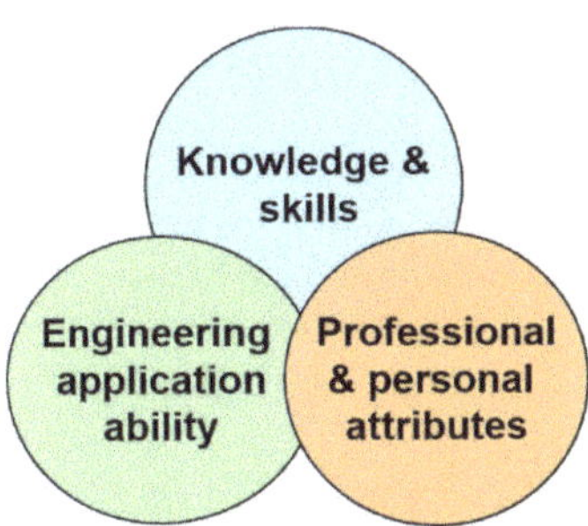

Figure 2. Engineers Australia, Stage 1 competency schema.

From this high-level schema, it is clearly necessary to develop more detailed outcomes at the level of the individual learning-unit level and show how integration across learning units addresses the outcomes within the *Knowing, Acting* and *Being* schema.

3.2. Building Blocks

Similar to flowsheet development that deals with processing units and the connected streams, curriculum development requires description of the learning units or modules. In some institutions, these are called "courses" or "subjects". They represent the lowest-level learning and teaching element in a degree program. The connections between units show required prior learnings (RPL). Besides these topological elements, the development of Intended Learning Outcomes (ILOs) for individual learning units, and subsequently the whole curriculum, require a wide range of other important elements. Figure 3 shows the key elements related to the process leading to the "curriculum as designed" stage.

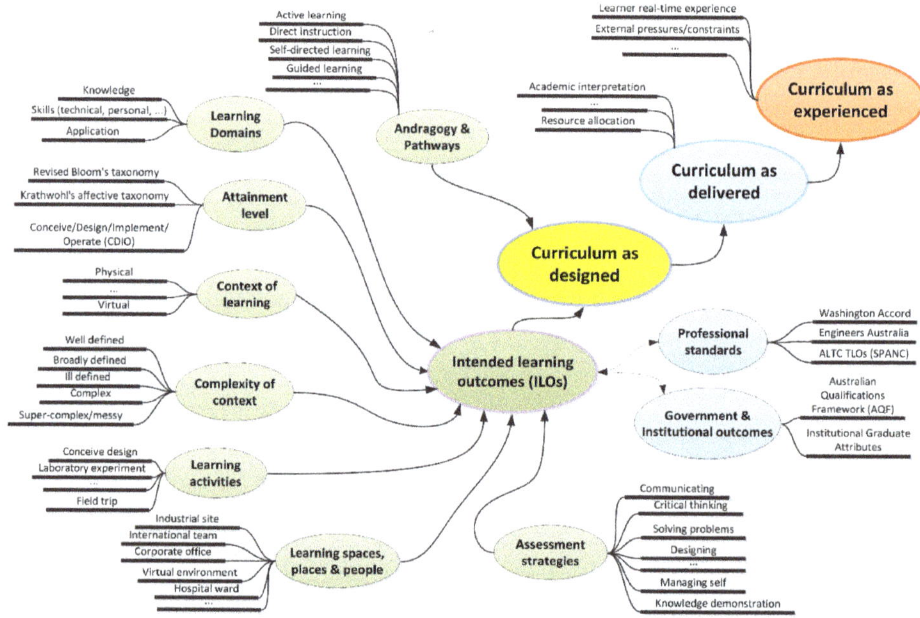

Figure 3. Curriculum design building blocks and information.

It should be noted that, beyond "curriculum as designed", there are two further concepts: "curriculum as delivered" and "curriculum as experienced". These concepts relate to the actual delivery of the curriculum by instructors and staff to student engineers, and the curriculum as actually experienced by the students. These are important concepts, but discussion of these is beyond the focus of this paper.

It should be noted that, in many higher education institutions, students have the ability to design their own personalized curriculum. Currently, for many students, it is difficult for them to see the connections between learning unit choices, sometimes leading to fragmentation in the overall curriculum. Moreover, over time, well-designed curricula can fragment, as learning units change with little consideration given to the implication of those changes on other integrated learning units.

3.2.1. Information Management

In order to gain maximum benefit from the development of a computer-aided environment for curriculum design, it is important that the information contained in these elements is both organized and extensible within data structures. We can classify the elements into two main categories:

1. Generic concepts;
2. Specific concepts.

Generic concepts are not discipline-specific, but need to be extensible and editable. They could include learning activities, attainment levels or assessment strategies. They are vital in developing the learning-outcome statements. Specific concepts relate to the discipline area considered in the design. Obvious specific areas would be discipline knowledge and application.

Taxonomies can be used to capture and utilize the important terms within an area. For example, the concept of "attainment level" can be organized as a multilevel taxonomy that is fully editable and extensible with the ability to select synonyms as needed. Figure 4 shows part of an attainment taxonomy within a design environment. It shows three key taxonomies often used in building learning outcomes.

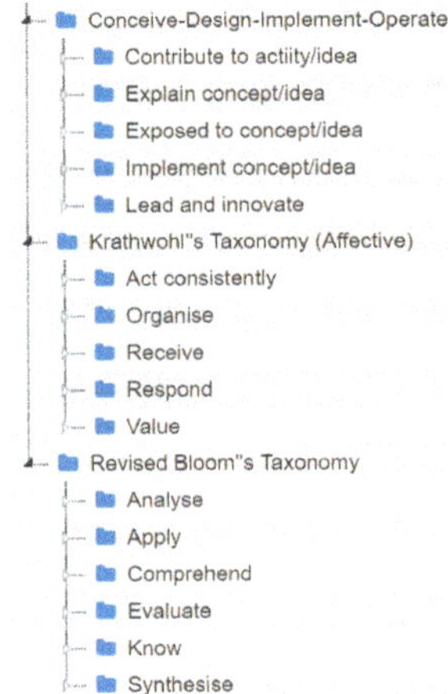

Figure 4. Part of an "attainment" taxonomy for building ILOs.

Similar approaches can lead to taxonomies for all the elements within Figure 3. Of particular interest are the discipline-specific knowledge taxonomies. Figure 5 shows a taxonomy for Chemical Engineering that is both extensible and editable within the design environment. It consists of levels that express a *domain, sub-domain* and *topic* taxonomy. This is clearly a multilevel view.

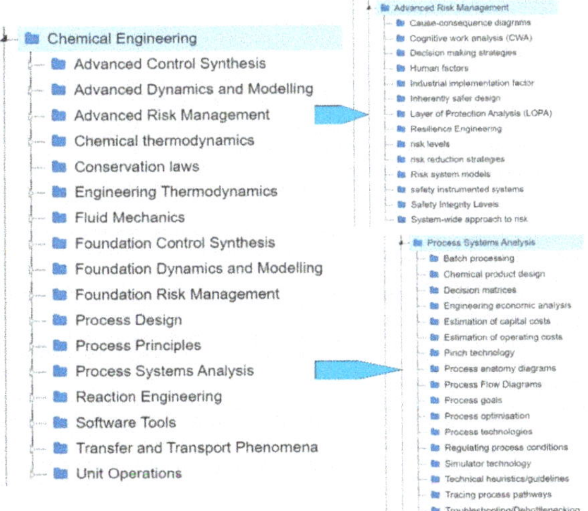

Figure 5. Knowledge taxonomies for Chemical Engineering (domain/sub-domain/topic).

3.2.2. Learning Unit Design

The key object in any curriculum design is the learning unit. This requires the integration of many elements in Figure 3:

1.   The intended learning outcomes (ILOs);
2.   The required prior learning (RPL);
3.   The assessment of ILOs: what, when, how and weightings;
4.   The attainment objects within the unit;
5.   The learning activities;
6.   The knowledge domain, sub-domain and specific topics;
7.   The complexity of the topics and tasks;
8.   The learning spaces or places to be utilized;
9.   The chosen andragogies.

This object is complex and could be regarded as analogous to a process engineering unit of a flowsheet. Learning outcomes can be built by using a defined but flexible syntax that uses the various taxonomies displayed in Figure 3. Structuring the ILOs provides significant power to reason over the completed curriculum and also visualize the characteristics of the curriculum. An example ILO from an advanced modeling course introducing the theory and practice of hybrid modeling is shown in Figure 6.

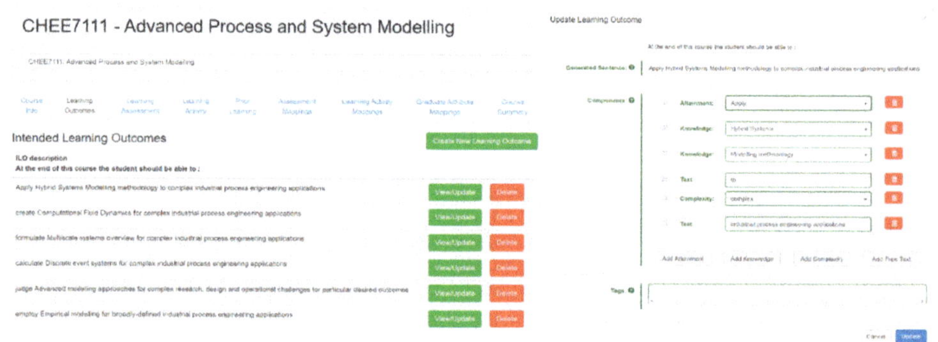

Figure 6. Example of intended learning outcomes (ILOs) (left) and the structured text ILO builder (right).

Moreover, the various tabs on the course profile show the other course-building options that are used in fully specifying the course characteristics. They include assessment items, learning activities, prior learning for the course and important mappings to professional competences.

3.3. Pathways

Curriculum designs ultimately lay down a series of learning pathways, professional attributes and competency development. As noted previously in the introduction, the program schema elements are developed incrementally across the length of the program. The representation and tracking of this incremental development are a crucial part of the design process.

For example, we often desire to understand the linkages within years and across years. This is similar to process streams passing through operating units. A simple visual illustration of course integration via learning outcomes is shown in Figure 7 [17].

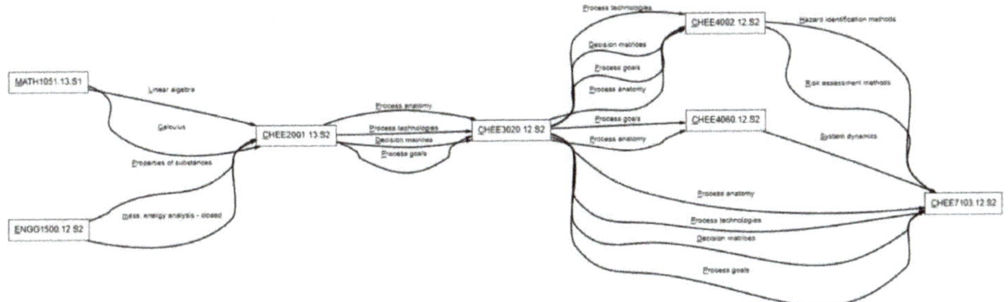

Figure 7. Example of knowledge domain outcome linkages for process system courses. (By permission of Engineers Australia [17]).

4. Curriculum Design: *The Journey Maker* Computer-Aided Design Environment

In the previous sections, we described the importance and challenges of complex curriculum design. The key goals were discussed, as well as the design information and processes required to build curricula. The development of a comprehensive web-based design tool which is adaptable, extensible and editable provides for design across all disciplines across higher education. Importantly, it was also developed for student use in building personalized curricula and visualizing the impact of choices. This is particularly important in the Arts and Humanities disciplines.

The functionality of this curriculum environment includes the following [17]:

- The ability to describe knowledge and skill domains and their interlinking across space and time;
- The ability to describe and track intended learning outcomes (ILOs) and attainment levels across space and time;
- The ability to describe the context and complexity of the planned learning;
- The ability to describe the spaces and places where learning is planned to happen;
- Linking assessment activities to ILOs and learning activities for promoting learning and proving outcomes;
- Ease of use by program coordinators and academic staff;
- Use by students to gain insight into the characteristics of a designed curriculum, and potentially to design and track the "experienced" curriculum;
- The ability to interface to other systems, such as institutional course and subject profiles;
- The ability to embed within the environment current "best practice" concepts and design processes as a vehicle for innovative curricula and better-informed academic use;
- The ability to adapt the environment across disciplines.

The majority of this functionality was developed in 2011/2012 as a standalone application [17]. The further development expanded activity across all university faculties, resulting in a web-based environment called *The Journey Maker*. The environment has two major components: a design component and a visualization component called "The Visual Journey". Figure 8 shows the entry page to the development environment, showing the design functions available for use in building and visualizing curricula.

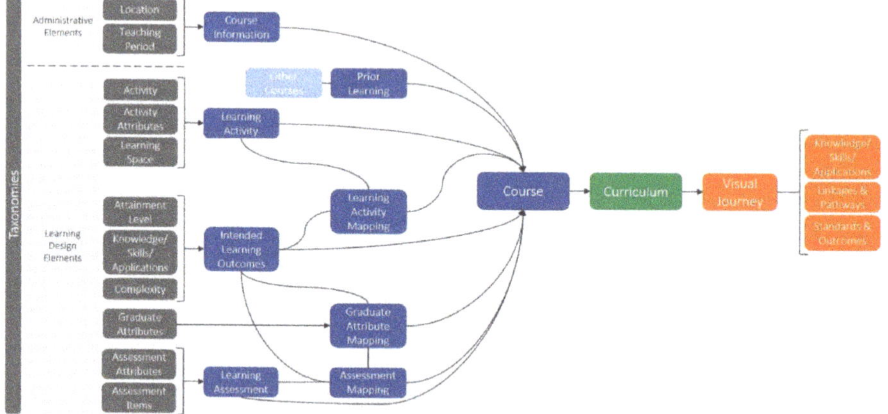

Figure 8. Overview of *The Journey Maker* design-environment functionality.

Elements of this environment are illustrated in the next section, using the design of some selected learning units, as well as showing an overall five-year curriculum design.

5. Case Study in Curriculum Design: A Five-Year Chemical Engineering Program

As a case study to display the development of a curriculum, we use the design of a five-year combined Bachelor and Masters Chemical Engineering program (BE/ME) at The University of Queensland. This is sufficiently complex, to demonstrate the design tools and the visualization tools that give deep insight into the design. The overall program structure is shown in Figure 9. This program typically has four learning units per semester, including elective options, which are shown as "+" options in each semester. These are chosen by students to fulfill their degree requirements, build specializations and address personal interests. There are 10 semesters in this program, with industry placements in Year 4, Semester 2.

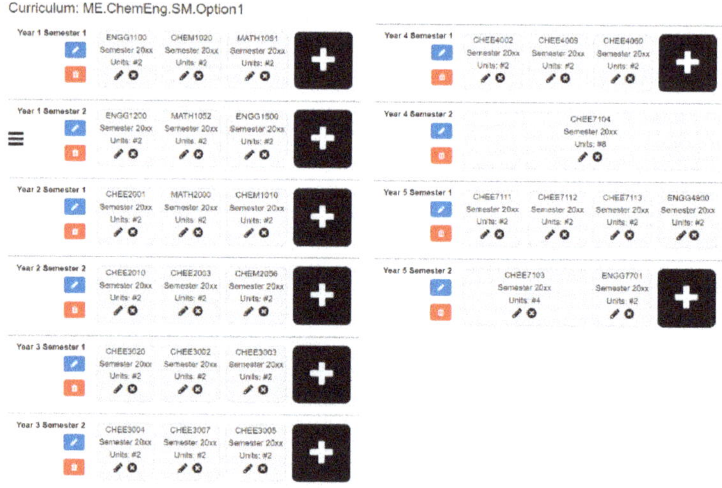

Figure 9. Compulsory course requirements for five-year BE/ME program.

As shown in Section 3.2.2, all the course profiles have been built by using the course builder, and so they have a rich description of the learning within each module and also the connectivity throughout the curriculum. It is possible to investigate this particular design through the use of the visualization facilities within "The Visual Journey" web environment.

Overall Curriculum Characteristics

The visualization is able to view the overall picture of the curriculum as seen through the ILOs. Figure 10 shows the relative frequency of ILOs across the principal domains of the program. Clearly, the knowledge domain of Chemical Engineering dominates, but we also see the relative emphases around professional competences and the basic sciences: mathematics and chemistry. These insights might lead to unit-level or curriculum-level redesigns, depending on overall outcomes requirements. The visualizations can be presented in many ways: frequency, weighted frequencies and the like.

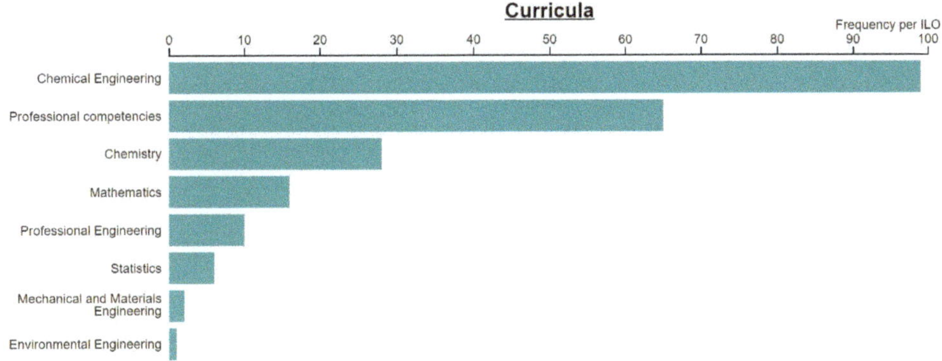

Figure 10. ILO frequencies for major knowledge domains across whole curriculum.

For further insights, the individual domains can be expanded, as seen in Figure 11, which focuses on the sub-domains within the Chemical Engineering domain. The thumbnail of the whole curriculum can be viewed to see where a specific focus might occur. In this case, process design can be seen within the ILOs across a series of courses starting in Year 1.

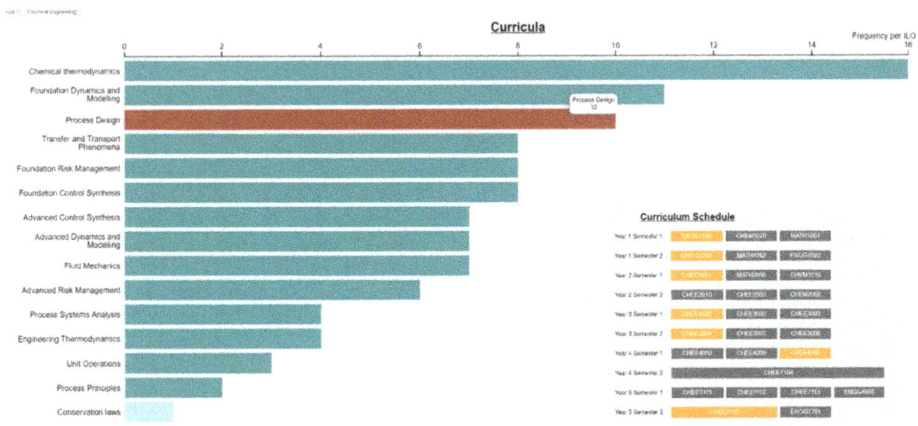

Figure 11. Lower-level domain knowledge visualization and location within the curriculum.

This form of visualization, and many others, can also be applied to other aspects of the learning outcomes, such as cognitive objectives as described by Bloom's taxonomy. Figure 12 shows the distribution of ILOs where "synthesis" is a major objective within courses. In particular, the issue of synthesis described by "design" can be seen in later years of the curriculum. If some redistribution of that attainment to earlier years was considered important, then the visualization helps in redesign.

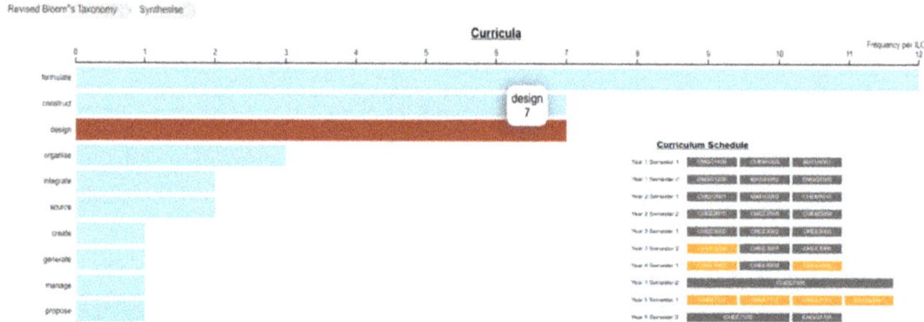

Figure 12. Attainment levels and distribution across the curriculum.

As well as considering issues around where knowledge, use of knowledge and personal attributes are developed, it is important to understand the linkages through the curriculum. This is particularly important as curriculum evolves, and in many cases, fragmentation occurs over time due to loss of integration. Figure 13 shows the connections of a particular ILO into following courses. Loss of that ILO due to changes in the learning unit could be important. The alluvial plots also allow unit instructors to see the position of the unit in relation to other units from a learning perspective. They see the required prior learning and also the future use of learning outcomes.

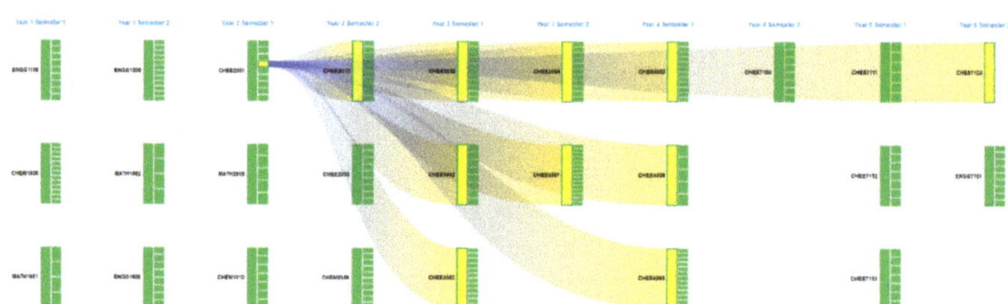

Figure 13. Tracking the linkages of a specific learning outcome through courses in the curriculum.

Figure 14 shows how course integration occurs across the curriculum. It shows key linkages along important learning pathways and identifies what might be "orphaned" courses within the curriculum.

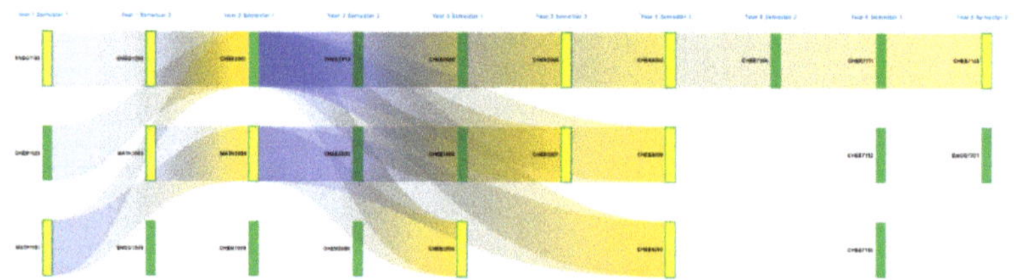

Figure 14. Tracking linkages between courses that have specified required prior learning.

Many other tabular and graphical views can be generated due to the structured descriptions and taxonomies embedded within the design environment. Within the approach, there is a very extensive subsystem for designing the assessment strategies. Figure 3 shows the extent of all the curriculum components, which cannot all be discussed in this paper.

These provide the ability for any discipline to develop curricula that is sharable, easily updated and extensible for the specific issues often dealt with in higher education programs.

The effects of using these formalized approaches, as described in these developments, are that now staff and students can clearly see the interconnections within curricula and are able to see the flow of knowledge and skills that are needed for future learning. It also allows curriculum changes to be easily mapped and tracked over time, providing easy access to professional accreditation organizations to see the impact of those changes. The ability to visualize learning outcomes and the forms of assessment used to provide evidence of attainment levels is an important aspect of the design.

Due the structured, extensible nature of all the curriculum building blocks outlined in Figure 3, it is relatively easy to export a wide range of reports, visualizations, curriculum data and statistics for numerous purposes.

6. Conclusions

This work has shown the use of certain PSE principles in developing complex curricula. The design of curricula resembles in many ways the basic ideas of process design and flow sheet development, since curricula design is an educational process. Through the understanding of the crucial building blocks for curricula and the structured manner of developing learning units, it is possible to produce whole curricula designs that capture the many characteristics that make up complex learning environments.

By using structured information management approaches, the final designs can be visualized to understand the whole integrated curriculum. The fundamental concepts, organization and deployment started within Chemical Engineering, but they have now been adopted and used in many other disciplines, including nursing, science, agriculture, pharmacy, veterinary science, medicine and philosophy. It is significant that the idea of the importance of an integrative approach rather than disparate designs was at the heart of the PSE efforts of Roger Sargent from the very beginning of his long and distinguished career.

The application of PSE ideas to bring an integrative approach to curriculum design now yields similar benefits to many other disciplines that have already been realized within the PSE community.

Author Contributions: Conceptualization, I.C. and G.B.; methodology, I.C. and G.B.; software, eLIPSE.; formal analysis, I.C. and G.B.; investigation, I.C. and G.B.; resources, I.C. and G.B.; data curation, I.C.; writing—Original draft preparation, I.C., G.B.; writing—Review and editing, I.C. and G.B.; project administration, I.C. and G.B.; funding acquisition, I.C. and G.B. All authors have read and agreed to the published version of the manuscript.

Funding: We acknowledge funding from The University of Queensland, Technology Enabled Learning (TEL) Grant scheme.

Acknowledgments: We acknowledge the web development work of the Centre for eLearning Innovations and Partnerships in Science and Engineering (eLIPSE). We thank Erzsébet Németh (E.N.) for her work on the visualization aspects of curricula.

Conflicts of Interest: The authors declare no conflict of interest. The funders had no role in the design of the study; in the collection, analyses or interpretation of data; in the writing of the manuscript; or in the decision to publish the results.

References

1. Sargent, R.W.H. Integrated Design and Optimization of Processes. *Chem. Eng. Prog.* **1967**, *63*, 71–78.
2. Williams, T.J. *Systems Engineering for the Process. Industries*; McGraw-Hill: New York, NY, USA, 1961.
3. Sargent, R.W.H. Chemical Engineering and Engineering Science. *Chem. Eng.* **1963**, *168*, 151–155.
4. Dewey, J. *Democracy and Education: An introduction to the Philosophy of Education*; The MacMillan Company: New York, NY, USA, 1916.
5. Tyler, R.W. *Basic Principles of Curriculum and Instruction*; University of Chicago Press: Chicago, IL, USA, 1949.
6. Barnett, R.; Coate, K. *Engaging the Curriculum in Higher Education*; Open University Press: Milton Keynes, UK, 2005.

7.   Hicks, O. Curriculum in Higher Education—Hello? in Enhancing Higher Education, Theory and Scholarship. In Proceedings of the 30th HERDSA Annual Conference [CD-ROM], Adelaide, Australia, 8–11 July 2007.

8.   Ambrose, S.; Bridges, M.W.; Lovett, M.C.; DiPietro, M.; Norman, M.K. *How Learning Works: Seven Research-Based Principles for Smart Teaching*; Jossey-Bass Publishers, John Wiley & Sons: San Francisco, CA, USA, 2010.

9.   National Research Council. *How People Learn: Brain, Mind, Experience and School, Expanded Edition*; National Academy Press: Washington, DC, USA, 2000. [CrossRef]

10.   International Engineering Alliance (IEA). Available online: https://www.ieagreements.org/ (accessed on 14 October 2019).

11.   ABET, Accreditation Board of Education & Technology. Available online: http://www.abet.org/accreditation (accessed on 14 October 2019).

12.   European Network for Accreditation of Engineering Education (ENAEE). Available online: https://www.enaee.eu (accessed on 14 October 2019).

13.   Oliver, B.; Ferns, S.; Whelan, B.; Lilly, L. Mapping the Curriculum for Quality Enhancement: Refining a Tool and Processes for the Purpose of Curriculum Renewal. In Proceedings of the Australian Quality Forum 2010, Gold Coast, Australia, 30 June–2 July 2010; pp. 80–88.

14.   Roy, G.; Armarego, J. *CCmapper: A Tool for Curriculum and Competency Mapping*; Edith Cowan University, School of Engineering: Perth, WA, Australia, 2012.

15.   JISC. The Design Studio. Available online: http://jiscdesignstudio.pbworks.com/w/page/12458422/Welcome-to-the-Design-Studio (accessed on 14 October 2019).

16.   SOFIA Interactive Curriculum Mapping. Available online: https://www.sofiacurriculum.com/ (accessed on 14 October 2019).

17.   Cameron, I.; Birkett, G. A Curriculum Design, Modelling and Visualization Environment. In Proceedings of the 2012 AAEE Conference, Melbourne, Australia, 3–5 December 2012; Session 2B: Curriculum Design II, paper 1.

18.   Requirements Engineering. Available online: https://en.wikipedia.org/wiki/Requirements_engineering (accessed on 14 October 2019).

19.   Engineers Australia. Available online: https://www.engineersaustralia.org.au/About-Us/Accreditation (accessed on 14 October 2019).

20.   UK Engineering Council. Available online: https://www.engc.org.uk/education-skills/accreditation-of-higher-education-programmes/information-for-higher-education-providers/ (accessed on 14 October 2019).

21.   EUR-ACE Framework Standards and Guidelines. Available online: https://www.enaee.eu/eur-ace-system/standards-and-guidelines/#standards-and-guidelines-for-accreditation-of-engineering-programmes (accessed on 14 October 2019).

MDPI
St. Alban-Anlage 66
4052 Basel
Switzerland
Tel. +41 61 683 77 34
Fax +41 61 302 89 18
www.mdpi.com

Processes Editorial Office
E-mail: processes@mdpi.com
www.mdpi.com/journal/processes

www.ingramcontent.com/pod-product-compliance
Lightning Source LLC
LaVergne TN
LVHW070116170726
843515LV00007B/1697